T0419185

CROWDASSET

Crowdfunding for Policymakers

CROWDASSET

Crowdfunding for Policymakers

Editors

Oliver Gajda
European Crowdfunding Network, Belgium

Dan Marom
Hebrew University of Jerusalem, Israel

Tim Wright
twintangibles, UK

World Scientific

NEW JERSEY · LONDON · SINGAPORE · BEIJING · SHANGHAI · HONG KONG · TAIPEI · CHENNAI · TOKYO

Published by

World Scientific Publishing Co. Pte. Ltd.

5 Toh Tuck Link, Singapore 596224

USA office: 27 Warren Street, Suite 401-402, Hackensack, NJ 07601

UK office: 57 Shelton Street, Covent Garden, London WC2H 9HE

Library of Congress Cataloging-in-Publication Data

Names: Gajda, Oliver, editor. | Marom, Dan, 1979–   editor. |
    Wright, Tim (Management consultant), editor.
Title: CrowdAsset : crowdfunding for policymakers / editors, Oliver Gajda,
    European Crowdfunding Network, Belgium, Dan Marom, Hebrew University of
    Jerusalem, Israel, Tim Wright, twintangibles, UK.
Description: New Jersey : World Scientific, [2020] | Includes bibliographical references and index.
Identifiers: LCCN 2020004326 | ISBN 9789811207815 (hardcover) | ISBN 9789811207822 (ebook) |
    ISBN 9789811207839 (ebook other)
Subjects: LCSH: Crowd funding--Political aspects. | Finance, Public. | Community development--
    Finance. | Political planning--Citizen participation.
Classification: LCC HG4751 .C783 2020 | DDC 352.4/4--dc23
LC record available at https://lccn.loc.gov/2020004326

British Library Cataloguing-in-Publication Data
A catalogue record for this book is available from the British Library.

For any available supplementary material, please visit
https://www.worldscientific.com/worldscibooks/10.1142/11485#t=suppl

Desk Editors: Balasubramanian Shanmugam/Yulin Jiang

Typeset by Stallion Press
Email: enquiries@stallionpress.com

To our wives and children,
and changemakers all over the world

Preface

The emergence of a practical and functioning crowd economy is the most significant change in the way we think about the process of creating value. Hyper-connected technologies have transformed our ability to engage with a wide, diverse, and highly distributed collection of collaborators and, simultaneously, driven down the incremental cost of each engagement. In tandem, these two factors present us with entirely new opportunities which we have only just begun to explore.

It is in the number and diversity of available collaborators in the crowd that we find resources, insight, and answers and it is what we refer to as the "crowdasset," a source of extraordinary potential but one which requires new thinking and new models to mine effectively. To articulate the idea that the crowd is inherently valuable is not a new or necessarily novel idea, but it is also a widely misunderstood concept in terms of the nature of its characteristics, and the practicalities of how best to harness or unleash it have bedevilled us for ages and have been the base for many principled and often hot disputes.

Our contention is that it is time to re-examine this topic in the context of the digital era. We suggest that perhaps we are on the verge of a transformation, where it is becoming possible to understand the crowd as the asset to radically change our future for the better. We also assert that this will emerge from active efforts to directly tap into the asset to create value and that the crowd itself, in line with the demonstrable empowerment delivered by digital technologies, now has the power to effect change by itself directly. We set out this position somewhat in the spirit of Elinor Ostrom's work in that we can recognise that things are happening even if our existing models don't comfortably account for it, so perhaps we should seek to explain them. In this book, we capture a number of powerful examples of what is actually happening in the collective or crowd action and offer a series of frameworks which can rationalise and explain aspects of that and also offer guidance and advice to those seeking to take on the opportunity.

The notion that the crowd has powerful capabilities and a profound inherent legitimacy underpins the philosophy of our democratic institutions. But, as the commercial and civic realms begin to explore the possibilities presented by the crowd economy in its many forms, and begin the process of accessing that crowdasset, it is a circumstance which presents both exceptional opportunities and formidable challenges for the policymaker.

On the one hand, the emergence of a connected and empowered "crowdasset" offers the opportunity to re-imagine and transform our civic, public, and commercial lives. New ways of engaging, consulting, informing, and transacting are within reach and already being used to shake up and improve tired and moribund models unsuited to the crowd-empowered era.

On the other hand, such profound changes can present considerable obstacles and risks in a world where gatekeepers, rule makers, and custodians find themselves side-lined, redundant, and even bypassed. To those policymakers tasked with shaping the environment and context under which these transactions take place, this is, at the very least, confusing. To those policymakers seeking to engage with and make use of these new opportunities, they can find that established models and thinking from a more intermediated world are unsuited for the task and they are unprepared.

As Sherry Arnstein so eloquently put it when introducing her Ladder of Citizen Participation "The idea of citizen participation is a little like eating spinach: no one is against it in principle because it is good for you," but the practicalities of doing it are much more challenging. This may have some foundation in the idea that so much of our policy agenda emerges from the top-down, whereas crowd empowerment is fundamentally bottom-up.

In this book, we hope to walk a middle path which can accommodate both ends of that telescope. We hope to help policymakers able to better understand, respond to, and build upon bottom-up crowdasset opportunities. We also hope that we can ensure that where their top-down approach is necessary these are well aligned and properly configured for best effect.

We do this in the unshakeable belief that the best prepared and least confused policymakers will be able to effectively manage these challenges and, better still, maximise the opportunities for themselves and others.

The purpose of this book then is to provide policymakers with an understanding of the scope of the opportunity, some perspective on its provenance, exemplars of those who have already begun to explore what crowdassets can deliver, and a comprehensive set of tools to equip them to act, respond to, and make the most of, the new circumstances that are emerging.

And when we say "policymakers," we mean that in the widest sense. For us, a policymaker has the ability to shape the terms and aims of transactions within a defined group. As such, we can include in our audience policymakers in a traditional sense in those that directly shape and inform legislation at a national or supranational level. But we can also think of many other groups whose purview perhaps does not range as far as a national government but can, nevertheless, have significant roles in shaping and informing what occurs within their remit. This might include local and regional governments, civic bodies, religious communities, NGOs, and many others. For all of these bodies, and more, their role in enabling and managing the context for crowd activity is wide and so too is their ability to harness the opportunity to develop and enhance the lives of those who are part of their community by harnessing the crowdassets open to them. We include them all, and we seek to address those aspects of both managing the ecosystems they shape and also the consideration of how best to construct a programme of interventions to maximise the benefit from the opportunity on offer.

The ability to engage huge numbers of citizens at a low cost, in a low-friction, and real-time manner is the realisation of a long-held desire to build a more equitable, democratic, empowered, and inclusive society and economy.

This is made possible by the widespread availability of personal technology devices that utilise the global common asset of the internet. Their use as a mechanism to facilitate and empower people's enduring desire to take control and ownership of things that affect them and the things which they care about is the most influencing phenomenon of our times.

The pace with which these changes have come about does, however, bring with it challenges. It is by definition new and so throws up situations and conundrums that are novel and not easily resolved by well-worn approaches or existing models. It is also the pace at which we are experiencing these changes that can see it outstrip legacy regulatory frameworks and conventions and, in so doing, unbalancing what has been a relatively stable, integrated, and well-understood architecture of rules and accepted protocols for policymakers and citizens alike. The new crowd-empowered mode both circumvents previous thinking by making the previously "undoable" suddenly "doable." At the same time, as existing structures gradually flex in response to this new environment, new gaps and openings appear to permit yet more disruptive activities to emerge.

The breadth and depth of the impacts of the crowd economy are still only in their very early stages, and while we cannot predict its development far into the future, we can say with certainty that it has enormous potential. What we can also say with confidence is that it is here to stay and it is already demonstrably causing disruption, challenging our thinking, outstripping our historically bound regulatory frameworks and mental models, while also offering very real and meaningful opportunities to do things differently and better. There is then the potential to resolve long-standing and once insurmountable challenges for the benefit of humanity if we can successfully harness the opportunity.

It is in this spirit of new collaboration and community we suggest that the crowd is the key partner and source of value in a way that it has not been before and, as such, we define it as the asset, the crowdasset which, if used wisely, will be the crucible of political, economic, cultural, and civic innovation and regeneration for years to come. Sebastian Thrun helpfully reminds us that searching far and wide in non-conventional sources is a sensible thing to do if you are in search of talent. In the foundation of Udacity, he taught a parallel course on Artificial Intelligence to

200 Stanford University students simultaneously with more than 100,000 non-traditional students accessing the same course virtually through the Udacity platform. A total of 23,000 non-traditional students completed the course successfully, and in the ranking of the course graduates, it was not until you reached the 413th ranked student that you found a Stanford student.[1] The UK Arctic survey vessel via crowdsourcing exercise resulted in the name Boaty McBoatface topping the poll.[2] This speaks of the fact that it is a process that requires more thought and structure than often imagined and also the power of influencers within such a poorly, managed exercise. But it also creates anxiety among policymakers as they understand that poorly managed crowd engagement can present significant perceived reputational risk.

The crowd economy and the value of crowdassets encompass many practices including crowdsourcing, open innovation, collaborative consumption, and many other collaborative and user-driven practices. This can include the emergence of the value derived from Big Data as those data are an expression of the crowds' activity.

There are two key dimensions which define and locate these crowd economy activities. Firstly, there is the changing power relationship between those seeking the value from the crowd and those providing it. This dynamic is founded in perhaps the most disruptive aspects of the crowd economy as it is the empowered nature of individuals as agents with real capacity to shape their individual and collective choices and experience which thus exemplifies the change brought about by digital empowerment.

The second dimension defines the source of the value within the crowd, which ranges from the collective being the source of value in its widest sense to an individual as a source of value who is otherwise hidden among the crowd and who, perhaps, has no connection or relationship whatsoever with those seeking the value.

Across this canvas, we can paint a range of different crowd engagement models and crowdasset types.

We have captured this in the past as the crowdasset framework (see Figure 1). It is a tool that seeks to simplify navigation across the diversity of the crowd economy.

Preface

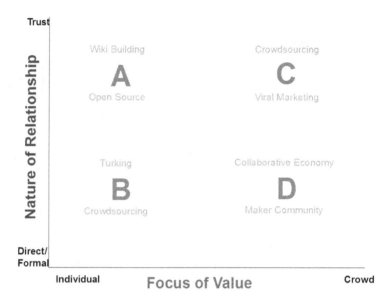

Figure 1: *The crowdasset framework.*

But we do not propose to explore all of these diverse aspects of the crowd economy in this book preferring a more focused approach, and we do this for two reasons.

Firstly, the principles we set out here are, we believe, widely applicable across much of the crowd economy and so a focused approach does not in and of itself exclude other crowd engagement models. We would contend with confidence, based on our experience across the sector, that the principles, ideas, and approaches we set out here are widely and thoroughly applicable in almost any crowd economy or crowdasset circumstance.

Secondly, it is, in our opinion, of greater value to focus on a more mature aspect of the wider crowd economy to demonstrate our thinking as it provides us with a richer and more developed source of material to draw on and perhaps provides indications of where other crowd engagement models might develop.

With that in mind, we will limit our examples to the financial expression of the crowd economy phenomenon as it is here that we have perhaps witnessed the most rapid and visible changes and the greatest maturity.

Namely, we focus on crowdfunding and we make no apology for giving our emphasis to that aspect of the crowd economy and particular expression of the crowdasset phenomenon as the most developed and established to date. In doing this, we do not overlook the other ways in which the crowd can be an exceptional asset beyond a source of funding. While this is a necessarily focused study of crowdfunding, it will become very apparent throughout the course of this work that the phrase itself fails to capture the breadth of return available in a crowdfunding campaign. We illustrate four key groupings of returns from crowdfunding, only one of which is directly finance related. We refer to these groups as Finance, Insight, Communication, and Networks. Our intention is to demonstrate the breadth of real value of the crowdasset in the context of an activity, which might be specifically termed crowdfunding. This should illustrate that it is both legitimate and proper to use crowdfunding campaign as a vehicle to deliver against other important outcomes. For example, a well-run campaign or series of curated campaigns could be exceptionally powerful vehicles to bring about engagement at the seventh and eight rungs of Arnstein's eponymous ladder.

Crowdfunding is quickly emerging as a significant and valuable part of the financial landscape. The financial crash of 2008 exposed a fragility in the system and sparked an explicit expression of brooding dissatisfaction, among a wider public, with a sector seen as insular, exclusive, aloof, and out of touch with the needs of much of the population. At the same time, it exposed the vulnerability of some of our core institutions and activities on which so many aspects of society are dependent when they are shown to be so reliant on a non-diversified model of supply. This, and the drive to find alternative approaches coming at a time of technological and societal change, was a combination of factors that rapidly accelerated the growth of crowdfunding, which, although not a wholly new concept, became one which had found its moment.

The practical and philosophical underpinnings of crowdfunding make it highly distinctive from traditional capital and a welcome extension to the financial markets not just in terms of supply but also in its real and demonstrable differences from what had gone before and in the nature and characteristics of the finance it provides. Through widened and inclusive participation in the financing process, by reducing the distance between investor and investee, and in the digital spirit of the "long tail" where many small contributions can be easily aggregated into significant impacts, crowd finance can demonstrably be shown to have a more diverse motivation, a broader set of expectations of return and, through that, a uniqueness of character and colour in an otherwise monochrome marketplace.

As it has grown, these advantages have become ever more apparent to the extent that crowdfunding is now increasingly seen as a finance of choice for many because of the additional and valuable benefits it can deliver over and above finance alone.

As its use continues to grow in terms of its breadth by range of application and deepens in terms of the size of funds raised and, as it normalises in terms of its perceived acceptability, and the general level of understanding of its availability and distinctive qualities is increasingly acknowledged, so too will the value it creates and the possibilities it presents. This process of growth and embedding can be an organic one, growing and solidifying slowly and organically over time, or it can be cultivated, nurtured, and encouraged by intervention so as to yield greater results and better outcomes sooner. In this respect, the role of the policymaker as a builder of the ecosystem for such innovation to grow as well as a user of the opportunities presented becomes very distinctive.

It is increasingly apparent that by virtue of the nature of the crowdfunding process of broad outreach and extended deepened communication the returns available by engaging with the crowdasset are much wider than simply financial. Its power to elicit insight, create extensive networks and connections of value, and drive communication and engagement through the process of reaching out to the crowd means that it is imperative for policymakers to understand, enable, and make use of it.

Policymakers will inevitably rub up against this rapidly emerging crowd phenomenon either by accident or by design. To our knowledge, and based on our experience, this presents two potentially overarching scenarios of this encounter. The first is an upward–driven, ill-prepared engagement initiated by citizens and clients. Being an inherently reactionary response on the part of the policymaker, it can be challenging, difficult for resources to be anticipated, already assuming a life of its own, which may or may not be aligned with, or supportive of, the policy agenda.

Alternatively, there is a proactive and visionary engagement shaped by policymakers to nurture, accelerate, and guide the process where the initiation is by the proactive policymaker, with leadership and with an intention to support, and align with, the policy agenda.

Both these scenarios can yield results, but we believe the latter is the scenario that is best suited for everybody involved. The purpose of this book is then to help prepare and develop that process into a valuable and informed approach, which will yield the best results.

In this book, we set out both the fundamentals of crowdfunding and the crowd economy and what makes it distinctive to what has gone before. Following this grounding and foundation, we introduce a framework of analysis applicable to all policymakers at both a macro- and micro-level. We also include a rich set of fascinating and diverse exemplars of policymakers' engagement with, and impact on, crowdfunding developments. Drawn from around the world and written by a range of contributors at the heart of this development, these cases offer perspectives on the possibilities and real stories of the practicalities of what the crowd can achieve when policymakers engage with them.

The purpose of the framework shared here is to bring rigour, structure, agility, and strategy to an otherwise challenging and potentially confusing circumstance as a policymaker considers what interventions might be best for their needs. It will permit the development of a holistic, thoughtful, and strategically aligned approach for utilising the potential of crowdfunding and the wider crowd economy to provide the best effects for the policymakers of the organisation and area of oversight, its strategy and aims and its impact. Instead of the response to crowdfunding opportunity being the simplistic one of "let's run a crowdfund," policymakers can use the framework to develop a more sophisticated and integrated approach bringing crowd-based approaches and opportunities to the heart of what they do and in tune with existing activities. It prevents crowd-based activities from becoming the agenda itself but rather integrates crowd-based approaches into the wider strategy for better outcomes.

The framework is simple and understandable, but it is this very simplicity which allows it to be used effectively in a variety of ways and in a wide range of circumstances and scenarios from the micro- to the macro-level.

Its use and application are based on a series of broad activities and outcomes common in an agenda-setting context. We describe these applications as follows:

- *planning*—for defining and choosing interventions,
- *analysis*—to better integrate crowdfunding into the existing activities by exploring other interventions and actions currently being undertaken for other objectives but which might be relevant to Crowdfunding,
- *audit*—tool to monitor progress and impact as the intervention plan and process is operationalised, and finally
- *benchmark*—a framework to evaluate what insight similar organisations can provide, for comparison purposes.

A policymaker may use one or all of these approaches and we do not mandate or encourage a mechanistic or slavish approach. Similarly, we expect that more approaches and ideas will emerge around the use of the framework as the range of circumstances in which it is tried and applied grows. That said, perhaps the most sought-after aspect of the framework is its ability to speed and define the planning process allowing policymakers to quickly develop clear and robust plans that can be justified and expounded in the structure of the framework.

Apart from its utility, the framework's main and original value proposition is its categorisation of intervention options into the four headings of Infrastructure, Leverage, Education, and Matching.

In our experience, all interventions can be accommodated in these four groups; however, there is endless and ongoing innovation in the way the interventions are developed and deployed. The nature of the framework means that it can be applied to an organisation or any defined frame of analysis regardless of size, i.e. from supranational bodies to the government and civic bodies to charities and larger corporates. Within and across these high-level groupings, there is immense scope for innovation and flexibility to accommodate and reflect local sensibilities and specific needs.

Let us deal briefly with each of those four high-level groupings and expand a little on what each encompasses.

Infrastructure—This aspect focuses on creating and strengthening crowdfunding activities by addressing the infrastructural or ecosystem needs. This might, for example, include supporting the creation and activities of crowdfunding platforms, encouraging and developing online communities for these platforms, or indeed ensuring a conducive environment by means of regulatory action, tax incentives, and other such interventions.

Matching—Matching concerns itself with offering support and benefits in steps with the activities or market participants. While this is most commonly expressed as some form of financial matching, there are many other areas of matching support that can be offered in terms of brand, social, and relationship capital and a range of other valuable resources and assets that a policymaker may have at their disposal.

Education—Education is a remarkably wide-ranging grouping that can vary widely in terms of course and mode of delivery. But typically and most commonly, the effort will address three groups. Firstly, educating the public and creating awareness on crowdfunding. Secondly, providing the entrepreneurs and innovators and activists with the proper tools to crowdfund efficiently. There may also be a need to educate "investors" as to the benefits and opportunities of crowdfunding

Leverage—The fourth pillar of intervention focuses on leveraging the success of crowd activity by developing their ability to build on their success. So in this scenario, you are explicitly seeking to build on already initiated activities to advance with them in and of themselves or as an adjunct or feeder to other strategic aims.

By providing such broad but clearly defined groupings, the framework offers sufficient space to explore, innovate, and consider the possibilities while providing sufficient structure to make the process manageable and approachable. This tool ensures that the policymaker has both rigour in approach and a structured and meaningful mechanism to describe and support decisions on how to engage with the crowd economy.

The key outcomes from using the framework are, in our experience, a series of interventions that are more aligned with the strategic aims of the organisation and more fully integrated. The organisations have a better understanding of the opportunity and are better able to plan the process of developing the interventions and monitoring their progress against a series of sensible benchmarks. Crucially, it ensures that the interventions are supportive of all your activities and not separate from them.

In addition to describing the theoretical and more abstract nature of the framework, we illustrate the real-life and highly practical applications of the technique by a variety of policymakers to more effectively build, use, and derive tangible results and outcomes from a considered structured and strategically aligned approach in order to engage with the new crowd-empowered economy and society.

The breadth of examples we include here is, in our view, impressive but can never be comprehensive as the sector continues to develop and innovate. But they do nevertheless present a range of diverse and fascinating accounts of policymakers getting to grips with the crowdasset opportunity through the vehicle of crowdfunding.

Examples of regulatory responses seeking to create a structured and safe environment are captured and accounted for from countries like Lithuania and Finland. More localised regulatory innovation and a more federal model is explored in the Intrastate funding in the USA. The power of engagement and community mobilisation are well illustrated in examples from the office of the Mayor of London and Goteo experiences. We have also given a nod to the future as we include the emerging field of blockchain and crypto-currencies as these innovations add a rich new layer of opportunity by helping, for example, the City of Berkeley in California re-imagine mini municipal bonds for a new generation.

It is apparent that we are on the threshold of one of the most remarkable changes as to how we can create value and resolve challenges and problems for the benefit of all. How quickly and effectively we can embrace and exploit that opportunity will, in large part, be driven by the attitude, actions, and responses of policymakers. It is our hope that by sharing the insights and approaches found this book, we will provide the best and most robust foundation for policymakers to be best prepared to thrive and prosper in these new and changed circumstances.

Endnotes

1. https://thenextweb.com/insider/2015/06/02/udacitys-sebastian-thrun-how-nanodegrees-can-democratize-tech-education/
2. https://www.bbc.co.uk/news/world-europe-jersey-35860760

About the Editors

Dan Marom is a Thought Leader in the FinTech and crowdfunding arenas and is currently focusing on his role as a faculty member at the Business School of the Hebrew University (Israel, Jerusalem), alongside business consultancy, venturing, and investment activities. In 2010, Dan co-authored a pioneering book in the field of crowdfunding— *The Crowdfunding Revolution*. The second edition was published in 2012 by McGraw-Hill. Dan has advised international organisations such as the World Bank, the European Commission, and other organisations and foundations that are operating in the alternative finance field. His second book, *Crowdfunding: The Corporate Era* (2015) was recently published by Elliott & Thompson. Dan has gained substantial personal experience in start-up venturing, impact investments in emerging markets, and venture capital, mostly in the various sub-domains of fintech. He works currently with leading corporates and entrepreneurs on various international projects. He lives in Tel Aviv, Israel, holds a PhD in Finance from the Hebrew University of Jerusalem, an MBA (Cum Laude), and a BSc in Electrical Engineering.

Oliver Gajda co-initiated the European Crowdfunding Network (ECN) and went on to establish ECN as an international NGO in 2013 as Founding Executive Director and Chairman. Previously, Oliver has worked with venture capital, microfinance, technology, and social entrepreneurship in both commercial and non-profit settings in Europe and the USA. Oliver is an Advisory Board Member at the SolarCoin Foundation and the Förderkreis Gründungs-Forschung e.V. (FGF). In 2016, he published with Springer International Publishing a first academic compendium on crowdfunding called *Crowdfunding in Europe: State of the Art in Theory and Practice*. A former journalist, he started his career in the early 1990s in the publishing and business information industries. Oliver holds Master's degrees from Solvay Business School and from the University of Hamburg and studied at SEESS (UCL) in London.

Tim Wright is a Director and Co-Founder of twintangibles, a management consultancy that helps organisations create value through accessing the crowd economy. A leading thinker and practitioner on crowdfunding, Tim provides expert insight to a wide range of organisations and his clients include the OECD, European Commission, the Scottish Government and Scottish Enterprise. Tim has worked for Egon Zehnder International and McKinsey & Co. and a number of other highly respected organisations. He is a Board member of the Scottish Fire and Rescue Service, and Registers of Scotland, member of the Scottish Internet Domain Policy Advisory Board, and an advisor on the Buckminster Fuller Challenge. Tim is co-author, with Daniela Castrataro, of *Crowdfunding—Come Finanziarsi Online and Ideas to Reality* co-authored with John Reid and presenter of the 5* rated online training course Crowdfunding to Win available on Udemy. A graduate of Durham University, he also has a Masters Degree from Robert Gordon University, and an MBA with Distinction from ULSM when he also won the Ashridge Guardian International Essay of the year competition.

Contents

Part 1

The Crowdfunding Opportunity

Chapter 1

Introduction to Crowdfunding

Tim Wright, Dan Marom and Oliver Gajda

1. Have We Entered an Era Where Innovation Could be Democratised?

You may answer this question positively with the reasoning that we have always been democratising innovation. As consumers, we have purchasing power. Everyday, we choose products or services we want to buy with our hard-earned money. The most innovative offerings will often result in the highest number of sales and will continue to be produced by businesses, and undesirable commodities will be eliminated from the market. Unfortunately, innovation requires more than that. Crossing the "valley of death" for ventures requires prior funding.

You may also approach the answer to this question in terms of voting power, choosing elected officials who will support innovation through their regulation and tax policies or governments spending on research and development. Indeed, policymakers can do much in order to encourage, support, and nurture innovation. Nowadays, policymakers and the public can work hand-in-hand, in order to make it happen, create jobs and utilise the potential impact of innovation for job creation and economic growth.

Crowdfunding has revolutionised the way innovative ideas get funded. In the past few years, we have experienced a significant paradigm shift, with the emergence of the internet, social media, and proliferation of crowdfunding platforms. Consumers no longer have to wait for commodities to be produced in order to voice their opinions about them; and residents can do more than vote for electing officials who will decide which programmes and projects get funded. Crowdfunding has placed power in the hands of the people, taking a bottom-up approach to decision-making and financing ventures.

Crowdfunding is a method of raising capital for a venture from a large number of individuals, each investing or offering small amounts of capital, commonly through online platforms or social media. Crowdfunding is disrupting traditional funding models, opening up capital opportunities for start-ups and small and medium-sized enterprises that were shut out of traditional sources of funding, such as bank loans, venture capital, and angel funds. In order to fully understand the definition of crowdfunding and its implications on innovation, we will look at its evolution—of how crowdfunding was formed, the different categories of crowdfunding, the benefits and risks associated with it, countries and sectors increasingly using crowdfunding—and discuss and define civic crowdfunding.

2. The Modern Evolution of Crowdfunding

When websites were first being built, in the 1990s, Hotmail was introduced, Google was launched and the World Wide Web connected people from around the globe with the click of a button. The modern form of crowdfunding as we know it today was launched in 1997. A British progressive rock band, Marillion, became the modern-day crowdfunding pioneer by financing their tour to the United States through online donations from their fans. Marillion, unlike many bands at the time, embraced the internet. They kept in frequent contact with their fan base, fondly known as the "Freaks," through their website and a mailing list. The mailing list provided the means for fans to talk to one another and the band. Ten years had passed since the band's latest hit record and the band was strapped for cash. Mark Kelly, the keyboardist in the band, closely monitored the mailing list and answered fans' questions. In January, a fan asked if they will be touring the US that same year, Kelly responded that unfortunately the band was not able to tour the United States, due to the lack of financing from a record company. At that point, one of his devoted fans, Jeff Pelletier, suggested that the Freaks could pitch in a few dollars each in order to raise the money for the tour. Kelly replied that it would cost more than a few dollars each and that the cost would be around US$60,000.

Pelletier's idea was met with overwhelming enthusiasm by the Freaks, so Kelly assigned Jeff Woods to oversee the campaign and collect the funds. Woods was well known by the band and the Freaks and previously proposed to his wife on one of the bands tours. Legitimacy and trustworthiness were instantly formed when a band member from Marillion handpicked Woods to manage the funds. Investors knew their money was going to a trusted member of their group, so they were willing to transfer money to him. After a few months, the campaign managed to raise the entire US$60,000 and funded the band's tour to the United States. Not only was Marillion was impressed that the money was raised so quickly but also the band was elated because the investors also bought tickets for their shows.

Most of the donations that Marillion received originated in the United States, but some of the donations came from other countries, since the band rewarded a limited-edition CD of a song recorded on the tour for fans donating US$10 or more to the campaign. This was the most successful tour that Marillion experienced, not due to concert attendance, but as a result of the media coverage centred on the story of its unique funding campaign.

The first documented crowdfunding platform or website in the United States was introduced in 2001. At the time, it was referred to as a "fan-funding platform." Brian Comelio launched ArtistShare in the hope that it would enable fans to fund the production costs for digital albums sold online, under favourable contract terms for the artists. ArtistShare served as a model for other crowdfunding websites that developed in subsequent years, such as Kickstarter, Indiegogo, and PledgeMusic. Crowdfunding continued to grow in popularity and received more traction in 2014 with 375 crowdfunding platforms formed in North America and 600 in Europe.

3. The Four Categories of Crowdfunding: Donation Based, Rewards Based, Lending Based, and Equity Based

Through crowdfunding, capital was easily able to flow from the smallest or largest of investors in the market, to nascent entrepreneurs or well-established organisations that propose innovative projects that need funding. These platforms simplified, sped up, and lowered the cost of the fund-seeking process. We will look at four widely accepted models of crowdfunding: donation based, rewards based, lending based, and equity based and discuss how each of them streamlined the funding process.

The first crowdfunding model is donation-based crowdfunding. In platforms using this model, a project is initiated by an entrepreneur or organisation seeking funding, a marketing profile is added to tell the "crowd" or the readers on the platform what the money will be used for, and a funding goal is set. The funding goal is carefully selected and strategised, in order to meet the project initiators' funding needs to accomplish the project and to ensure that the campaign will be successful. Some platforms incentivise the project initiators to be conservative while picking their funding goal, by adding a penalty fee for projects that fall short of their funding goals in the time allotted. Some platforms establish an all-or-nothing funding policy, where the project initiator will not receive the funds raised if the funding goal is not reached. In this case, the investors or "backers" are not charged till the goal is met, instead they "commit" to or "pledge" a certain amount of funds. The reason behind the all-or-nothing policy is to protect the backers, ensuring that the project initiators have enough budget to fulfil the project scoped out.

Donation-based crowdfunding is commonly used in the public sector, it involves investors donating funds to a cause, charity, or person, without expecting a return on their investment. This crowdfunding tactic is regularly employed in funding projects for disaster relief, politics, or educational grants or to assist sick or poor individuals. The incentive for the charitable giving is usually a sense of pride, showing solidarity or support. A few well-known donation-based platforms are Kickstarter, Indiegogo, Crowdfunder, GoFundMe, and JustGiving.

The donation model is also common in the political realm. The first time that a donation-based crowdfunding campaign was used in the United States presidential elections was in 2008, when Barak Obama raised a remarkable US$500 million from 6.5 million online donations and his Facebook page was followed by nearly 2.4 million fans. Approximately half of the overall donation sum consisted of donations under US$200. Political crowdfunding not only resulted in him raising a record amount in private donations but also inspired individuals and communities to work together towards a common goal to get him elected. This has created a blueprint for a range of political activism and is now a common mechanism to fund a political campaign. Its popularity as a model is not simply for the funding. The outreach and communications inherent in a crowdfunding campaign are fundamental in driving awareness for a cause, policy, or issue. It is highly inclusive, giving the most minor backer a sense of direct involvement. It is highly insightful as the conversation around the campaign can create deep insight into supporters and activist's ideas. But, perhaps most importantly, being directly involved in the process through even a minor financial investment will typically translate into a turnout which is a currency as valuable as funding in a political context. This feature of a sense of involvement of ownership of a process is very much aligned with a great deal of tone and offer of wider digital disruptions, and it is one of the most important and enduring characteristics of crowdfunding in any context where much of its value is derived from.

The second model in crowdfunding is rewards-based crowdfunding. This model is very similar to donation-based crowdfunding, except in this model the investor or backer expects a return, reward, or perk for their investment. The reward may be in the form of an exclusive gift or being able to buy the product before its release date to the public, for a discounted rate. Popular platforms utilising this crowdfunding model are Kickstarter and Indiegogo.

One of the most memorable rewards-based campaign seen in the early days on Kickstarter was in 2012, for the Pebble smartwatch company. Eric Migicovsky ran the campaign to raise funds for the Pebble smartwatch before Apple watches were released. Migicovsky came up with the idea for the smartwatch, while riding his bicycle in the Netherlands and attempting to use his smartphone. With the assistance of a small team, he developed early prototypes, but the product was not tested in the market and finding investors for hardware products was challenging. He decided to turn to the crowdfunding platform Kickstarter. At the time, Kickstarter was only 3 years old and Migicovsky was unsure if his campaign would generate enough funds, because most projects were focused on artistic works.

Migicovsky placed a convincing video for his project and rewarded backers who pledged US$115 with a Black Pebble Watch. Early adopters were very enthusiastic about the reward, and the campaign managed to secure approximately 40,900 backers who received the reward. Within a mere 2 hours of Pebble's first campaign on Kickstarter, the company raised US$100,000, followed by US$1 million in its first 28 hours, reaching a record-breaking US$10,266,845 with 68,929 backers. Pebble's second campaign, Pebble Time, secured an even more remarkable amount of funding, US$20.3 million from 78,471 backers. Both these campaigns shattered the records for the most funded projects on Kickstarter and served as a model for other projects of how to create successful campaigns. The reward model is perhaps the most popular model, being relatively lightly regulated and highly accessible.

The third crowdfunding model is the lending-based or peer-to-peer crowdfunding model. This model is a direct competitor to financial intermediaries, such as banks. In these platforms, individuals or businesses seeking loans fill out an application, then after the crowdfunding platforms verifies and approves the loan, the loan gets categorised with a certain risk level. The loans are then funded by numerous investors who can invest small sums of money, for example, US$25, which will get repaid in periodic payments with interest. One of the largest lending-based platforms in the United States is LendingClub, with Prosper being the second largest. Lending club provides personal loans up to US$40,000, business loans up to US$300,000, auto refinancing, and patient solutions to finance healthcare costs.

Certain lending-based crowdfunding platforms target social causes, such as Kiva, providing microloans for low-income entrepreneurs and students in over 80 countries. In Kiva, investors do not receive interest on the funds given, they only receive a series of payments that repay the principal of the loan and then choose whether to reinvest those funds in another venture.

The fourth crowdfunding model is equity-based crowdfunding or crowd equity. In this model, the investor purchases private company securities through the platform, owning a small piece of the company. This is the most regulated type of crowdfunding, since almost every country around the world regulates the selling and issuance of securities to protect investors. Proponents of equity-based crowdfunding believe it provides greater access to capital, reduces geographic barriers, and enables more people to invest in small businesses through trusted online platforms. The model was pioneered in the UK where a common law system provided sufficient leeway for innovation in this field to make initial progress and platforms like Crowdcube and Seedrs exemplify the model. Jumpstart Our Business Startups Act (JOBS Act), signed by President Obama in April 2012, loosened long-standing federal restrictions on equity-based crowd-funding in the United States. Title III of the JOBS Act, which went into effect in May 16, 2016, allowed both accredited and non-accredited inves-tors to purchase equity in early-stage businesses, but limited how much can be invested during a 12-month period. Other types of financial instru-ments can be traded through these platforms and even civic projects utilise this model with the issuance of withdrawable community shares.

4. Benefits and Risks of Crowdfunding

Crowdfunding platforms do much more than fund innovative offerings. They build a global environment where innovative ideas can be cultivated from a diverse population and a variety of socio-economic backgrounds and can be applied in all industry sectors. Let us examine the benefits of crowdfunding for entrepreneurs or project initiators more closely. The strongest benefit is that crowdfunding platforms provide access to low-cost capital that can be processed quickly and efficiently online, connect-ing project initiators and backers across the globe. Budding entrepreneurs who would not be able to get funding from traditional bank loans, venture capitalists, or angel investors now have a way to access funding for their ventures. These projects span across all sectors, where small start-ups in industries that are less interesting for a venture capitalist may be of particular interest to other investors in the market who prefer to invest in projects that correspond with their expertise.

A second benefit to crowdfunding platforms is that they raise aware-ness to innovative commodities before they have been put on the market. Potential investors are able to see a marketing profile about the company and the planned commodities and decide if these are ideas that are worth funding. By placing the marketing profile up on the crowdfunding platform, project initiators can gain a deeper understanding of their consumers' needs and which demographic is most interested in the offerings. The companies' offerings can be enhanced and marketing campaigns can be further targeted. Collecting information and suggestions to improve the product or service from the backers and the general population on a crowdfunding platform is part of crowdsourcing.

The third benefit of to the project initiators is that crowdfunding has been shown to create jobs and open other traditional funding opportunities with venture capitalists, within 3 months of the campaign. Companies often use the funds raised on hiring new staff in order to accomplish the project.

Crowdfunding like other funding options also possesses its own inherent risks. For example, the project initiator risks lowering the business' reputation if the project fails to meet the campaign's fundraising goals. Additionally, the project initiator may think the offering is not attractive enough to consumers and stop developing it, when the marketing profile may have been the cause of low investor engagement. In the case of having multiple campaigns, donor exhaustion or fatigue may occur, where backers withdraw from funding the projects that they committed funds to in the past. Another risk that innovative businesses face is the risk of intellectual property being stolen, so they protect their products, services, and business ideas through patent filings, using copyrights, and trademarks.

5. Crowdfunding Offers Numerous Benefits to the Backers of the Projects

Crowdfunding platforms are easily accessible and invest in an early-stage company that has a low search cost. Having the projects listed on the platforms saves the investor time and energy of seeking out start-ups in the market that are looking for funding. Also, it is possible to invest in early-stage ventures in a specific sector where the investor is most familiar and comfortable. For example, if an investor comes from an agriculture background, they may want to invest in a company selling innovative farming equipment instead of a high-tech company. Another benefit is transparency of results in most crowdfunding platforms. For instance, in lending-based platforms, investors have access to the consumer credit asset class information, which in traditional financing was only available to banks and large institutions.

Even though there are numerous benefits to investors using crowdfunding, there are also a few risks. Information asymmetry can occur with equity-based crowdfunding in early-stage investments. To mitigate this risk, investors turn to "lead investors" with more information and connections to lead the crowd of backers. Information asymmetry can also exist in rewards-based platforms. To alleviate this risk, investors monitor the amount of capital raised as a signal of project performance.

6. Crowdfunding Utilised in a Variety of Industries

Crowdfunding offers opportunities for investors to fund ventures in a wide array of sectors. This is a great benefit to investors who may have had to rely solely on the choices presented by venture capital companies. These venture capital companies may have inadvertently or purposefully focused only on specific sectors and types of businesses to invest in, such as a focus in high-tech companies. Much like social media transformed the way the public consumes news from a variety of sources, investors have more options presented to them in terms of ventures to invest in. With the growing popularity of crowdfunding, in May 2017, the sectors receiving the highest total capital crowdfunding commitments from backers in the United States were Wine and Spirits, Food and Beverage, Sport and Fitness, and Entertainment and Media.

7. Global Crowdfunding

Billions of dollars are invested through crowdfunding worldwide in a variety of sectors. China by far has the largest cumulated transaction value in crowdfunding and holds the largest FinTech market in the world. Other countries leading in crowdfunding are the United States, Hong Kong, the United Kingdom, Australia, Japan, Korea, and Israel. So what led China to its rapid adoption of crowdfunding? Traditional state-based banking firms were unable to keep up with the demand for capital by businesses in

China, and with very few regulations, there was a rapid growth in lending-based crowdfunding platforms. Currently, with tightening of regulations in China, the industry is going through a consolidation process, where the number of lending-based crowdfunding platforms is being reduced.

8. Civic Crowdfunding

Crowdfunding is used worldwide, and another way that the concept of crowdfunding has been implemented is through civic crowdfunding by municipalities and governmental organisations. Civic crowdfunding is a subset of crowdfunding campaigns that produce shared goods, which are valued by the lenders and the community. Residents have commonly relied solely on municipalities or property developers to raise the funds needed for local municipal and community spaces, but civic crowd-funding facilitates individuals in the community to easily stay abreast of proposed developments as well as support those developments financially.

To understand the beginnings of civic crowdfunding, we can review the most commonly cited instances of early civic crowdfunding in the United States, which can be traced back to 1885. The French Ship Isère delivered over 200 crates of sculpted metal to construct the Statue of Liberty. Surprisingly the statue sat unassembled for over a year, due to the inability of government sources' to raise enough funds to build a granite pedestal required to display it.

Press magnate Joseph Pulitzer, who owned and operated *The New York World*, launched a fundraising campaign urging readers to donate to the construction of the pedestal and in exchange their name would be printed on the front page of the newspaper, no matter how small the contribution. The campaign was an extraordinary rescue effort. In just 5 months, over 160,000 people donated more than US$100,000 (worth approximately US$2.6 million today), which was enough to complete the pedestal.

Recently, civic crowdfunding platforms have gathered speed through-out major cities. Governments and individuals are able to propose projects like parks, libraries, or festivals through civic crowdfunding platforms, such as Spacehive, Citizinvestor, Neighborly, and IOBY. Spacehive is one of the largest civic crowdfunding platforms in Europe that has financed £4.9 million in projects, partnering with over 68 towns and local municipalities. One of the successful projects on Spacehive was The Peckham Coal Line project in 2015, where local residents Nick and Louise identified an unused coal line that could be transformed into a beautiful elevated park that would unify two communities separated by the coal line. Nick and Louise, with the help from other residents, outlined clear steps, timelines, and costs for a feasibility study, marketing campaign, and events on Spacehive. Their campaign was met with great interest from over 500 residents who attended their events. Support continued to grow quickly for the campaign through social media and press coverage, securing backers beyond local residents. They surpassed their funding goal of £64,132 and raised £75,757, through 928 backers.

9. Crowdfunding and Policy

For policymakers, a key question in the search for economic growth and stability is how to foster the creation and funding of innovation, how to help small and medium-sized businesses, individuals, or other small and also non-profit organisations to access adequate financing when pursuing ventures of positive social, technical, or economic impact. In the very early stages, innovative ideas usually used to be fully financed by the entrepreneurs themselves, their families, and their friends. Crowdfunding has proven to be able to provide risk diversification to innovators, start-ups, and micro-entrepreneurs.

Crowdfunding has been by now shown to be an adequate tool to fund capital requirements of established businesses, real estate and renewable energy projects, civic developments, as well as art and culture. It can be used before and as a supplement to investment from government funds, business angels, venture capital funds, and bank finance, while enabling entrepreneurs and innovators to either grow their idea organically or to scale it fast. Both the financial and non-financial benefits have increasingly attracted the attention of not only a wide range of policymakers but also local and regional authorities, development agencies, and supranational organisations. Some are partnering with crowdfunding platforms, others are still exploring opportunities, while yet others remain sceptical.

The flexibility of crowdfunding allows for an array of engagement options for policymakers.

In general, one can differentiate non-coercive (soft) and coercive (hard) policy instruments. Soft policy instruments might include public acknowledgement or event participation, topical publications, and other general supportive actions. The hard engagement forms would be more tangible, for example, in the form of financial involvement through co-funding or financial support to the sector. And they might include law-making and regulatory oversight. The right choice of policy and the appropriate balance between soft and hard instruments is key for any intervention to develop into a local or national crowdfunding market.

Policymakers seeking to make the right choices of instruments at regional, domestic, or international levels need to understand which measure, for example, regulatory, voluntary (self-regulatory), economic, or information instruments, will deliver the policy goals. Of course, there has not yet been a method developed for simply defining and applying "soft" or "hard" policy instruments within the crowdfunding sector. In practice of course, we are likely to see the application of mixed policy instruments combining both soft and hard approaches.

The right approach is always dependent on the state of the market, the overall responsibilities of the individual government organisations, regional development or cohesion policies, and, of course, overall doctrine of individual governments. In this book, we cannot answer every possibility that a potential reader might face, but we can provide a mix of examples and a tool box for the interested party to delve into and take from.

To provide another view, we can also categorise potential policy instruments between macro-, meso-, and micro-level measures. In such a scenario, macro-level engagement would try to shape the overall environment of the crowdfunding sector within a given country or union of countries, for example, by creating tax incentives, applying regulatory frameworks, and incentivising voluntary self-regulation and industry best practices. It might also include soft actions that are aimed at the development of an overall market by generating knowledge and trust within investors and the financial markets to welcome crowdfunding.

As an example, we can point to the European Commission proposal for regulating crowdfunding under the European Crowdfunding Service

Provider Regulation for businesses from 2018, which would address legislations across all European Union member states alike. On the national level, we can point out regulatory sandboxes created by the UK government for FinTech development, which also contributed to the testing of crowdfunding models or relevant tax incentives by the French government for community-linked renewable energy funding. On top of this, there have been numerous regulatory approaches by national lawmakers over the past few years across the world.

The meso-level policy instruments would encompass measures that would aim to address the market via a general framework, which would generally be designed to cover a defined aspect of the market. Examples would be hard instruments such as co-investing or match-funding schemes, in which government bodies, such as development banks, provide financial involvement to market players in the crowdfunding sector via a defined set of parameters. This would likely either happen as a loan guarantee, an investment into a loan, or as an equity portfolio or grants and also support the operational financial instruments to crowdfunding platforms. Soft measures could also include specific capacity building or educational efforts focused on partial aspects of the market, such as efforts to improve reporting or transparency within the industry.

The European Investment Bank has already actively made asset allocations to lending-based crowdfunding platforms with the European Fund for Strategic Investments (EFSI) providing loan guarantees. The bank is also seeking to expand existing support mechanisms for other forms of alternative finance, such as business angels or microfinance, to include crowdfunding. There are also regional activities throughout the world, where national or regional development banks and other government organisations actively engage the sector with some form of financial support or incentives. An example of a soft intervention is the *Assises de la Finance Participative*, the French crowdfunding sector's key policy event, hosted in the years 2013, 2014, and 2015 by the French Ministry of Finance in Bercy, Paris. In 2015, Emmanuel Macron, the current President of France, then minister for economy, industry, and digitalisation, joined the discussion in support for the crowdfunding sector.

Lastly, on the micro-level, policy instruments will be either regionally limited in reach or highly specific in application. Here, a mix of soft and hard instruments can be applied, including capacity building and training, awareness creation through events and publications, topic or region-specific micro-grants to support individuals in creating crowdfunding campaigns or investment-ready programmes for entrepreneurs. Overall, we notice increasing regional cooperation between government organisations, often development related, and crowdfunding platforms in support of specific regional sectors, such as the creative industries or start-ups and entrepreneurs.

Examples include development agencies and development banks working in cooperation with crowdfunding platforms, such as the business development banks NRW Bank and IFB Hamburg, both based in Germany with crowdfunding platforms in support of the creative and arts sector, or Investitionsbank Berlin and L-Bank, both based in Germany with a loan programme for start-ups and small and medium-sized enterprises based on market proof via crowdfunding on a dedicated platform.

For the above discussed areas of policy engagement, there are plenty of examples that can be found across the world, and a good selection is found within this book. We have categorised the book in a different way, focusing on local, national, and pan-national examples. However, sometimes a clear distinction is not possible, and we had to make judgements that may not always seem to fit a reader's view. Therefore, we ask the readers to approach the categorisation we have made here with an open mind and to explore other possibilities on their own, by cross-reading.

Part 2

Our Framework

Chapter 2

The Framework

Tim Wright, Dan Marom and Oliver Gajda

1. The Origins

In the past 10 years, crowdfunding has developed beyond what even some of its most ardent advocates might have expected. While making public calls for funding is not a totally new idea, and always part of the public discourse, it has grown to become an accepted, courted, and valued part of the financial markets across the world and in ever-increasingly diverse models and sectors.

This outcome is fundamentally a result of a combination of societal, technological, and economic circumstances. The desire for individuals to empower themselves and take ownership of activities that were once permission based has changed our view of what is possible in many areas of life with crowdfunding being the financial manifestation of that yearning. The technology that allows us to quickly, cheaply, and directly strike up conversations and transact with an extraordinarily dispersed crowd opens both the practicality and capability of vast numbers of people to become directly engaged in support and investment in a way that was previously unimaginable. And while crowdfunding, in its modern incarnation, did exist before 2007, it is undoubtedly the case that the financial discontinuity of 2007 transformed our collective appetite and need to explore new options in finance. Traditional sources of capital dried up building a demand-side impetus, while the search for returns as crashing interest rates drove down retail investment opportunities, primed a supply-side effect.

The ecosystem resulting from this disruptive period has created a fertile and energised environment for financial innovation, civic engagement, and collective action.

The growth and sense of its permanence has meant that crowdfunding, in its broadest sense, has gone beyond it being perceived as simply a novel quirk used in exceptional and uncommon circumstances to being increasingly regarded as a mainstream, recognised, acknowledged and, in some cases, the preferred finance option for a range of individuals and organisations. It has introduced a new and novel opportunity for engagement and inclusivity. The growing recognition that, in common with many other digital disruptions, the principles and ethos that underpin crowdfunding indicate that it brings not only finance with different characteristics to the marketplace but also the process of a highly public and accessible call for funding also yields a range of immensely valuable additional benefits including insight, communications, and network development.

This can create extraordinary and significant additional value from well-run crowdfunding campaigns and offer a range of opportunities for governmental, public, and civic partners and policymakers who are increasingly aware of this.

The result of this remarkable growth in scale, popularity, and impact is that crowdfunding is now a topic that policymakers can no longer afford to ignore or put off to another day. The empowering nature of digital technologies has prompted a groundswell of activity from grass-roots activists to create a circumstance where the policy agenda, and strategic imperatives that go with it, cannot ignore this crowd-powered activism. Indeed, in some cases there is a growing recognition that, far from being a disruptive threat, it can be harnessed for strategic and disruptive advantage and provide a powerful mechanism to address long-standing and entrenched challenges.

The role of the policymaker to both nurture and benefit from this opportunity is now too compelling to ignore.

But still, the question for many policymakers is—How best to respond to this opportunity?

The necessity to engage is driven by a number of factors as follows:

1.1. The Opportunity

The emergence of such a common and widely used mechanism is an extraordinary opportunity for policymakers to use this "ground up" grass-roots and empowering activity to advance the wider aims and main strategic goals of the policymakers' organisation, remit, and polity. As crowdfunding becomes part of the established ecosystem, it is incumbent upon policy-makers tasked with the delivery of a range of outcomes from service delivery, civic engagement, economic development, and social innovation, among many others responsibilities, to explore opportunities as to where they can make use of, and to intervene in, this growing activity to further those aims and to do so in the most effective and efficient manner.

1.2. The Regulatory Responsibility

In keeping with many other digital disruptions of the recent times, much of the enduring and profoundly challenging aspects of crowdfunding are the way in which it confronts norms and behaviours that have persisted over long periods of time. Digital technologies have "empowered" individuals and groups to take ownership of activities that once required either "permission" to be granted by gatekeepers, some form of intermediated processes, or were constrained by limited geographical reach. Many of these behavioural and operating norms are reflected in, and supported by, policy, regulatory, and legal statutes. With the integration of allied services and processes, there is tendency for a well-understood and, to a large extent, ossified framework to emerge over a period of time, which rigidly defines how certain activities operate and, in so doing, to determine the art of the possible. This can both intentionally and unintentionally limit innovation and constrain ambitions to only the understood and recognised models and can create long-standing inequality of opportunity. It leads to the development of entrenched challenges that appear to be insurmountable in the "cage" of existing norms.

The emergence of digital technologies has significantly challenged many of these established norms through their ability to empower grass-roots and bottom-up activity and bypass, or step outside of, the established norms and, in so doing, change the art of what is possible. We can see this in many sectors where digital technologies have disrupted convention, and crowdfunding is really just the financial manifestation of other similar changes which have transformed our understanding of retail, distribution, and other sectors which had for many years seen little or no significant change.

By running ahead of these frameworks, the pioneers of digital disruptions, be it in the music industry, publishing, and transport sectors, have forced a reexamination of the regulatory landscapes to take account of the new art of the possible and, with that, a reexamination of what is an appropriate or even relevant framework for such a changed environment.

It has also begun to teardown frameworks that had developed into barriers for entry, defending the position of established market players and making it difficult for new entrants to offer novel thinking.

For many policymakers, the regulatory responsibility can run from a local, regional, and national remit to a pan-national dimension.

The approach and range of intervention are wide and varied but can be broadly divided between the extremes of those seeking to review, renew, reform, and reimagine their environment to those who tend to constrain, define, or protect their environment.

But the underlying imperative is to be seen to "respond" to the changed circumstances even if that results in a considered declared decision to not act or to reinforce the status quo. Simply ignoring the reality of a changed circumstance is no longer a sustainable response.

Crowdfunding is no different than any of these other disruptive activities in that it does impinge on, is subject to, or forces the reconsideration of, regulatory frameworks, codes of practice, and established norms.

Regulation always runs behind innovation, but its impact on innovation can be profound in both positive and negative ways, but for policymakers with a responsibility for shaping these matters, it is a task which in most cases cannot be ignored.

The regulatory response will define the operating regime of these changed circumstances and the range of options available can simply reflect that no change is necessary, in order to create a circumstance to provide adequate controls and safeguards to those operating in the sector or even to seize an opportunity by amending or changing frameworks to embrace and encourage the emergence of the new models which are made possible by digital innovation. For us, the option to seek change and renewal is the most attractive in almost every case.

1.3. The Need for Integration

The temptation in the early days of crowdfunding was for many established actors, be they policymakers or otherwise, to regard the crowdfunding phenomenon as an aberration which sat outside of the norm and was pursued by a small group of odd balls in highly specific circumstances. In many cases, the implicit thinking was as follows: it is a phase that will pass and it is best to wait and see as, chances are, it will go away. As crowdfunding has permeated so many aspects of society, it has established itself as part of the mainstream. In much the same way as Social Media was originally viewed as a separate, distinctive, and niche activity, it is now recognised as part of the fabric of our existence and so touches on and integrates with a range of other related and parallel activities and actions. This has created much debate, for example, on the impact that it has had on politics, news, public discourse, and marketing. Crowdfunding is much the same in that it is having an impact in areas and aspects of public life beyond where it might once have been seen to reside. Entrepreneurship, innovation, inclusion and equality, and creative activity are, to name but a few, all affected and invigorated through the emergence of crowdfunding and the changes it brings.

The time for considering it as a potential passing and transitory presence is over. Similarly, the temptation to treat it in isolation, apart and distinct from the portfolio of activities and responsibilities in the policymakers remit, is a mistaken one and the necessity must now be to properly integrate it to complement and mesh with this nexus to make it most effective and properly aligned.

For all these reasons then it has become increasingly apparent that as crowdfunding has both grown and become an accepted part of normal daily life that there is a need for policymakers in their widest of forms to act.

But we restate it, the common and abiding challenge for many has been "how best to respond?"

In our work as crowdfunding experts and consultants engaging with a wide range of organisations and individuals across the world, this reality has become ever more apparent and the need for guidance, mechanisms of analysis, and exemplars to guide policymakers in this crucial question has never been more urgent.

Governments, pan-national and supranational organisations, local governments, social and civic bodies, even larger charities, and NGOs are increasingly aware that there is work to be done. But, for many, it can be challenging to undertake that process and finding approaches that work in a structured and clear way has proven elusive for many.

In some cases, the challenge for policymakers is founded in a lack of awareness or insufficiently detailed understanding of what the crowdfunding opportunity looks like to take adequate action. This can mean that the opportunity that is present in the unique characteristics of crowdfunding can be missed or misunderstood. It can also mean that the lens of analysis used is one that is founded in what has gone before and so lacks the necessary calibration to be appropriate in a changed circumstance.

In other cases, there is a necessity to try to retrofit this new phenomenon into an established and functioning strategic approach, and to do this without the new ideas distorting or overwhelming the original plans becomes the imperative. How to avoid, what we might call, "the tail wagging the dog."

Often, the policymaker is also faced with the need to act with some urgency in a context of an established and demanding programme of work and eternal insufficiency of bandwidth. It is also common that action is required but that the supporting resources of time and funding are limited.

In helping our clients to answer these challenges, we developed a series of approaches to work with them and help them to find elegant, effective, and aligned solutions. Our aim has always been to assist them to understand, embrace, and integrate the crowdfunding opportunity for maximum benefit and for the most positive outcome within their areas of responsibility, whatever they might be, and to do that in a timely and managed fashion.

We have condensed the insight gained from these experiences into a series of frameworks and techniques, which have served to make this process easier, quicker, and better for us and for a wide range of clients.

It is these core frameworks and their application that we are sharing in this book in order that you can use them to undertake this process of exploring planning for, integrating, and making the most of the crowd-funding opportunity for yourselves.

If you have already begun that journey of exploring how crowdfunding and crowdassets can advance your policy agenda, you will almost certainly still find value in applying the lens of the frameworks and the accompanying techniques found in this book to analyse what you have done to date and to search for additional opportunities and values.

The strength and appeal of frameworks to consulting generally is in their ability to quickly bring structure to an often-complex set of circumstances and make that situation navigable and actionable. In so doing, they create a situation where a set of possibilities and options can be translated into activities that can be quickly acted upon.

We have sought to bring that same ethic to these frameworks and techniques. This can range from a quick set of structured interventions to a one-off audit and evaluation process or a complete comprehensive strategic and interconnected set of activities that allow the organisation to develop a much more comprehensive solution. They are then approaches that can operate at a macro- and micro-level as required. But at its heart, the intention is to provide a concrete and reusable lens of analysis that is comprehensive in its scope yet simple in its application.

By having a toolset as approachable and as flexible as this, we find that its application is very wide and, as such, is relevant and useful to a wide range of organisations. By extension, the range of what and who we are able to consider policymakers is a broad one. The default standpoint of what constitutes a policymaker tends to point to a more governmental, political, and legalistic one. Many people, when presented with the word "policymakers" would typically imagine we mean only politicians or those they task to carry out on their policy commitments. But, we would challenge this notion.

The truth is that a responsibility for, or influence on, policy is a much more widely found responsibility than at first sight. The ability to shape the circumstances and operation of an activity from the international to the local level is a wide one, and the ability to influence the effectiveness and use of activities in a range of different contexts is much more widespread than often first imagined. The key characteristic is the realm over which the policy gift and decisions run. With that in mind, we can imagine a huge diversity of circumstances where this might fall within the purview of an organisation, team, or individual.

It is then, in most cases, the setting of strategic aims and setting of the context where the tactical activities of others can integrate to advance that strategic ambition.

To all of these communities and individuals with a responsibility to create the circumstance where crowdfunding can take place or its operation can advance your strategic ambition, we find the application of the framework has brought value and clarity of action and that this breadth of application has been demonstrated in our work with clients across a wide and varied portfolio.

The purpose of the framework is to provide a mechanism for analysis, sense-making that leads to action. The actions are not prescribed or defined beyond high-level thematics thus leaving tremendous flexibility for interpretation and accommodation of local circumstances. Indeed, the cases in this book, collected from across the world, are testament to the diversity of activity that is already taking place as policymakers in many guises respond post the opportunity. But the framework is robust and repeatable and of great assistance when considering the question we began with—How to respond?

2. Using the Framework

The strength of the framework is, in many ways, its simplicity complimented by its rigour and completeness. It is this openness which gives it wide applicability and flexibility. But, in order to operate it successfully in its myriad of options, it is important to have a sufficient base level understanding of crowdfunding and its context to make the most informed choices when using the tools.

There are in fact two stages of understanding. Firstly, understanding the nature of the opportunity, and secondly, the scope and actions that might apply to the circumstance.

Primarily, it is important to grasp what we might refer to as "The Opportunity of Crowdfunding" By this we mean that, at a high level, how we articulate the broad areas of value that crowdfunding, and by extension crowdassets, can create.

3. The Crowdfunding Opportunity

A phrase that we have used in all the time we have been engaged with Crowdfunding in its many forms and one which becomes apparent once anyone has ever undertaken a crowdfunding campaign is that "crowdfunding is about so much more than money!" It is this proposition which sits central to the crowdasset idea. The crowd is the source of value in a wide and diverse sense and the activity of crowdfunding is merely a mechanism to operationalise the engagement.

It is a crucially important truism and extremely important in the context of considering why and how policymakers should want to intervene in facilitating the crowdfunding opportunity for achieving the best effect. Without this basic knowledge, the interventions are likely to be limited in scope and, quite possibly, overlook opportunities to yield significant additional value and return. By the same token, it may be difficult to plan, monitor, and value the impact of interventions to the crowdfunding sector without a sound and comprehensive understanding of the direct and indirect impact crowdfunding is already having on both those who undertake it and the economies they operate within.

To address this, we will introduce the mechanism we use to expose the opportunity and value of a crowdfunding campaign to those considering running a crowdfund and simply adjust it to serve the needs of the policymaker. We have found that we can use this same structure in the context of explaining what crowdfunding can deliver to the policymakers' aims as well. After all, what we are seeking to do in both circumstances is align a crowdfunding effort to deliver value against a strategic plan and hence make a crowdfunding campaign return maximum strategic and tactical value to a crowdfunder as well as to the policymaker.

The aim of the exercise when used with a crowdfunding campaign manager is to ensure that they place the campaign centrally and strategically for the organisation and that they ensure it addresses the key needs of the organisation. This is intended to subordinate the campaign to the larger and wider aims of the organisation so as to avoid the common problem of a resource-intensive and demanding process, like running a crowdfunding campaign, overtaking, and distorting the aims of the organisation. It is also an essential step if the significant investment of time, effort, and resources required to run a successful crowd engagement campaign achieves the maximum return on investment.

The simple schema we use to unpack and illustrate the opportunities for value creation via a crowdfunding campaign suggests five key zones or groups of returns for putative crowdfunders.

We describe these as Funding, Insight, Networks, and Communications with a central fifth aim of Growth. It is worth noting immediately that only one heading deals with that aspect which most would commonly associate with the crowdfunding process, namely funding. By ranking each of the five categories as being of equal potential importance, we are emphasising the diversity of return available.

It is not so hard to understand these categories of return when you reflect on the diversity and size of the addressable crowd on offer to you in a crowdfunding campaign. But the process of opening the eyes to the range of values available also simultaneously exposes the enduring value of the crowd as an asset and the importance of taking a longer term view of how to nurture, protect, and develop that crowdasset.

As with many successful analysis tools, it is the simplicity and openness of these categories that ensures their widest applicability. We can utilise that openness in the context of assisting policymakers to understand their opportunities in precisely the same way that we explore outcomes for any other individual or group considering crowdfunding.

As the understanding of crowdfunding matures in the marketplace, we can thus see more examples of crowdfunders running very strategically aligned campaigns that are actually intended to yield very specific and valuable returns that are not financial, and the financial total sought is a much more notional target and a simple vehicle to operationalise the quest for the other benefits of such a campaign. This is, in our opinion, neither wrong nor ill-advised. On the contrary, it is a tacit acknowledgement of the power of crowd engagement and the importance of crowdasset.

Policymakers are presented with the same opportunities, and their engagement with it should be informed by the same clarity of outcome.

By using the same lens of analysis, they can bring greater understanding of why they are engaging and what they wish to gather from it prior to their decisions on the mode of intervention in the sector.

For each of the five headings, there are a wide range of additional areas of focus and specificity to be developed. Each detail will be of immense importance or value specifically to the unique circumstances of each crowdfunder. The analysis and specification process for each individual campaign is an important, early, and necessary step if the plan and structure of a crowdfunding campaign is to be made to reflect and deliver on the specific objectives and needs of each unique campaign.

It is all too common to encounter crowdfunders to express the idea that they are merely "seeking funding." But, even at an elementary level, the questions of what scale of fund and the characteristics of the funds sought are clearly likely to vary significantly depending on the individual business' point in its life cycle. Also, importantly, this will not be the sole source of funding for any meaningful initiative, so how might this funding round fit to best effect into a wider funding strategy? These are just scratching the surface of the legitimate questions to be asked at this point in order to progressively drive out the specifics, accuracy, and rigour necessary for any credible campaign. But it demonstrates how we can begin by using the framework as a simple sorting mechanism to progressively drill into successive layers of detail and specificity.

But, for all of the importance and often centrality of the funding aspect to many campaigns, it is worth noting that there are four other headings which can, and often do, yield as valuable or in some cases more valuable returns than the finance alone.

For example, it might be that the organisation is seeking to validate some figures in a business plan like cost of customer acquisition, acceptable product price point, or what is the optimum format for their new product? All exceptionally valuable pieces of Insight which can all be delivered via a well-run and targeted crowdfunding campaign. Similarly, they might also wish to develop aspects of their connectivity and network by sourcing new partners, distributors, and manufacturing options and in so doing develop the Network aspect of their crowdfunding campaign. Many examples of exactly such outcomes can be cited to validate the advantages of such an approach, and it is both legitimate and appropriate to target these aspects in a campaign as not only is it one of the best arenas to find the answers but also it would be an oversight not to take advantage of the opportunity when investing such a significant quantity of time, effort, and resource in a campaign.

The final and perhaps more obvious communications opportunity of a highly visible and public activity like crowdfunding almost goes without saying. But it is the breadth of communication and engagement that can be undertaken and achieved in the course of a campaign that should be properly explored. The options are diverse, and each crowdfunding project owner can specify what they wish to achieve, be it public relations, visibility, campaigning, or other outcome and to plan for it accordingly to meet the specific needs and context of their campaign.

If we were to refer to the Arnstein's Ladder, we can certainly consider crowdfunding being a vehicle that can be a means of driving engagement on the upper rungs. Similarly, we can understand how as a campaigning and consulting tool it has enormous possibilities as illustrated in a number of our case studies included in the book.

It is not our intention to set out here all of the potential approaches to driving out details for a crowdfunding campaign as its execution is not the thrust or focus of the book. However, we argue that the same framework can, and should, be used as a precursor to developing intervention options as it is possible to use these same areas of potential return as a mechanism of developing clarity about how crowd engagement activities can deliver value and how policymaker interventions can best enhance the most effective and optimal outcomes of the crowdfunding activity to advance strategic policy aims.

This frame of reference is also useful in helping the policymaker understand both the broad returns the crowdfunding community might be seeking and to develop the specificity within that. This can be extremely helpful when we reach the exercise of "auditing" based on the Policy Intervention framework, as it can surface other opportunities to align existing offerings and policy to meet, address, and enhance these broader outcomes. But this will be explained in greater detail below.

In the context of a policymaker, it might be that we adjust the semantics a little more to appropriately capture the typical remit or include subheadings more appropriate to that group. Funding might be adjusted to

Economic Value for example, and Communications might include Civic Engagement as a specific subheading.

But the same framework is of use in helping the policymaker frame specifically what crowdfunding might also be capable of delivering to them in a practical and operational manner. For example, it is entirely reasonable to apply the principles and use the same five headings to describe a broad level of potential avenues of value creation for the policymaker and their areas of responsibility and polity more generally.

For example, and in its simplest and crudest form, in the context of constrained funding, crowdfunding can indeed provide an alternative, additional, and distinctive source of funding. In terms of insight, a campaign can be a powerful vehicle to drive understanding of perception and sentiment analysis and co-creation. From a communications point of view, crowdfunding can be a powerful mechanism of civic engagement and narrative delivery. Networks are similarly powerful in terms of finding like-minded delivery partners and collaborators. And to round off the five, the possibility for Growth to emerge from these combined "zones of return" is significant. And we can consider these emerging returns as part of a collective outcome from many campaigns run by third parties encouraged or facilitated by policymaker interventions or indeed where a crowdfund was directly run or facilitated by policymakers (see Figure 1).

The same principle of gradual drilling down to develop specificity and opportunity works in a civic and public policy context as well as it does

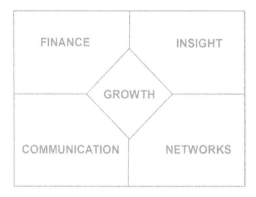

Figure 1: *The crowdfunding zones of return.*

for a commercial venture seeking to use crowdfunding. At both a broad policy level and at a tactical level, it is perfectly possible to see how this starting point helps to clarify where it might be useful to engage with crowdfunding in a very direct sense by defining its contribution and at a tactical level for more target possibilities and thus to act accordingly. It also allows the richness of understanding to consider the facilitation and nurturing aspect of policy intervention intended to facilitate others to undertake crowdfunding more effectively but with the express intention that by so doing they are contributing to a broader objective of the policymaker. This can be identified through this process of analysis and be knowingly and specifically targeted.

4. The Intervention Framework

The core policymakers framework, which we describe as the Intervention Framework, is similarly straightforward. As in our previous model, it is intended to act as a frame of reference, a facilitator, and not as a mechanistic answer provider. Founded firmly in the sense-making tradition, it is intended to offer a starting point to segment and shape thinking and, as a holistic framework, to begin and complete an analysis process, working towards a series of actionable steps based on sound and demonstrable reasoning.

At its heart, the core framework assumes that the starting position is that an organisation and the policymakers who operate within it have become aware of crowdfunding and wish to determine how crowdfunding can either directly advance their wider aims or how the crowdfunding opportunity can be managed or enhanced for the best and widest benefit of those who fall within the purview of the policymaker.

In our experience, the default positions for many policymakers fall into two main camps. In many cases, it is that this emerging practice, being novel, is necessarily regarded with suspicion and the imperative is to constrain its operation through regulation for safety and to chase it back into an established regulatory framework which makes it suddenly an "understood" phenomenon. This is in our view highly detrimental to the development of novelty and will lead to missed opportunity. In this case, we would seek to try to encourage a more positive and innovative understanding and open up the possibilities that a more supportive approach might yield.

The alternative and, in our view, more positive stance is to be enthusiastic, open, and impatient to embrace the potential and opportunity it represents within the context of the strategic policy agenda.

This more enthusiastic response, while welcome, can, however, sometimes lead to what can be described as "lets put the show on here!" approach, whereby organisations enthusiastically embrace the opportunity by imagining that they themselves should in some way run their own crowdfunding campaigns. This is particularly common among organisations that already had some responsibility for a sector or activity where raising funds forms part of their remit. This type of approach may well be exemplified by many larger charities and NGOs for example.

While these may be entirely understandable and obvious responses, as they play to their strengths and understood competences, it is also a highly tactical, narrow, and somewhat unsophisticated interpretation and application of what can be done. To be clear, the results can be valuable, useful, and demonstrable, so our argument is not that this approach should never be adopted. Indeed, many examples in our collection of inspiring cases certainly include examples of exactly such an approach to a lesser or greater extent. Our assertion is, however, that while these are all good initiatives which demonstrate the willingness and enthusiasm, there is often a wider and more sophisticated agenda to be developed from these foundations and adopting a more strategic and holistic approach will yield much greater value in the longer term. At its very least, any direct intervention to actively run crowdfunds should always be informed by the Zones of Return which we set out above and that these are considered in the widest possible context for the community in the policymakers realm and its main strategic aims.

Generally, the focus of running a single crowdfund is simply far too narrow and is based too much on the crowdfunding activity itself. It is the wider context within which crowdfunding is taking place and, perhaps most importantly, the breadth of impact and return crowdfunding can actually yield where the transformative opportunities lie.

The narrow approach is not dissimilar to the first time crowdfunders' response of focusing solely on funding to the exclusion of the wider returns to be won from a crowdfunding campaign and so the crowdfunding campaign should be run as an isolated project instead of an integrated, strategic, problem-solving, and value-creating activity for their organisation.

In the same way that we attempt to counter that narrowness of thought through the use of a framework which presents a much wider canvas to develop a richer picture, so is the intervention framework an effort to broaden the thinking around the options to intervene for the policymaker (see Figure 2).

It is important then that the policymaker has the same comprehensive and holistic approach to analysing and defining what the best interventions might be, and thus, this framework sets out the four main pillars of intervention. Each pillar is a placeholder for a diverse range of approaches and potential interventions but still effectively compartmentalises them

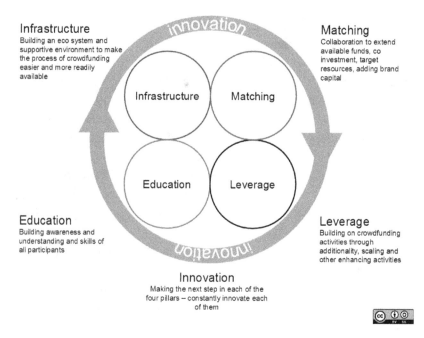

Infrastructure
Building an eco system and supportive environment to make the process of crowdfunding easier and more readily available

Matching
Collaboration to extend available funds, co investment, target resources, adding brand capital

Education
Building awareness and understanding and skills of all participants

Leverage
Building on crowdfunding activities through additionality, scaling and other enhancing activities

Innovation
Making the next step in each of the four pillars – constantly innovate each of them

Figure 2:   *The intervention framework.*

as being of one of four distinctive types and, in so doing, is a highly accessible and understandable starting point. We call these four pillars:

- Education
- Matching
- Ecosystem
- Leverage

Under each of these four broad pillars, we can list examples of a wide range of distinctive and specific actions and interventions each of which can individually or in combination have an impact on the development, growth, and operation of crowdfunding and, through the strategic alignment process with the use of the framework, advance the overarching aims of the policymakers by creating a circumstance where crowdfunding is most effective and most possible for their community. The context in which these four groups of interventions and activities can be developed and undertaken is one of constant evolution and development as a result of ongoing and necessary innovation in the intervention method, its aim, and its mode of delivery. The opportunity for creative innovation is expanding in many ways and is doing so constantly, not least, as new technological tools allow us to do more. Again, a cursory glance at the many case studies included here speaks to the diversity of approach in what is only a limited set of exemplars from a rapidly growing field. It is interesting then that among our examples we have examples of very well recognised interventions from a traditional policy perspective, and some more contemporary ideas, for example, embracing new technologies like BlockChain to further develop the range of possible interventions.

By creating these four distinct and comprehensive groups, we can begin to bring elements of order and analysis to an otherwise complex and disordered landscape of possibility.

This "sense-making" approach is distinct from many other frameworks which set out to define the approach and actions for you. Such frameworks, in our view, can be too prescriptive and limiting. Our approach is to help users to navigate a complex environment allowing the actors to respond to all of the contextual circumstances only they are aware of and, in so doing, arrive at a course of action that is appropriate and specific to them. Clearly, skilled facilitation can make this process quicker and better, but it is a framework which is in the best knowledge management tradition in that it is a self-help tool and it assists the actors to synthesise a plan to act based on their own unique circumstances and draws on their own information and understanding.

The application of the framework is also broad and not limited to simply defining and planning a set of actions and interventions, although it is this challenge which compelled us to create it in the first place. What we have found is that the framework has, to our knowledge, at least three other forms of use and yet more may emerge over time. But each is of considerable value, and in each case, it is the simplicity of the model which allows it to be used effectively to bring structure to an often otherwise messy circumstance.

The main uses of the framework are as follows:

- Analysis—Through its broad categorisation model, it can be used as a tool to better integrate crowdfunding into existing activities and interventions already being undertaken by the policymakers or their agents for other strategic objectives but which might be relevant to or aligned with encouraging crowdfunding,
- Audit—Again, using the broad categorisation to set and monitor progress and impact as the intervention plan and process is operationalised,
- Benchmarking—As a lens to evaluate what similar organisations are doing in the crowdfunding space and to use that insight for comparative purposes.
- Planning tool—This is in fact its primary use, as a mechanism for creating a series of linked and considered interventions by defining the four main options for intervention. This is in many ways a combined analysis and planning mode whereby a gap analysis could be undertaken as to what is missing or what could be improved under each heading and within the gift of the policymakers remit in order to develop a plan of defined interventions.

What is increasingly apparent as the use of the model matures is that each of these activities can be supportive of each other and can be aligned into a series of interconnected and supporting phases in developing interventions.

The sequence and completeness of use of the framework is entirely the choice of the client organisation, and while again a good consultant can look to optimise the process, it is entirely reasonable for the organisation to use the framework in what order and completeness they might wish to, in order to satisfy their own needs. It might, for example, be only used as a benchmarking tool to analyse what similar bodies are doing. Similarly, it might be that an organisation may wish to use just the planning and analysis applications of the framework and, if so, then this is fine. There is no mandated or rigid approach to its use, and its strength and utility is on its robustness, simplicity, and breadth of application (see Figure 3).

Let us deal with each of these four methods of application in a little more detail.

4.1. Analysis

While the origins of the framework lay in the need to define a simple to consider set of intervention groups, we have found that its application has grown over time and one popular model of use is that of an analysis tool.

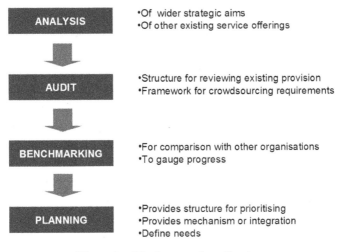

Figure 3: *The framework applications.*

The great strength and value of this step is that it can be a significant aid in placing crowdfunding within an existing set of provisions and interventions. There is a common tendency to regard any novel or emerging activity or approach as being somehow out-with the existing order of things. Its very novelty tends to isolate it and so it remains an external, vestigial, and vulnerable initiative. Crowdfunding is not immune from this. In many cases, it is "othered" as having a certain uniqueness that can prevent approaches to it being integrated or aligned with other activities. Quite apart from the many other issues with such a dislocated approach, it misses the opportunities for great efficiency, speed, and cost savings to be won by the optimisation and re-purposing of existing interventions and other support efforts and services towards support of crowdfunding.

Crowdfunding is not something that sits out with other enterprise support for example. In fact, it is an approach which is well aligned and draws on many existing business practices and, as such, can benefit from the integration with any existing provision.

Running a successful crowdfunding campaign will require its champions and managers to utilise many well-recognised skills and approaches which are common in any entrepreneurial and funding activity. The distinctiveness and novelty of crowdfunding in large part is contained in its capacity to undertake these activities in a faster, cheaper, and a more disintermediated way which can yield significant and unique benefits. But the processes of executing these activities are in many cases commonly understood business skills. Of course, it does have sufficient distinctiveness and novelty to require new thinking and new skills to be developed in order to operate effectively in the sector, but this does to detract from the fact that it requires skills and competencies which are well understood and supported elsewhere in many a standard business intervention programme.

Consequently, in this usage of the framework, it provides a structure to review the existing circumstances for what might be of use in a crowdfunding context. Let us illustrate this with a simple example. A local government with responsibility for economic growth and development in their area will already operate a range of interventions of typical business support. These can all be drawn into the framework's four headers and then included and signposted as part of a provision that seeks to address the crowdfunding marketplace.

For example, there is likely to be existing training and education for, let's say, digital marketing, business plan preparation, or intellectual property protection. Each of these three topic areas are all relevant to many crowdfunding schemes and could, with little or no adjustment, become part of a package of provision under the Education and Training banner for crowdfunders.

So the thrust of the analysis application is to ask "What do we already do?" and to create a comprehensive overview of the existing provision and consider how applicable it is and valuable it might be to a crowdfunding audience. This might then lead to a new set of signposting approaches to highlight these existing resources to the new audience. It might also lead to extensions and amendments to these existing provisions to tweak them to make them more relevant, useful, or applicable in a crowdfunding context. Alternatively, the same principle applies when undertaking a gap analysis to disclose where there is a lack of existing provision and consideration can be given to addressing and closing that gap.

It is not difficult to imagine how a thorough analysis under each of the four framework headings can often create a surprisingly good existing provision which, with only minor alterations, is very suitable for a crowdfunding sector. By adopting this process, it draws crowdfunding inside the wider initiatives to quickly become part of the existing provision and so overcome the challenge of crowdfunding being out on a limb or poorly aligned.

It can also be a useful mechanism of reducing apprehension and caution towards crowdfunding by demonstrating that many aspects of it are entirely mainstream and not to be feared! This can be as effective to an internally facing audience for a body or organisation as it can for an external audience.

One of the outcomes we are looking to achieve at this analysis phase is to actually progress beyond an evaluation of what is actually happening but to achieve a level of alignment. It is entirely reasonable and probable that as the framework is used and developed even further, the alignment step becomes a separate activity with a series of activities and best practise underpinning, which define how best to approach that particular challenge. But for now, we will simply describe it as one of the benefits and outcomes available through the analysis phase.

4.2. Audit

The audit function of the framework like all of the others is founded in its ability to compartmentalise the four distinctive groups of intervention types. But in this case the intention is to provide a framework for monitoring activity, progress (value-added or created), and calls on interventions under each heading. It is then a mechanism of use for ongoing assessment purposes and development of key performance indicators.

The key question at this phase is "How are we doing?"

For a more externally focused application, we could simply use the four pillars to filter requests, discussions, feedback, and enquiries on the subject of crowdfunding or to get a bottom-up sense of what the actors in the crowdfunding sector are most exercised by or are lacking. In this situation, it is also possible to use the Zones of Return as an initial filter to broaden out the sense of crowdfundings' generic capabilities to deliver returns.

As with all of the other applications of the framework, it is possible to use audit to drive others into other uses of the framework so perhaps moving from audit to say benchmarking. But there is no defined or mandated approach. One can use it as best suited or as needed for any specific circumstance. For example, it is perfectly possible to imagine that a series of interesting interventions emerging from a planning phase could be monitored and audited in a range of ways through further use of the framework and indeed it is the commonality of categorisation across the stages which makes them easier to integrate.

4.3. Benchmarking

As with all benchmarking exercises, the intention of this application is to either establish a baseline from which future actions and outcomes can be judged or, alternatively, to compare oneself against other comparable bodies or individuals. And so, it is perhaps reasonable to suggest that the most common question in this phase is "What are others doing?"

Again, the strength of the framework is to simplify the process of undertaking either of these exercises by providing a lens through which one can group and classify in a consistent and comparable manner.

The purposes of such a process can be diverse. Comparison exercises can, of course, simply be seeking to identify and mirror what is perceived as best practice. Alternatively, it could be used for undertaking a gap analysis to identify opportunities to bring oneself in alignment with that perceived exemplar and all of the steps and activities necessary to do that. But it can also look for nuance and divergence which can uncover subtlety and localisation of approaches. This can be an important feature of exploring crowdfunding as there is a tendency to imagine that the industry has a certain singularity of approach, whereas a deeper dive into the sociology of crowdfunding will quickly illustrate some quite marked differences which can be in part because of the circumstance—which may be changed by policymakers—or reflect specific cultural and immutable characteristics which should be emphasised or at least retained and responded to sensitively. Indeed, we are pleased that our case studies cover a wide range of cultures and locations and have some apparently similar examples which on closer examination demonstrate considerable divergence in approach and emphasis.

Baseline benchmarking is an important step for many policymakers not least because it is often very relevant to the typical KPIs used in the Governmental and Civic sectors. Establishing where we currently are on the basis of some useful comparators and then developing monitoring and metrics which can consistently report on the evolving situation over time is a significant requirement and demand made upon those working in these sectors so, again, having a simple but robust framework to compartmentalise key aspects of that can be a valuable shortcut.

4.4. Planning

As you might expect, a central and essential component of the framework is its use as a planning aid. The ability to act upon findings from other framework-driven activities is essential, and, once again, through the use of the four pillars we have a structure which can define areas of activity and focus and a mechanism to monitor progress against that plan. We have already pointed out that the four areas for intervention are not unique to crowdfunding. They have applicability elsewhere and the framework can be used to analyse other active interventions by a policymaking body.

And so it can be seen that, in a planning context, it is also possible to use these broad pillars of intervention to build more coordinated and holistic plans by integrating different sets of policy interventions designed to address discrete issues but which, when considered more holistically and through closer alignment, may yield better results.

The many applications of the framework have clear synergies and links with each other and there is remarkable flexibility to apply it to meet a range of micro-/macro- and national/local circumstances. Logical linear paths through it can easily be defined with analysis, audit, benchmarking, and planning being examples of a clear, obvious, and increasingly well-trodden path through the various applications of the framework. But it is the very absence of any specific mandated or solitary approach which maintains its flexibility and wide applicability. Our expectation is that the application of the framework will continue to grow as its use expands and specific techniques to facilitate particular activities or aspects of each stage and phase will emerge and develop over time and we both welcome and encourage this ongoing process. As such, we invite you to develop your own approach, combining the stages and applications as you see fit to meet your own particular circumstances.

We hope that by sharing the models we will, over time, learn of further techniques, applications, exercises, and specific approaches that will be developed and similarly be shared to provide greater richness to the tool.

Clearly, the cases we include in this book are examples of existing and established interventions in a wide set of circumstances and we do not make claims that any of them used the framework. But it is entirely reasonable and proper for you to use the frameworks to run the rule over each to get your own sense of what their thinking was and is, how it can be de-constructed using the tools set out here and perhaps even, dare we suggest, identify opportunities to do even better for the exemplars set out here but, and most importantly, to ensure your journey into building on the crowdasset opportunity is the most effective and is as rewarding as possible.

Part 3

Crowdfunding in Action

In this section, we have collected a series of examples of policymakers in action, entering the crowdasset arena. In some cases, these have been written by the policymakers themselves and in others they are reflections and analysis of the policymakers' actions written by third parties.

What they demonstrated is a huge diversity of activities utilising all of the intervention options we have set out above and demonstrating the full range of returns we assert are available.

We wish to make clear that to our knowledge none of these activities were directly influenced by our framework and we claim no credit for their excellent achievements. We thank our contributors for sharing such valuable insight with us and through us to you.

We have grouped the cases into a series of very high-level collections to aid in your access to them but they are simply notional groupings and many others would have been valuable and possible.

We encourage you to explore the diversity of the story told here and to consider, as you go, how you might act and develop an approach for your policy remit for each of the circumstances described and based on the tools we have supplied you with.

Enjoy!

Part 3.1

Societal Impact

Chapter 3

Civic Crowdfunding, Equity, and the Role of Government

David J. Weinberger

Abstract

The number of city governments looking to promote civic crowdfunding as a tool for organisations and residents to fundraise for public goods and services is poised to increase. This chapter examines how city governments may leverage civic crowdfunding in ways that ensure resources are distributed equitably across neighbourhoods, build strong and authentic relationships with residents, and increase the capacity of local organisations to plan and deliver innovative projects that benefit their neighbourhoods. Relying on case studies from two civic crowdfunding programmes—the El-Space Challenge in New York City and the Great Streets Challenge in Los Angeles—the author measures the extent to which crowdfunding may help to achieve these goals. This chapter culminates in a series of recommendations and conditions for governments interested in using civic crowdfunding as a tool for financing projects. Ultimately, the author finds that, to effectively leverage crowdfunding to build relationships in a community, a government must first demonstrate its willingness to invest in and trust the people of that community. With these requisite conditions met, and with a government's genuine commitment to pursuing meaningful partnerships with residents, civic crowdfunding has the potential to fundamentally shift the power dynamics in a government's relationship with a community.

1. Introduction

Informing, consulting, and placating the public results in policies and plans that lack community ownership, weakens communities' trust in their governments to do what is right. For governments looking to build trusting, productive relationships among residents, the co-production model of public administration remains a helpful framework. Rather than make spending decisions unilaterally, co-production requires governments to consider residents as active participants in the creation of public goods. Tony Bovaird argues that public administrators should overcome any reluctance to sharing power with communities and place more importance on supporting, encouraging, and improving the capabilities of citizen groups.[1] The co-production model requires that governments and communities commit similar amounts of resources to a project.[2] The resources committed to a project in a co-production scheme vary depending on the actor, but may include time, fundraising efforts, feedback, and technical expertise. According to Bovaird, the co-production model hinges on the notions that: (1) no one actor—community or government—should have the power to dominate outcomes and (2) both the government and the community must assume some risk.[3] Bovaird posits that community stakeholders need to trust the "advice and support" of the government and that the government must trust and accept the "decisions and behaviours" of communities, rather than prescribe solutions unilaterally.[4]

2. Bringing Planning Online

The advent of digital tools offers new opportunities for planners to pursue the kinds of genuine and mutual community partnerships that Arnstein prescribes. As technology has advanced, since John Forester first discussed his model of participatory planning, so have the tools that planners use to listen to and mediate between communities' varied and sometimes conflicting concerns. Daren Brabham argues that new online crowdsourcing mechanisms are appropriate channels for enabling citizen participation in public planning projects.[5] Joanna Saad-Sulonen considers the most widely used online community engagement tools, including websites to help citizens understand existing plans, questionnaires that allow citizens to post their solutions to specific problems presented to them on a map, and websites that allow citizens to respond to entries in planning competitions.[6]

Saad-Sulonen argues that these new tools limit citizen interaction with government and, because planners who employ them ultimately retain complete decision-making power, they are insufficient to ensure that citizen participation in a planning process reaches its potential impact.[7] Unlike Saad-Sulonen, Brabham believes in the power of digital crowdsourcing tools to harness collective intellect or "crowd wisdom" and spur ideation in a planning process.[8]

2.1. The Emergence of Crowdsourcing Online

Whereas Marschall focuses on the role of government in creating space for meaningful citizen participation, in 2009 Albert Meijer examines the role of what he calls citizen-to-citizen (C2C) interactions in facilitating effective governance.[9] Given the great extent to which public discourse now occurs online, Meijer suggests that governments should be strategic in their online engagement with constituents.[10] He notes that, while online tools do enable governments to engage in conversations with citizens, there is also value in learning from conversations among citizens in a digital environment.[11] In 2011, Meijer considers the impacts of new, digital media on the successes of co-production and introduces a new concept of "networked co-production."[12] Meijer argues that digital forums for people to find like-minded individuals who depend on similar services enable citizens to receive timely information and assistance (i.e. public services) from peers, without needing to seek the help of government.[13]

This form of co-production saves government effort because citizens are giving each other the answers that they would historically have expected their government to provide.[14] Further, when citizens offer each other answers to a problem on an online forum, they are able to offer each other emotional support that government employees are not always able to provide.[15] A theoretical line may be drawn between Meijer's model of co-production, which depends on meaningful citizen-to-citizen interaction, and online crowdsourcing tools, which offer mechanisms for citizens to respond to and build upon each other's ideas for public goods and services.

2.2. Conditions for Successful Civic Crowdsourcing

In 2013, Brabham distils the conditions for successful crowdsourcing, writing that the most effective crowdsourcing efforts have been the result of an organisation needing a discrete task performed or question answered, a community that is willing to perform the task voluntarily, and an online space for the work to be done.[16] Importantly, for crowdsourcing to be successful there must be a mutual benefit for the organisation and the community participating in the effort.[17] Brabham argues that control over the process should reside somewhere between the crowd and the government asking the question, so that the process blends the benefits of a top-down process with the openness and creativity encouraged by a bottom-up approach.[18]

2.3. Resetting Power Dynamics between City and Citizenry

Similar to Saad-Sulonen's critique of planners' use of digital tools to bolster engagement, Anttiroiko argues that the introduction of digital tools alone to the planning process will not do much for governments looking to promote and deepen citizen participation.[19] In other words, true participatory planning will require governments to allow spontaneous, citizen-led projects to inform the direction of plans.[20] Unless and until governments shift their thinking about planning processes to be bottom-up—where the lines between citizen and practitioner blur and planners follow the direction of the people they serve—citizens will simply be engaging with planners at the same moments in a plan's development, using new tools to communicate their feedback to a plan that will ultimately be decided by the government's team of experts.[21] In contrast to Brabham, who argues that ultimate control of crowdsourcing processes should be shared equally by government and the community but that the roles of government and citizen should remain the same, Anttiroiko argues that actors in a planning process should be permitted to spontaneously swap roles and that power structures should be less rigid to allow for situational reasoning to guide decision-making.[22]

For Jane Fountain, applying the co-production model in a digital environment raises the same concerns about rigidity in planning hierarchies that Anttiroiko raises in his examination of participatory planning. Fountain posits that technology-driven tools are not replacements for meaningful civic engagement; planners using online tools must strive for a deep understanding of citizens' lives.[23] To avoid moving towards systems of technocratic management that do not appropriately respond to the needs of citizens, there needs to be close, consistent interactions between government and neighbourhood residents.[24]

For co-production to work in a digital environment, planners must prioritize human relationships over their use of technology to make data-driven decisions.[25] Examples of processes that have been successful in this regard include those that dismantled bureaucracies that separate stakeholders, fragment problems, and rely on a few professionals' expertise, in favour of systems of rapid prototyping that make constant improvements to processes.[26]

2.4. Participatory Budgeting

Bovaird offers participatory budgeting (PB) as a powerful and effective model of co-production.[27] In a PB process, a government asks residents to volunteer ideas for expenditures in a district and then calls for a vote to rank the community's spending priorities. Residents lead fiscal decision-making, and the funds for the projects come from the government budget. The relevant city agencies are then responsible for the implementation of the projects, and communities may or may not be involved in delivering the project.

According to the PB Project, the mechanism has been a successful tool for putting residents at the helm of capital budgeting and building trust between residents and their elected officials.[28] Jurisdictions that have adopted PB have seen measurable increases in civic participation, political participation, and the number of community leaders.[29] Since the completion of the first PB process in Porto Alegre, Brazil in 1989, the practice has spread to over 3,000 cities worldwide.[30] In 2009, Chicago became the first city in the United States to successfully run a PB process.[31]

Governments have reported that PB processes result in stronger relationships between government, civil society, and residents.[32] While this process enables the public to make spending decisions, governments ultimately accept responsibility for implementing communities' ideas. Given that municipalities in the United States have only recently adopted the practice, it is too early to determine an American municipality's ability to maintain close contact with community participants following a PB process. Future research may examine the extent to which governments work with community participants in PB processes to implement and maintain civic projects.

2.5. Crowdfunding's Relationship to Tactical Urbanism

Tactical urbanism is the practice of making temporary and low-cost improvements to the built environment in order to draw attention and propose a solution to a problem in a public space.[33] In some ways, tactical urbanism is the theoretical successor to the co-production model of delivering public goods and services. Recently coined by Mike Lydon and Anthony Garcia, tactical urbanism refers to the dual notions that citizens know what public space interventions would work best for their blocks and neighbourhoods and that cities should enable citizens to experiment with creating public amenities without government interference.[34] Like co-produced public goods, tactical urbanism projects require inputs from citizens and government.

Rather than apply new tools to open existing planning processes up to greater participation, Lydon and Garcia see tactical urbanism as a planning model that disrupts traditional hierarchies and separations between citizens and their government.[35] The installation of temporary amenities, such as pop-up bus shelters, parklets, and chalk-painted curb extensions, permits passive citizens to engage with a proposed use of a space and share their preferences without needing to invest time attending a more traditional planning conversation. Lydon and Garcia's tactical urbanism model aligns neatly with Anttiroiko's proposal that traditional roles in planning should be allowed to adapt based on a situation, as well as Fountain's embrace of rapid prototyping as a vehicle for systemic change.

A theoretical cousin to tactical urbanism and co-production, civic crowdfunding similarly upends the traditional roles of government-provider and citizen-consumer. In 2014, in what seems to have been the first academic exploration of the topic, Rodrigo Davies surveys the field and proposes that institutions use civic crowdfunding to reconsider their roles in facilitating a community's participation in its decision-making.[36] While Davies does not explicitly draw a connection between co-production, tactical urbanism, and civic crowdfunding, he does suggest that governments looking to engage in civic crowdfunding should work in partnership with community groups to create shared resources in public space.[37]

In his 2016 PhD dissertation, Martin Mayer adds that, while civic crowdfunding boasts abundant historical precedent, what sets it apart in its newest iteration is its capacity to create strong, new relationships between citizens and government.[38] Mayer discusses the theoretical roots of civic crowdfunding in the area of co-production, finding that the practice should be considered to be a "complementary tool," rather than a "replacement for traditional budgeting and project development structures."[39] When used in addition to governments' standard modes of investing in neighbourhoods, Mayer argues that civic crowdfunding may benefit the government and communities.[40] Specifically, Mayer offers that civic crowdfunding gives governments the opportunity to finance untested project types from which decision-makers are able to learn but that risk-averse bureaucracies are unlikely to fund.[41]

2.6. Looking Ahead

Future thoughts on civic crowdfunding may be characterised by arguments for or against the institutionalisation of the practice. In *Democracy Reinvented*, Hollie Gilman argues that innovations that seek to make systemic changes to how decisions are made within an institution should ultimately be embedded in the institutions themselves.[42] Just as Marschall calls for a programme in which government provides meaningful direction for citizens who are asked to co-produce a public good,[43] others may argue that civic crowdfunding will require similar, government-ordered structure and direction. In what is perhaps a signal of more theories to come, Rodrigo Davies offers four roles for governments to play in leveraging civic crowdfunding: promoter (i.e. government raises and receives its own funds), curator (i.e. government draws attention to crowdfunding projects led by citizens or partner organisations), facilitator (i.e. government invites constituents or community-based organisations to use crowdfunding for their projects), or platform (i.e. a government creates its own crowdfunding platform on its site).[44]

Any new discourse related to civic crowdfunding should be informed by the evolution in discussions of co-production, which has moved away from imagining the practice as a government-driven model of partnership (e.g. Ostrom and Clary) and has now seemed to embrace collective co-production as a means of authentic, flexible, and tactical collaboration between actors (e.g. Marschall, Meijer, Fountain, and Lydon and Garcia). Questions for possible further inquiry include those related to (1) citizens' motivations for participating in civic crowdfunding, (2) the extent to which a government-led crowdfunding programme should involve citizens in the implementation of projects, (3) questions related to citizens' technical expertise to meet government's expectations in project delivery, and (4) barriers to participation in civic crowdfunding and the equity implications requiring citizens to pay for and create a good or service that might otherwise be provided by government.

1. *Motivations*: While citizens' reasons for engaging in the co-production of a pubic good or service are well documented,[45] less has been written about why citizens might be willing to shoulder the monetary cost of providing that good or service. That is, existing research does not adequately explain why citizens would decide to ask people in their own networks to contribute their own disposable income to a service that had previously been provided by the government. Empirical evidence from successful civic crowdfunding projects led and funded by citizens might provide a closer look at their motivations and the possible returns on an investment of volunteer time and resources.

2. *Citizen Involvement in Project Implementation*: As previously discussed, fundamental questions about a government's rightful responsibility to its constituents underscore much of the scholarly and public debate regarding civic crowdfunding. The notion of co-production suggests that involving citizens in the creation of public goods is an appropriate mechanism for a government to adhere to its responsibility to act in the public interest. Even proponents of co-production may disagree about the extent to which governments

should involve citizens in creating projects, including funding and delivering them. Few would dispute government's duty to commit (usually contracted) labour and funding to projects that improve neighbourhood outcomes; however, there is an unclear distinction between inviting participation in governance and asking citizens to pay for their government to fulfil its basic responsibilities. This study, and future research, may help governments determine how to involve citizens in co-production without shifting its responsibilities to the people they serve.

3. *Citizens' Technical Expertise*: Because residents looking to crowd-fund and co-produce a good may lack the technical expertise of a professional planner or designer, governments may need to relax their standards.[46] To what extent are governments willing to do this? Interviews with planners who have participated in the co-production of public goods—particularly design interventions in public spaces— may be helpful in understanding governments' perception of the design or performance trade-offs involved in asking or enabling non-technical citizens to implement their own projects.

4. *Equity Implications*: Davies and Mayer each surface questions about the equity implications of civic crowdfunding. Davies worries that, because civic crowdfunding in its purest form requires a resident to make a monetary contribution in order to contribute to civic life, the model may only entrench the systems of inequality that civic crowd-funding is often intended to dismantle.[47] Mayer's argument aligns with Davies's assessment but he adds that civic crowdfunding plat-forms should be equipped with safeguards to ensure equitable access to the benefits of using the tool.[48] These arguments are based more in reasonable conjecture and anecdote than in empirical evidence. In their 2017 working paper, Daniel A. Brent and Katie Lorah find that neighbourhood characteristics such as median income and racial composition are not strong determinants of the total amount that a crowdfunding campaign might receive.[49] In comparing communities' capacities to crowdfund and finding that there is no correlation between income and crowdfunding success, Brent and Lorah begin to address Davies' concern that wealthier neighbourhoods might be better equipped to run successful crowdfunding campaigns. Still, there

remains a question of how crowdfunding projects are distributed geographically across neighbourhoods with variable socioeconomic conditions. Future research may examine the degree to which projects are funded equitably in cities, and the degree to which civic crowdfunding projects are concentrated in neighbourhoods with abundant resources. With this information, the field might move towards a shared understanding of what interventions or safeguards might be needed to protect equity. This information might also give governments a better understanding of the role that they may play in ensuring the equitable distribution of crowdfunded projects across neighbourhoods.

2.7. Civic Crowdfunding and Participatory Democracy

Civic crowdfunding, as a model for a productive partnership between governments and community, should be considered among other tools for community planning, such as PB and participatory planning. Civic crowdfunding is also closely aligned with an asset-based approach to community development, an approach that requires anyone looking to make a change—residents, community-based organisations, and decision-makers—to first consider the assets of a particular community before looking at its weaknesses and opportunities for improvement. Proponents of asset-based community development also ask the creators of civic projects to leverage the skills and resources of the community members they are aiming to serve. Like asset-based community development, civic crowdfunding makes use of capital—creative, social, financial, and political—from the community where a project will ultimately be implemented. Proponents of both models hold that civic projects are strongest when their development actively engages stakeholders in the affected community.

In recent years, civic crowdfunding has become a tool that governments use to fund public goods. Several American cities, including St. Louis, New Orleans, and Jersey City, have used the practice to seek funding for small projects like sidewalk improvements, civic building restorations, and bike racks. In these cases, employees of the city lead the fundraising effort, turning to private stakeholders to contribute dedicated funding for projects that fall within government's typical responsibility but fall outside the limits of their budget.

Proponents such as Katie Lorah, Communications and Creative Strategy Director at ioby, say that civic crowdfunding offers residents an ownership stake in a public good, offering a new way for communities to organise for change.[50] Lorah argues that crowdfunding offers a new opportunity for residents to communicate with their governments. By leading or contributing to a crowdfunding campaign, residents are essentially expressing their demands, desires, and preferences for their communities in an open forum. Lorah calls on governments to value datasets generated by civic crowdfunding campaigns—including the types, locations, and relative successes of campaigns—and use them to inform larger decisions about where and how to invest in larger programmes and capital projects. A successful civic crowdfunding campaign led by a neighbourhood group for a small or temporary installation in a public space, Lorah's argument suggests, might improve the City's understanding of the neighbourhood's needs and priorities. With this information, the City may choose to invest its resources in a neighbourhood's infrastructure, transit, amenities, and social services in ways that are more responsive to local demand.

Despite the relatively small portion of the crowdfunding industry that civic crowdfunding platforms occupy, the model remains heavily scrutinized. Opponents argue that crowdfunding amounts to an additional tax on residents in communities that have not received their fair share of public investment. To critics of civic crowdfunding, this practice amounts to an additional tax on residents. In his 2016 piece in *Wired*, entitled "Crowdfunding for the Public Good is Evil," journalist Peter Moskowitz claims that, when governments ask residents to contribute to a crowdfunding campaign, they are essentially requiring residents to purchase enhancements to their own quality of life. If applied at the scale of an entire city, this model may pose a considerable threat to low-income communities that already struggle with a scarcity of public resources.

Governments, critics like Moskowitz say, should not be in the business of fundraising for amenities that should unquestionably be included in their budgets. Using the example of the city government of Central Falls, Rhode Island's crowdfunding for improvements to parks, Moskowitz claims that city agencies have turned to crowdfunding when their budgets were slashed. It is true that the rise of civic crowdfunding in the last decade has coincided with a period of fiscal austerity in many American cities following the Great Recession; however, there are no known examples of governments admitting to using civic crowdfunding to release their burdens of responsibility.

Conversely, there are documented examples of governments turning to crowdfunding as an extension of their approaches to community partnership and promoting civic participation. Rather than use crowdfunding to fund essential services (e.g. healthcare, roads, and education), local governments commonly use crowdfunding as a tool to encourage residents and civil society to create and fund their own ideas. In each of these cases, rather than fundraise for a project that falls within its own purview, a government asks constituents and community-based organisations to scope and help to implement projects that are ancillary to an existing government project or that are new and untested. In many cases, the city will commit matching funds to projects with demonstrated support from local communities. In 2015, the New York City Department of Transportation (DOT) asked residents to create and fund ideas to utilise the dark and unoccupied spaces underneath elevated train tracks in Brownsville, Brooklyn. The DOT, aiming to learn from the ideas of residents who are most deeply affected by the derelict spaces, offered matching funds to alleviate the burden of having to fundraise for these projects. The projects that were funded included a healthy food festival and block party under the train tracks, as well as a series of kiosks showcasing the work of young local artists. The city then worked in partnership with the residents who led these campaigns, in the form of technical assistance, permits and approvals, and logistical support. These partnerships built trust in city government among community leaders in Brownsville and forged stronger community bonds in the process. In this case, the city did not do its own fundraising, and it is unlikely that residents would have reasonably expected the DOT to have funded these types of projects with tax revenues.

This strategy is employed heavily among municipalities in the United Kingdom, where 45 (or 10%) city councils have invited a civic crowdfunding platform to help fund projects led by residents and community-based organisations. As in the case of the New York City Department of Transportation, some UK municipalities—including London and

Essex—have secured matching funds to accelerate the creation of civic projects. For example, in 2015, a collective of residents in the London Borough of Southwark used Spacehive to crowdfund for an urban park on the abandoned Peckham Coal Line. With the support of the Mayor of London and £10,000 in matching funds from the Southwark Council, organizers of the project were ultimately successful in raising more than the £64,132 that they needed to complete the feasibility study for the project and purchase basic asset protection.[51]

In some cases, residents or community-based organisations decide to fundraise to close a gap in a capital budget. For example, in Memphis, a team of community groups—including the city's community development council, a local arts district, a community development corporation (CDC), and a small business owner—came together to close the funding gap for the Hampline, a two-way protected bike lane through a disinvested neighbourhood. While a bike lane can reasonably be expected to fall within the responsibilities of the government, the team decided that they did not want to wait for the City to fully fund this project and attempted to use the crowdfunding campaign to increase and showcase public support for bicycle infrastructure in the city.

3. Comparison of Civic Crowdfunding Platforms

Most crowdfunding platforms, including those built to support creative and entrepreneurial endeavours (e.g. Kickstarter and Indiegogo), allow project creators to fundraise for charitable or civic projects. This study examines those crowdfunding platforms that are explicitly and solely designed to invite and support projects with public benefits (see Table 1). Though each of these civic crowdfunding platforms is aimed at creating public goods or services, they vary significantly in their approaches and models. The primary differences between these platforms are found in the roles that community and government actors play in delivering the project (i.e. who initiates the fundraising campaign, who leads the implementation of the project) and the types of projects that the platform supports. Patronicity and ioby are the only donation-based civic crowdfunding platforms based in the United States that engage residents and community-based organisations as leaders of their own fundraising efforts. On Neighbourly, donors give with

Table 1: Comparison of civic crowdfunding platforms discussed in Section 2.

Platform	Platform's incorporation status	Primary service area	Project risk level	Who fundraises?	Contributors receive	Who receives and spends the funds?
Neighbourly	For-profit	United States (all)	Low	Government	Financial return via municipal bonds	Government
Citizinvestor (defunct as of October 2017)	For-profit	United States (all)	Low	Government (or official partner)	Tax deduction	Government
Spacehive	Non-profit	UK	High	Community members	N/a	Project creators
Patronicity	For-profit	United States (MI, MA, IN)	High	Community members	Tax deduction if the project is led by a registered non-profit organisation	Project creators
Ioby	Non-profit	United States (all)	High	Community members	Tax deduction	Project creators

an expectation of financial return through municipal bonds. Positing that the platform more closely resembles an open market for municipal bonds, its founders do not consider Neighbourly to be a civic crowdfunding platform.

Spacehive, Patronicity, and ioby ask residents and community-based organisations to fundraise their own projects. While they all claim to offer different levels of government support in the form of matching funds and implementation assistance, by centring the community as the creator, originator, and fundraiser for these projects, these platforms effectively enable communities to create projects that carry higher risk and fall beyond the scope of what the government can typically fund. Projects need not have government support or matching funds to operate a crowd-funding campaign on any of these platforms. Neighbourly works with governments to fundraise for relatively low-risk projects, or projects that taxpayers might reasonably expect to be funded by their governments. Neighbourly, which calls itself the platform for "modern public finance," stands out among this array of websites as the only platform to offer financial return to campaign contributors. As in the more traditional model of public finance, governments issue bonds for discrete projects and investors make contributions with an expectation of return. Neighbourly's novelty is found in its ability to open public finance up to uncertified investors, enabling more people to take advantage in invest-ment opportunities and large-scale infrastructure projects to be funded more quickly.

Although governments provide funding and attention to communities' ideas presented on Patronicity, Spacehive, and ioby, the nature and depth of a government's engagement with affected communities varies with each platform. Each of these three platforms works with government part-ners to commit matching funds for community-led projects that meet partners' funding criteria. Any resident or community-based organisation may start a project on Patronicity, Spacehive, or ioby, and the degree to which a government is willing to release matching funds depends on the project's scope and objectives. The important differences between these platforms' approach to government partnerships are in their varied approaches to enabling government's involvement in the implementa-tion of projects. There is no evidence that the government partner plays a consistent role in working with community groups to implement projects after they are funded on the Patronicity and Spacehive platforms. Different from Patronicity and Spacehive, ioby explicitly asks governments to work closely with the leaders of projects that receive matching funds throughout the implementation process. In most cases, ioby then monitors the implementation of the crowdfunded projects to encourage continued collaboration. As evidenced by the sudden and unannounced closure of Citizinvestor in October 2017, the field of civic crowdfunding platforms is competitive and evolving. Those listed here have endured, at least in part because each has employed a different approach to civic crowdfunding (e.g. in the types of projects and originators of ideas they have served and the rewards they offer to contributors).

Among the civic crowdfunding platforms that are built to enable communities to innovate and create high-risk projects outside of the normal scope of government, ioby's approach to government partnership is most explicitly aligned with the model of co-production. The New York City Department of Transportation and the Office of the Mayor of the City of Los Angeles has each used ioby's platform and services to work collaboratively with communities to fund and deliver projects in public spaces. Extending their support for community-led projects beyond matching funds and committing explicitly to invest time and resources to assisting in the realisation of their ideas, these governments were the first in the United States to use civic crowdfunding intentionally as a mechanism for co-production.

4. The City's Motivations

A city government might consider leading a crowdfunding programme for any combination of the following reasons:

- *Deficit of trust*: When NYC DOT was developing plans to implement strategies included in its Under the Elevated report, the agency began by scoping possible interventions in Brownsville. Without any previous experience working with community leaders in the neighbourhood, NYC DOT's Urban Art and Design unit hoped that a crowdfunding programme would help them to build trust with local stakeholders who might ultimately help to steward the City's plans to install additional el-space projects in the future.

- *Supplement to existing funds*: The Los Angeles Great Streets Challenge stemmed from Mayor Garcetti's directive that the City identify corridors to receive sustained, interagency funding for safety interventions and place-making projects. The Mayor's Office had already set aside funding for grants to community-led projects, and they decided to use crowdfunding to build local support and unlock new capital from communities. With the Great Streets Challenge, the City successfully leveraged grant funding for additional funding from communities.

- *Fiscal austerity*: Especially during periods of fiscal austerity, municipalities may look to crowdfunding as a way to eliminate budget items without reducing overall spending on civic projects. Much of the resistance to crowdfunding in the fields of public administration and city planning stems from critics' fear that crowdfunding will be used to justify and encourage reductions in governments' expenditures. To date, there is no evidence that suggests that crowdfunding is a viable replacement for government investment in public goods and services.

- *Lack of inspiration*: The NYC DOT Urban Design Unit hoped that a crowdfunding challenge would inspire local designers and residents to create solutions for the el-space that would exceed their team's own creative capacities. They hoped that, because ideas would originate with non-professionals outside of the agency's bureaucracy, their ideas would push the boundaries of what is possible for the space. From photo kiosks to an urban agriculture festival, NYC DOT successfully used crowdfunding to generate new ideas to activate an el-space in Brownsville.

5. Community/Government Relationships in a Civic Crowdfunding Model

When governments invite residents to lead civic crowdfunding campaigns for projects that are important to them, they are effectively offering residents some degree of control over the outcomes of public decision-making. This is especially true when governments commit to working with residents to implement the projects and/or offer matching funds to projects that demonstrate significant community support in the form of donations.

In a civic crowdfunding model, platforms may either disburse to government or directly to the group that led the campaign. When governments raise and receive funds, the funds are directed to the agency's coffers and are spent pursuant to the agency's typical procurement procedure. Governments, whose procurement systems are set up to award contracts rather than to leverage volunteer labour, are not often equipped and inclined to meaningfully involve communities in the implementation of projects. Because communities are able to contribute funds but not to spend them, they have no outlet for shaping the creation of civic projects. These platforms offer a superficial model of co-production that resembles a tax scheme through which citizens with disposable incomes have outsized impact on the distribution and delivery of public resources.

When funds are disbursed to communities, citizens are placed at the centre of service delivery. Government may offer technical assistance but cannot dominate outcomes. Because it necessitates the inclusion of communities in decision-making process and empowers citizens to take leadership roles, this model of crowdfunding enables communities and governments to pursue true co-production. When communities receive crowdfunded money for a civic project, governments may choose from several approaches to working with the community, ranging from hostile to deeply collaborative:

- *Ban*: When a community crowdfunds a civic project, a government with relevant jurisdiction may decide to prohibit the implementation of the project. In these cases, communities typically do not seek government approval prior to fundraising. When the community approaches the government, the government bans the community from following through on the project.
- *Stall*: In some cases, the government may decide to put a temporary hold on the community's project and delay making a decision on whether it will approve a community-led implementation of a project. In these instances, a long waiting period may slow the momentum that the community had begun to build during the crowdfunding

campaign and reduce the likelihood that the community's project will be effective.

- *Wait and see*: Some governments decide not to take any action on the project until after they have a chance to measure outcomes. In each of these cases, the government does not act as a partner in the implementation of the project, but instead serves as a passive observer who is allowing the project to happen while it develops its formal stance.

- *Remove and reinstall*: Particularly in cases when a community is aiming to iterate on an existing government-produced good or service (e.g. street furniture, homeless services, or transit), a government may decide to ban or remove the project and deliver its own version with its own funds. Governments may do this because they doubt the community's ability to provide the good or service properly and safely.

- *Encourage and support*: In the strongest possible demonstration of a commitment to co-producing goods and services with the communities it serves, a government may decide that communities are well equipped to spend funds that they raise without strict government oversight. In these cases, governments may advise but not control communities in the spending of crowdfunded money.

6. Equity Implications

Critics of the co-production model argue that it would be irresponsible or unethical to shift the burden of public service delivery from government to citizens. Especially in low-income and historically disenfranchised communities, critics contend, citizens may only want to be minimally involved in public decision-making.[52] Marilyn Taylor argues that, in communities where government has disinvested or shirked its responsibilities to provide services, government has an obligation to invest without asking citizens to devote any time or resources in return.[53]

Seemingly discrediting any criticism that civic crowdfunding might deepen existing inequities in investment across neighbourhoods, Brent and Lorah find that groups in lower income neighbourhoods are no less able to successfully fund projects than groups in more affluent areas. Civic crowdfunding, they suggest, may give governments usable data on residents' preferences for civic projects and offer a "non-market valuation tool to help guide public funding."[54] These findings align neatly with the messaging that ioby uses to persuade governments to participate in a crowdfunding programme. On its website, ioby suggests that residents may turn to crowdfunding in order to do the following:

- bring attention to a problem in the public realm;
- demonstrate and pilot a solution in which larger funders may ultimately invest in bringing to scale;
- demonstrate a quick victory that restores a community's confidence in its own ability to create change.[55]

There are three potentially significant variables missing from Brent and Lorah's analysis: (1) the proportion of the original fundraising goal that each group successfully crowdfunded, (2) whether a project was success-fully implemented, and (3) the degree to which a group's local govern-ment supported or opposed the project.

Importantly, Brent and Lorah compared the amount that groups were ultimately able to fundraise without also looking at their original fundrais-ing goals. Without this information, it is impossible to ascertain the extent to which a neighbourhood's income impacts a group's ability to meet their fundraising goal. Using the amount fundraised as a proxy for crowdfund-ing success, Brent and Lorah exclude the degree to which projects have been implemented from their analysis. While the income of a neighbour-hood may not determine the amount that a group is able to crowdfund, there may still be unrevealed inequities in the distribution of finished projects across neighbourhoods.

In their analysis, Brent and Lorah include data from projects that were solicited or supported by the local government in some way and those that were more spontaneously generated by communities (i.e. without govern-ment involvement). By excluding this variable from their consideration, Brent and Lorah leave unexplored the relationship between governments' attitudes towards crowdfunded projects and the crowdfunded projects' success. Their study does not address the degree to which the support or opposition of government decision-makers impacts a group's ability to fundraise in a low-income area.

With these variables missing from their analysis, Brent and Lorah leave unanswered the important question of whether civic crowdfunding results in an equitable distribution of civic goods across neighbourhoods. ioby, the crowdfunding platform that supplied all project data used for Brent and Lorah's analysis, does not offer data regarding projects' completion and government involvement in their implementation. Without access to this information, questions regarding distributional equity remain largely open to conjecture. Lacking substantial empirical evidence upon which to base claims, arguments for and against government participation in civic crowdfunding programmes have been driven by opinions about the appropriate role of government in the provision of public goods.

7. Case Studies of Government-Led Crowdfunding Programmes

7.1. New York City Department of Transportation: El-Space Challenge, Brooklyn, NY (2016)

With the release of their 2015 publication, *Under the Elevated*, the New York City Department of Transportation (NYC DOT) and the Design Trust for Public Space set out to understand and begin planning for the city's hundreds of miles of elevated infrastructure—highways, bridges, and rails—that leave stretches of the public right-of-way dark, noisy, desolate, and uninteresting. NYC DOT calls these stretches "el-spaces" and claims that they serve as barriers that divide several of the city's most heavily disinvested communities.

NYC DOT recognised a particular urgency for intervention in Brownsville, a majority Black neighbourhood that has experienced decades of public and private disinvestment. They decided to focus their efforts in the short-term on improving public spaces along Livonia Avenue, where the subway tracks above cast a long shadow that bisects the neighbourhood (see Figure 1).

7.1.1. Intentions

In early 2015, NYC DOT planned to begin their transformation of el-space in Brownsville by investing in new lighting beneath the elevated train tracks along Livonia Avenue. The NYC DOT Borough Commissioner's office asked the agency's Urban Design Unit to address

Figure 1:   *Existing conditions on Livonia Avenue and Herzl Street.*
Source: Kelly, Quinn, NYC Department of Transportation, 2016.

the poor lighting conditions in response to demands from community groups. Without many direct connections to residents in Brownsville, the NYC DOT Urban Design Unit felt that they needed to meaningfully engage with community leaders and better understand local needs and demands.

In a fall 2017 interview with Patrick Smith, a Project Manager with NYC DOT's Urban Design Unit, he explained that his office felt that Brownsville is "politically complex; you had to be careful about engaging everybody, or at least the right people, or else you come off as a negative agent." At the time, East New York, the neighbourhood immediately to the east of Brownsville was enduring a contentious rezoning process. The Urban Design Unit wanted to be sure not to make assumptions about what the neighbourhood would want underneath the elevated train tracks.

At about the same time, the Urban Design Unit began to discuss the possibility of crowdfunding resident-led el-space interventions with ioby.

Smith felt that the concept of crowdfunding might be useful to soliciting and investing in communities' ideas for the el-space on Livonia Avenue, as a way of understanding the changes that community leaders wanted to see in their neighbourhood and beginning to build trusting relationships with residents. Non-profit organisations, unincorporated community groups, and independent residents would submit their ideas to NYC DOT for review and, if approved, would create crowdfunding campaign pages on ioby.org. According to Smith, "crowdfunding acted as [a] proxy for how much the community liked the idea."[56] NYC DOT would consider a donation from a resident of Brownsville to be a vote of confidence in that idea. To reduce the burden on the group leading the fundraising campaign, NYC DOT would offer matching funds and implementation assistance. Importantly, the group would retain ownership of the project through the end of the implementation process.

7.1.2. The city's existing relationship with community members

With the crowdfunding programme, NYC DOT set out to better understand the neighbourhood's priorities and build trusting relationships with residents. Patrick Smith reports that his office expected the Brownsville community's perceptions of the programme to be affected by three important dynamics as follows:

- *The many and complex relationships of the city with local community groups*: NYC DOT was not the first agency in recent history to decide to do work in Brownsville. Several agencies, including the Department of Health and Mental Hygiene, the New York City Housing Authority (NYCHA), and the Department of City Planning, were already focusing resources on developing programmes in the neighbourhood. At the same time, residents were eyeing the fight over rezoning in neighbouring East New York and voiced their concerns about gentrification and displacement.
- *The relationship between NYC DOT and the Brownsville community*: In his interview, Smith offered, "NYC DOT is more neutral than some [other agencies]. We do relatively uncontroversial things other than bike lanes, which weren't associated with this project." Smith also says

that residents' interactions with NYC DOT tend to be related to traffic and road conditions: "we're the pothole people."[57] NYC DOT's Public Space Unit may have had existing relationships with community members through their recent work on Osborne Plaza.[58]

- *The relationship between the Urban Design Unit and the Brownsville community*: The Urban Design Unit did not have existing relationships with residents and were not able to meaningfully assess which organisations and residents would be able to deliver on projects.

The levels of interactions that Smith describes are less meaningful to residents, particularly to those with limited knowledge of government operations and structures. Most residents who participated in the crowdfunding programme perceived the staff of NYC DOT as agents of the city. Any distinction between the Urban Design Unit and the Public Space Unit—or the differences in approaches between NYC DOT and the Department of Health and Mental Hygiene—may be difficult to understand or unimportant to residents whose experience working on civic projects is limited.

To residents who do not trust the government to serve their best interests, their fundamental distrust extends beyond the lines between agencies and offices. At least one participant in NYC DOT's crowdfunding programme began working with the agency despite feeling that the City has responded to their needs as quickly as they should. In a phone interview in November 2017, this participant pointed out that primarily white neighbourhoods with higher incomes have historically received more funding for civic projects. Residents of colour are acutely aware of the continued impacts that overtly and implicitly racist policies have had on the neighbourhood: redlining that has left Brownsville chronically disinvested, legacies of urban renewal projects that bisected communities and displaced long-time residents in the mid-to-late 20th century, a sustained police presence in the area that continues to result in the disproportionate incarceration of people of colour, and an affordable housing crisis that has largely gone unaddressed. In 2015, when NYC DOT came to Brownsville with the *El-Space Challenge*, a massive and controversial rezoning effort was well underway in the adjacent East New York neighbourhood. Although the Urban Design Unit had virtually no involvement in these policies and initiatives, residents were hesitant to trust their government to serve the neighbourhood's best interest.

7.1.3. Execution

In March 2016, NYC DOT officially announced their call for residents' ideas. To promote the opportunity, staff from NYC DOT and ioby presented to a meeting of the Brooklyn Community Board 16 and convened their own information session that was attended by about 15 prospective applicants. NYC DOT and ioby presented this opportunity as a three-part challenge to residents of Brownsville: (1) tell NYC DOT what you think should happen under the tracks on Livonia, (2) fundraise quickly for your project on ioby's crowdfunding platform, and (3) work with NYC DOT to bring your idea to life. From the pool of applicants, NYC DOT would choose a smaller group of participating groups and match every dollar they raised with three additional dollars from the City. NYC DOT did not require that each group meet its original fundraising goal in order to receive City funds. The Urban Design Unit planned to work with any group that did not reach its fundraising goal to reduce the scope of the project according to what was raised. Smith's team expected that the amount raised through crowdfunding would be marginal in comparison to the several millions of dollars in NYC DOT's annual budget, but that this process would create a new framework for groups to come to government with their ideas and receive investment based solely on merit. Notably, in their outreach materials and presentations to the community, the agency did not acknowledge the City's troubled history of investment in Brownsville.

By the deadline of April 1, 2016, six groups had submitted their applications to participate in NYC DOT's *El-Space Challenge*. To select participants from among this pool of applicants, NYC DOT's Urban Design Unit convened an advisory committee composed of representatives from ioby, the Department of City Planning, the Department of Housing Preservation and Development New York City, the Brooklyn Borough Commissioner of NYC DOT, and several other NYC DOT units, including the Urban Art, Public Spaces, Sidewalks and Inspection Management (SIM) Units. It was important to the Urban Design Unit that the relevant permitting agencies and offices represented in the advisory committee be familiar with the scopes of these projects before they were approved, so that approvals for implementation might ultimately be expedited.

The advisory committee recommended that the following four projects receive access to NYC DOT matching funds in the crowdfunding challenge:

- "Make Music Brownsville," led by Friends of Brownsville Parks. The organisation aimed to raise a total of $4,810 to host an event that would attract residents to the el-space by activating a series of locations along Livonia Avenue with live musical performances.
- "Best of Brownsville Photo Project," led by the Brownsville Community Justice Center (BCJC) and Brownsville Partnership. The two organisations aimed to raise a total of $14,102 to create and install kiosks under the elevated tracks on Livonia Avenue that would display images and information curated by local youth.
- "Isabahlia Healthy Food Festival," led by the Isabahlia Ladies of Elegance Foundation. The organisation aimed to raise a total of about $12,600 to host a celebration of fresh and healthy food under the elevated tracks on Livonia Avenue. Gardeners from the Powell Garden, which abuts Livonia Avenue, would donate fresh fruits and vegetables to residents while educating them about the importance of eating nutritious foods.
- "Grand St. Settlement's Brownsville Block Party," led by Grand St. Settlement. The organisation aimed to raise a total of $4,000 to host an event on Livonia Avenue that would showcase the work of local community-based organisations and start-ups.

When notifying the organisations that their projects had been selected, NYC DOT asked each of the projects' leaders whether they might be interested in combining efforts. Rather than commit their team's time to supporting the execution of three distinct events, NYC DOT preferred to work with the groups to produce a single event. The groups fundraised separately, but agreed to work together to produce a multi-site event that they ultimately called "Live! On Livonia."

Leaders of each of the groups were required to attend two 1-hour grass-roots fundraising trainings with ioby and work one-on-one with a

fundraising coach on ioby's staff. By the June 3, 2016 fundraising deadline, groups had raised nearly $11,000.[59] Of the 72 donations to the proj-
ects, 56 were under $100.[60] A small number of large donations to projects,
including a $5,000 donation from New Yorkers for Parks to the "Best of
Brownsville Photo Project," skewed the average donation to about $112.[61]
NYC DOT matched the amount raised on ioby's platform with about
$33,000 and the Urban Design Unit assigned a member of the team to
work closely with each leader to plan for implementation. Because the
groups would be working in partnership on the event, NYC DOT did not
expect to spend significant time and resources to working with the leaders
of each group.

   ioby released the funds to the groups within 2 weeks of the fundraising deadline, and NYC DOT disbursed their matching funds more slowly,
reimbursing each group as they made purchases. This is consistent with
the City's typical processes for funding a non-profit organisation, including the process through which discretionary funding from members of
City Council is disbursed.[62] The Urban Design Unit committed more staff
time and resources to making and tracking these disbursements, pulling
permits, and coordinating efforts across teams than they had originally
hoped. According to Smith, in an effort to ensure that the groups would be
successful, NYC DOT played an active role in planning the event: "we
couldn't leave it up to the groups and say, 'this is your thing and we'll
check back in.'"[63] This account suggests that the Urban Design Unit
determined that substantial oversight might have rendered the risk of the
crowdfunded projects' failure more acceptable to the agency.

   In phone interviews conducted in fall 2017, participants expressed
varying degrees of satisfaction with the degree to which NYC DOT
assisted in the development and execution of the event. Most participants
feel that NYC DOT offered at least some assistance in navigating permitting processes and coordinating the joint effort of the participating groups.
To one participant, NYC DOT's heavy involvement in the implementation
of the project seemed to signal the City's fundamental distrust of residents. This participant claims that NYC DOT withheld reimbursement for
purchases made that fell outside of what the agency deemed to be an
appropriate scope of work for the project. The participant felt pressured
into partnering with the other groups participating in the challenge and

that she would have preferred to have led her own event and to have spent her funds without being asked to comply with NYC DOT's burdensome compliance requirements. Importantly, the participant laments a lack of transparency earlier in the programme that would have enabled her to make an educated decision about whether to participate. Had she known the amount of spending oversight that NYC DOT would exercise, she might not have decided to participate in the challenge. In a different interview, another participant suggested that some of the participating organisations lacked the capacity and experience to navigate the City's procurement procedures. This participant believes that there is a positive correlation between a participant's availability, willingness, and technical ability to comply with NYC DOT's processes and his or her satisfaction with the programme.

7.1.4. Community outcomes

Ultimately, all four projects funded through the *El-Space Challenge* were fully completed (see Figures 2 and 3). In their interviews, all participants said that they do not feel as though this programme shifted the Brownsville community's existing perceptions of government. Following the "Live! On Livonia" event, NYC DOT did not invest in continuing to cultivate the relationships that they built through this programme. One participant claims that, if the City intends to build relationships with community members, agencies should work with residents in equal partnership, trust residents to make purchases that are in the neighbourhood's best interest, and continue to cultivate their relationships with community members.

Still, community groups report benefiting from the unrestricted capital that came to them from their crowdfunding efforts. Following the completion of their crowdfunding campaign, the leaders of the Isabahlia Healthy Food Festival planned to purchase premium t-shirts to distribute to attendees. NYC DOT would not allow the group to spend the agency's matching funds on items that did not explicitly align with the objectives of the *El-Space Challenge*. Because the funds raised on ioby's platform were unrestricted, the group was able to purchase and distribute t-shirts without interference from NYC DOT. The Best of Brownsville Photo Project aligned closely with the types of projects that the Urban Design

Figure 2: *A finished photo kiosk, funded through the "Best of Brownsville Photo Project" campaign and installed by volunteers and staff of the organisations that led the campaign.*

Source: Edmonds, Katie, September 2016.

Unit was accustomed to funding (i.e. public art installations on sidewalks), and the agency raised few concerns about the appropriateness of the project's expenditures.[64] As such, the organisations that led the Brownsville Photo Project may have received less utility from the absence of restrictions on the crowdfunded monies than the participating groups whose projects were less clearly in line with NYC DOT's typical categories of expenditure.

7.1.5. Government outcomes

Through this partnership with ioby, NYC DOT reports that they built trusting relationships with local leaders in the area and that they arrived at a better understanding of residents' needs and priorities.

Figure 3: *Volunteers distribute fresh foods and information about the local community garden to Brownsville residents at the Isabahlia Food Festival, funded through the El-Space Challenge and included in the "Live! On Livonia" event.*

Source: Ward, Liz, September 10, 2016.

Even so, the changes that this partnership made to NYC DOT agency culture were modest. The agency's attorneys initially raised concerns about the structure of the match challenge, which resembled a small grants programme. In his interview, Smith said, "Originally, our Legal Department said, 'We don't do grants. We do purchases and procurements.'" By writing NYC DOT's contract with ioby to exclude any language comparing the challenge to a "grant" to a community group or organisation, the attorneys approved a framework that enabled the Urban Design Unit to execute the programme as planned. Smith says that, now that the agency has a legal structure to accommodate a matching challenge, any of the agency's units would have no trouble repeating the programme in the future.[65] The Urban Design Unit is considering another round of the programme and is currently working out questions about whether the programme can be scaled to include other neighbourhoods without significantly increasing their staff's capacity to support it.

The NYC DOT Urban Design Unit cites "the optics of asking Brownsville residents to fundraise for their own projects" as their chief obstacle to moving forward with another round of a crowdfunding programme.[66] This hesitation may be rooted in the Unit's understanding of the history of disinvestment in Brownsville and the degree to which residents may feel that there is a historical deficit in basic funding for public goods and services. While the agency has funding available to invest in Brownsville, Smith claims that—without a deep knowledge and understanding of the problems in public spaces in the neighbourhood—his team does not always feel comfortable investing in a specific intervention in an El-Space. When interested groups fundraise for a project, his team can more confidently evaluate the popularity of an idea and direct investment to it appropriately. The "Best of Brownsville Photo Project" and "Make Music Brownsville" completed their crowdfunding campaigns primarily through large donations from major donors and businesses. Smith says that, "while it wasn't originally a strict read on the democratisation of choice here, I do still think that that these crowdfunding campaigns showed a level of support and buy-in that we don't usually seek in our projects. It was valuable." In the future, Smith hopes to find an additional way to have residents demonstrate interest in a project without needing to commit financial resources. This may help NYC DOT correct for any outsized influence from major donors and businesses.

According to Smith, his team benefited equally from sharing the project's workload and sharing creative responsibility. Because they were responsible for creating their projects, participants in the programme shared some of the risk that would otherwise be assumed by NYC DOT, had the agency been the only entity leading the implementation of the project. After the "Best of Brownsville Photo Project" kiosks were installed, a community member who had not been aware of the project when it was in its planning phase publicly criticized the project. Not knowing that two resident-led organisations had been responsible for their design, the community member received local news attention for his criticism of these kiosks, which he called a signal of gentrification in the neighbourhood. The Brownsville Community Justice Center (BCJC) publicly defended the project and met with the resident to discuss the community engagement that had driven the organisation to design the project. Now convinced that the project was in the neighbourhood's best interest, the resident appeared on a local television news channel to announce his support for the kiosks. NYC DOT points to BCJC's response to the controversy as an important reason that government should produce public goods and assets in partnership with community-based organisations. Smith claims that this controversy would not have been as easily resolved if NYC DOT were primarily responsible for the project's implementation.

Though NYC DOT has continued to work in Brownsville since the culmination of its *El-Space Challenge* in 2015, the Urban Design Unit has not reengaged any of the participating groups to co-produce additional events and projects. Groups that participated in the match challenge report that, after NYC DOT exerted considerable oversight over purchasing for their projects, the agency reverted to their typical project delivery protocol in Brownsville. It is not clear that the *El-Space Challenge* changed NYC DOT's approach to community engagement or that it opened up new programmes or procedures for the agency's decision-makers to more easily and adequately support resident-led initiatives. As a result, the relationship between government and the community remains largely unchanged.

7.2. Mayor's Office, City of Los Angeles: Great Streets Challenge, Los Angeles, CA (2015)

In 2014, in his first executive directive as mayor of Los Angeles, Mayor Eric Garcetti identified 15 "Great Streets" corridors where his administration would focus funds and other resources to creating safe and fun spaces for pedestrians and bicyclists.[67] To arrive at these corridors, the Mayor's Office asked City employees and elected officials to submit corridors to be considered by a working group of about 50 employees of City agencies.[68] After reviewing submissions, the working group assembled a penultimate list of 40 corridors and asked each of the 15 members of the Los Angeles City Council to pick a street in their district that they wanted to be included in the first iteration of the Great Streets Initiative.[69]

Figure 4: *Residents cross Crenshaw Boulevard and Florence Avenue using a temporary pedestrian scramble installed on February 20, 2016. This project was funded through the "Street Beats" campaign.*

Source: Los Angeles Great Streets, "Street Beats," LAGreatStreets.org, http://lagreatstreets.org/street-beats.

The 15 Great Streets corridors that the council members chose are shown in Figure 4.

According to Mayor Garcetti, the Great Streets Initiative would leverage available City resources to complete fixes to the public right-of-way that would return these corridors to a state of good repair, including replacing broken signs, filling potholes, and repainting curbs.[70] The Mayor's Office also considered the Great Streets Initiative to be an opportunity to build stronger relationships with community members and invite residents to create their own innovative place-making interventions on the corridors.

7.2.1. The city's existing relationship with community members

While some community groups in the Great Streets neighbourhoods had worked with the City of Los Angeles in the past, for most residents, much of the City's work in public spaces remained difficult to understand. Lilly

O'Brien, who led the 2015 *Great Streets Challenge* for the Los Angeles Mayor's Office, says that, to most residents, "government is healthcare, taxes, and infrastructure."[71] Most residents would neither know that the City is involved in investing in innovative solutions in public spaces nor know that the City had set out to intentionally invest resources in the city's priority corridors through the Great Streets initiative.[72]

According to Sarah Auerswald, President of the Mar Vista Chamber of Commerce and the leader of the "Make Mar Vista" campaign through the 2015 *Great Streets Challenge,* residents trust the City to varying degrees: "There are broken things about LA that we all kind of know, and that LA will never get around to fixing. There were pockets of people, like me, who said we had to try. There were pockets on the other end who said that there's no point in trying."[73] In the "vast middle," Auerswald says, are residents who do not often consider the role of government in their lives.[74]

7.2.2. Government intentions

When the Los Angeles Great Streets team began to design the *Great Streets Challenge* in 2015, they considered the potential benefits of coproducing place-making projects with community-based organisations. According to O'Brien, the team wanted this to be an opportunity to build trusting relationship with residents.[75] The team expected that any community-based non-profit—regardless of size, previous experience, or capacity—would be capable of delivering place-making projects. They believed that the only barrier to community-led projects in public spaces is the City's willingness to trust organisations. The *Great Streets Challenge* was originally conceived as a small grants programme, without any crowdfunding component.

The Great Streets team worried about their capacity to manage a small-grants programme. A conversation with the staff of ioby convinced the team that adding a crowdfunding element would help organisations to build the fundraising and organising capacities that they would need to do similar projects in the future. According to O'Brien, "the idea was that, not only would ioby help to coach applicants and recipients through crowdfunding, but they would also advise on putting projects together and building a groundswell of volunteers."[76]

In the *Great Streets Challenge Grant Application Manual* that accompanied the application to participate in the programme on ioby's website, the Great Streets team lay out basic criteria by which they would judge applications. The City required that projects met the following criteria:

- have clear public benefits;
- include a community engagement component;
- demonstrate creativity and innovation;
- contain a plan for evaluating the project's success and collecting data;
- offer an assessment of long-term impacts and implementation.[77]

In an attempt to solicit the largest pool of applications possible, the Great Streets team intentionally offered applicants only a few examples: "Successful applications will propose creative programs, interventions or other imaginative uses of public space that foster a sense of community. This may include community events or longer-term demonstrations."[78] In the *Application Manual*, the Great Streets team also informs applicants that they will be judged based on their "organisational capacity to complete projects" as well as the "feasibility of their proposals."[79]

7.2.3. Community intentions

Most applicants to the *Great Streets Challenge* had already planned to build some versions of their place-making projects before hearing about the City's programme. When they heard about the matching funds being made available to community-based organisations through the Challenge, Sarah Auerswald's team reported making small modifications to their plans to meet the City's requirements. According to Auerswald, "My producing partner said, 'If we're going do [the project] anyway, we might as well get grant money and more support and attention to make it bigger.'"[80]

7.2.4. Execution

In the City's call for ideas for the Challenge Grant on May 12, 2015, they intentionally asked applicants to share only the most important details of their projects.[81] Rather than create barriers for lower capacity and less experienced groups by including questions about the specifics of their projects' implementation, the team decided to "award applications for projects that felt relevant and were exciting, innovative, and creative, and we would sort out details along the way."[82] To get the word out about the opportunity, the Great Streets team attended community events, reached out to organisations with expansive networks in the 15 Great Streets neighbourhoods, and held an in-person "application workshop" where they could review the specifics of the match programme and all criteria by which they would be selecting participants from the pool of applicants.[83] Ultimately, 36 community-based organisations and local designers applied to participate in the *Great Streets Challenge*.

In July 2015, the Great Streets team convened an advisory committee composed of ioby staff and representatives from city agencies that work in public spaces, to review the plans of what groups were looking to accomplish. The advisory committee judged each applicant based on their community engagement strategy, demonstrated creativity and innovation, plan for evaluation and data collection, projected long-term impacts, organisational capacity, and feasibility.[84] From the 36 applicants, the following 8 were chosen to participate in the programme:

1. "Connect the Dots | Van Nuys," led by the Arid Lands Institute, Pacoima Beautiful, and the L.A. County Bicycle Coalition. The organisations raised a total of $13,192 to host a series of design workshops, focus groups, and bicycle festivals on Van Nuys Boulevard.

2. "YES (Youth Envisioned Streets) for a Healthier South LA," led by the National Health Foundation, A Place Called Home, Los Angeles Walks, Los Angeles County Bicycle Coalition, and the Coalition for Responsible Community Development. The organisations raised a total of $10,060 for local youth to create temporary street redesigns and host a one-day event on Central Avenue.

3. "Nuestra Avenida: Cesar Chavez Reimagined," led by Multicultural Communities for Mobility (a project of Community Partners), From Lot to Spot, and CALO YouthBuild. The organisations raised a total of $4,279 to host a series of sidewalk charrettes where they worked with residents to visualize new pedestrian, transit, and bicycle infrastructure on Cesar Chavez Avenue.

4. "Fig Jam," led by the North Figueroa Association. The organisation raised a total of $15,170 to host discussions, tours, and performances on North Figueroa Street, where attendees generated ideas for improving the corridor's health, mobility, prosperity, and social connectivity.
5. "Make It Mar Vista," led by the Mar Vista Chamber of Commerce. The organisation raised a total of $10,909 for a one-day festival that included a parade and the installation of temporary, pop-up protected bike lane, a mural, and wayfinding signs on Venice Boulevard (see Figure 5).
6. "REvisit Reseda Blvd," led by the Northridge Chamber of Commerce, the Northridge South Neighbourhood Council, Northridge Vision, Northridge Sparkle, California State University Northridge, the Museum of the San Fernando Valley, and Council District 12. The organizers raised a total of $6,182 to install flexible and collapsible street furniture to be used for performances, art displays, pop-up businesses, and product demonstrations at events on Reseda Boulevard.
7. "Street Beats Fundraising Campaign," led by Community Health Councils, Trust South LA, Ride On! Bike Co-Op, and Studio MMD. The organisations raised $10,536 to host a one-day community festival—with musical performance, food, art, and temporary pedestrian amenities—to encourage residents reimagine the design of the intersection of Crenshaw Boulevard and Florence Avenue.
8. "Pacoima Street Values," led by Pacoima Beautiful. The organisation raised $1,264 to create a kit of parts, community workshops, and a "CicLAvia" event where they proposed and received community feedback on new street activations on Van Nuys Boulevard.

After these groups were notified on July 13, 2015, O'Brien worked closely with groups to identify their anticipated barriers to implementation and make plans to engage local city council members and city agencies that would need to provide approvals. Based in the mayor's office, the Great Streets team was uniquely positioned to work across city agencies. O'Brien, wanting participants to feel supported by the City as they planned and implemented their projects, committed to speaking with each participant at least twice a week during and after the crowdfunding period.

Figure 5: *Residents use the pop-up bike lane and recreational spaces funded through the "Make It Mar Vista" campaign and created by community members with assistance from the City in November 2015.*

Source: Los Angeles Great Streets, "Make It Mar Vista," LAGreatStreets.org, http://lagreatstreets.org/make-it-mar-vista.

Virtually every leader of an organisation that participated in the *Great Streets Challenge* reports that the City's help was critical to the success of their project. Auerswald claims that, without the matching funds and other assistance from the Great Streets team, her project's implementation would have been far more difficult: "They were always there if we had questions. They were instrumental in getting us through roadblocks." Every group reached its fundraising goal, either through ioby's crowd-funding platform or in the form of offline (i.e. check and cash) donations. Of the 506 donations to the projects, 333 were under $100.[85] Every project received at least one donation greater than $500, and all received support from a combination of neighbourhood residents and local businesses.[86] Despite some large contributions from local businesses, funders, and elected officials, the average donation to projects in the *Great Streets Challenge* was $97.[87] Ultimately, every group successfully implemented their project.[88]

For example, Auerswald recalls a time when the California Depart-ment of Transportation ("Caltrans") tried to prevent her team from stopping traffic for her pop-up demonstration on Venice Boulevard. She was asked to present her project's scope to Caltrans executives, and O'Brien came along to defend the project. Caltrans ultimately allowed the project to progress, but asked that Auerswald obtain $25 million in liability insurance. When it became clear two days before the demonstration on Venice Boulevard that Auerswald's team would not be able to purchase an insurance policy of this size, the Great Streets team decided to have the City assume liability.

Among the people to whom the *Great Streets Challenge* participants appealed for donations, there were some who questioned the City's intentions. Some residents felt that the Challenge was a sign of government shirking its responsibility and moving the burden of delivering civic goods to the public. Auerswald says that these critics of the programme were the same residents who have historically bemoaned what they have perceived to be exorbitant government waste. To these residents, a government-led crowdfunding programme like the *Great Streets Challenge* was a mechanism for the government to withhold full funding for neighbourhood projects, masquerading as a platform for deepening community participation.[89]

The teams participating in the *Great Streets Challenge* successfully leveraged the promise of the City's matching funds for donations from residents and local businesses. Auerswald and her team took their fundraising asks from door-to-door and sent out a mailer to every address within a one-mile radius of the Great Streets project. They spoke to representatives and owners of over 120 businesses, so that they could address every stakeholder's concern and convey the objectives of the demonstration. At the time of the demonstration, Auerswald says that residents and business owners were generally supportive of the project.

7.2.5. Community outcomes

In phone interviews conducted in fall 2017, participants in the *Great Streets Challenge* reported that their understanding of the City's structures and operations—including an understanding of agencies' jurisdictions and the names of the most helpful agency representatives—had improved by the culmination of the Challenge in 2015. Since then, some say that it has been difficult to maintain relationships with City staff in light of turnover and shifting agency priorities and funding.

7.2.6. Government outcomes

According to O'Brien, her team's failure to build long-term relationships with residents through the *Great Streets Challenge* stems from a lack of long-term investment in community partnership. O'Brien says that partnerships are important to starting to lay foundations of trust with community-based organisations, but that the City should be concerned with "sticking around." For some participants, local neighbourhood council offices proved to be extremely valuable assets. Citing Los Angeles' sizable geographic footprint, O'Brien claims that neighbourhood council offices are better equipped than City Hall to build relationships with residents and respond to their needs quickly.

The *Great Streets Challenge* illuminated several problems with the City's permitting process for resident-led projects in public spaces. The Mayor's Office of Budget and Innovation hired a consultant to look at special events permitting reforms, who came back to the City with several discrete modifications that were to be made to the permitting protocol. The City has not yet adopted these changes.

Having initially assumed that the City's unwillingness to partner was the most important barrier to resident-led place-making interventions, the Great Streets team was surprised to learn that nonprofits are not always equipped to manage grant funds. O'Brien says that, despite their good intentions and the promises that they made to the City, "there are so many factions in communities. It can become so political. It was naive to think that politics only happen in City Hall. We were naive about what it meant to give communities financial tools and technical support to do projects of this nature." Despite the technical assistance that the City provided to the participants to execute their projects, some organisations are bound by internal agreements and opposition from the community.

To ensure that the role of the City remained clear to attendees of each project's launch event, the Great Streets team asked each organisation to commit to distributing an exact amount of printed collateral at their event. O'Brien regrets prescribing metrics of success for the *Great*

Streets Challenge participants after they had fundraised for their projects. According to O'Brien, organisations are far more effective at building awareness and consensus when government grants them autonomy to work in their own communities: "If you trust people enough to give them challenge grants, you shouldn't give them number goals late in the game." Despite the many challenges in executing and following up on the promises made to the participating organisations, O'Brien maintains that these partnerships are still worth pursuing because they changed and expanded communities' conversations about the safety and vitality of streets.

8. Synthesis and Conclusions

Both the *El-Space Challenge* and the *Great Streets Challenge* were built on the premise that investing in the co-production of public goods and services with community-based organisations would give the City:

- *A deeper understanding of communities' demands*: Neither NYC DOT nor the Great Streets team began their crowdfunding challenges with a thorough and nuanced understanding of residents' lived experiences in their target neighbourhoods. By committing matching funds to bolster residents' and local organisations' projects, rather than investing in projects designed and scoped by planners or policymakers with the City, NYC DOT and the Great Streets team hoped to learn more about the needs of the people whom they intended to serve.
- *Lasting relationships with community members*: NYC DOT focused their work in Brownsville in part because they did not maintain many existing relationships with community groups or residents in the neighbourhood and hoped that the programme would help to spread good will among residents with negative perceptions and low expectations of government. The Great Streets team, a small team working across a large geography, intended to build trusting relationships with residents in the 15 neighbourhoods where the City was investing in Great Streets programmes.
- *Buy-in to a larger planning effort that the city was leading*: NYC DOT hoped to generate interest in el-space interventions after the release of

their Under the Elevated report, while the City of Los Angeles hoped to build consensus around the mayor's Great Streets initiative.

- *Opportunities to fund new and experimental projects that might be considered risky*: Both the Los Angeles Mayor's Office and NYC DOT's Urban Design Unit had hoped to only commit matching funds to projects with elements that would fall outside of the City's categories of expenditure, effectively sharing the risk of their projects' failures with the groups that had crowdfunded.

Table 2 offers a side-by-side comparison of the programmes. They resemble each other in several ways, including a considerable amount of oversight that each government exercised over participants' spending and a shared intention of creating long-term, trusting relationships with residents. Even so, the structure and contexts of the *El-Space Challenge* and the *Great Streets Challenge* differ in important ways as follows:

- *Openness to risk*: The City of Los Angeles was considerably more comfortable with accepting risk in the execution of projects as evidenced by the City's willingness to purchase liability insurance on behalf of Sarah Auerswald's team.
- *Community's existing perception of government*: Residents of Brownsville, a neighbourhood with a long history of disinvestment and the sight of several failed urban renewal projects, were wary of the City's intentions in working in the neighbourhood. In many of the Great Streets neighbourhoods in Los Angeles, many residents had never considered the role of government in planning for public spaces. In these neighbourhoods, apathy had been an important barrier to building relationships with residents.
- *Number of applicants to the programme*: Despite spending considerable resources and time attempting to publicize their call for applications and to identify organisations with ideas for activating the El-Space on Livonia Avenue, NYC DOT was only able to recruit six applications. The City of Los Angeles was more successful in identifying groups with Great Streets projects and funding needs, ultimately bringing in 36 applications to the programme. This discrepancy between the

Table 2:   *Characteristics of the El-Space Challenge and the Great Streets Challenge.*

Characteristic of programme	El-Space Challenge (NYC DOT)	Great Streets Challenge (Los Angeles)
Government oversight	Significant	
Openness to accepting risk	Not open	Moderately open
Community's existing perception of the government's responsiveness to their needs	Mostly negative	Mostly uninformed/ unconcerned
Number of applicants	6	36
Percentage of applicants that government chose to fundraise through the programme	67%	22%
Breadth of eligible project types and scopes	Limited by geography and project type	Limited by geography and project type
Relative strength of government's existing relationships with community-based organisations	Weak	Mixed
Amount of match funding available for resident-led campaigns	Up to $60,000	Up to $200,000
Timeline of the crowdfunding programme (i.e. whether the programme was time-bound or is ongoing)	Time-bound	
Amount and nature of technical assistance provided by the crowdfunding platform operating the programme	Required grass-roots fundraising trainings, 1:1 coaching	

programmes may be due, at least in part, to a difference in scale: the NYC DOT programme targeted only one corridor in Brownsville, whereas the Los Angeles programme was open to applicants in 15 corridors around the city.

• *Percentage of applicants that the government chose to fundraise through the programme*: NYC DOT accepted four participants from an applicant pool of six (67%), whereas the City of Los Angeles accepted only 8 from a pool of 36 (22%). The Great Streets team was more selective, in part because their applicant pool was significantly larger.

- *Breadth of eligible project types and scopes*: Both NYC DOT and the Great Streets team required applicants to meet a set of criteria based on location and the type of project being proposed.

- *Relative strength of government's existing relationships with community-based organisations*: The Urban Design Unit of the NYC DOT had very few, weak relationships with community-based organisations before this programme began. The Great Streets team had previously worked with organisations in some—but not all—Great Streets neighbourhoods.

- *Amount of match funding available for resident-led campaigns*: NYC DOT had set aside up to $60,000 in the agency's expense budget for participating projects, whereas the City of Los Angeles had committed up to $200,000 in matching funds for projects in the 15 Great Streets neighbourhoods.

- *Rules of matching donations*: In Los Angeles, the Mayor's Office gave $1 for every $1 raised. To account for the systemic inequities in Brownsville, NYC DOT contributed $3 for every $1 raised by every group working through the El-Space Challenge.

- *Timeline of the crowdfunding programme*: Rather than give groups access to matching funds into perpetuity, both NYC DOT and the Great Streets team required groups to fundraise within a discrete time frame.

- *Amount and nature of technical assistance by the crowdfunding platform operating the programme*: Both NYC DOT and the City of Los Angeles used ioby as the civic crowdfunding platform to operate their programmes. In each programme, ioby paired every leader of a crowdfunding campaign with a grass-roots fundraising coach—called a "Leader Success Strategist"—on ioby's staff. These Leader Success Strategists helped each group develop its campaign messaging, approach, timeline, and budget. ioby also led two mandatory grass-roots fundraising and digital communications trainings for all community leaders in each programme and followed up with each leader to offer additional coaching throughout the fundraising period.

8.1. Trust and Expectations

In his 2007 piece, Tony Bovaird argues that, for governments and communities to successfully co-produce public goods and services, community stakeholders need to trust the "advice and support" of government.[90] Given long-time residents' experiences enduring city policies that disenfranchised, divided, and isolated the community, groups that participated in the *El-Space Challenge* entered the programme without fully trusting in the good intentions of NYC DOT. In part because they initiated the programme in a neighbourhood with such a profound deficit of community trust, the Urban Design Unit ultimately failed to build trusting and sustainable relationships with long-time residents.

NYC DOT's Urban Design Unit came to Brownsville with their *El-Space Challenge* without first having attempted to earn the community's trust. It is not clear whether the Urban Design Unit would have been more successful in building these relationships if they had begun the programme by explicitly acknowledging the City's role in the neighbourhood's unique history of disinvestment. Had NYC DOT and other agencies began their approach to engaging leaders and long-time residents in Brownsville by demonstrating a genuine effort to invest in their health, safety, and quality of life over the long term, the City might have found it easier to lay the foundations of successful, productive partnerships.

In Los Angeles, a pervasive apathy enabled the Mayor's Office to exceed residents' extremely limited expectations of government.[91] In a city where many residents had not previously considered the role of government in creating functioning public spaces, the *Great Streets Challenge* stoked an interest in civic life and promoted a deeper understanding of government operations. In Los Angeles, the Mayor's Office was able to set the public's expectations for government involvement in civic life. In Brownsville, NYC DOT failed to adequately challenge the public's pre-existing distrust in the good will of the government.

While both programmes succeeded in helping community-based organisations create projects in public space in the short term, the NYC DOT and the Great Streets teams failed to build a mutual trust that would last beyond the completion of the programme. Table 3 offers a side-by-side comparison of outcomes across programmes:

Table 3:   Outcomes of the El-Space Challenge and the Great Streets Challenge.

Outcome of programme	El-Space Challenge (DOT)	Great Streets Challenge (LA)
Amount spent in matching funds	$32,751	$70,261
Amount raised through crowdfunding	$10,917	$71,592
Success rate (% of groups who completed projects as planned)	100%	100%
Improved community's trust in City?	No	No

9. Findings and Recommendations

9.1. Questions to Ask before Initiating a Crowdfunding Programme

A city government looking to enable a crowdfunding programme similar to New York's *El-Space Challenge* or the *Great Streets Challenge* in Los Angeles should consider the following questions upfront:

- In which neighbourhood(s) might a crowdfunding programme help a city build trust with residents? In which neighbourhood(s) might such a programme harm a city's ability to create trusting relationships?
- How might a city design the rules and eligibility criteria of a crowdfunding programme to ensure that public resources are being spent equitably and that community groups hold at least some responsibility for delivering their crowdfunded projects?
- How will a city approach partnerships with communities to deliver civic projects? How does that city intend to maintain these partnerships with community groups beyond the completion of the crowdfunding initiative?

9.2. A Stepwise Approach to Launching a Crowdfunding Programme

9.2.1. Choosing a neighbourhood

The difficulties that the Great Streets team in Los Angeles and the NYC DOT Urban Design Unit faced when attempting to use crowdfunding to strengthen their ties to communities may offer some guidance to municipalities looking to initiate their own civic crowdfunding programmes. Firstly, governments should locate their programmes in neighbourhoods that have strong—or, at least minimally contentious or controversial—relationships to government. Because successful co-production requires a foundation of trust, a match challenge that leverages civic crowdfunding may not be a good fit for a neighbourhood such as Brownsville, where deeper and deliberate investment is needed to improve residents' quality of life. Until and unless a city demonstrates an interagency commitment to reversing the harm that the government's policies have caused, any of that government's agencies may be unsuccessful in building trusting relationships through one-off programmes such as the *El-Space Challenge*. Conversely, crowdfunding may work well to build trusting relationships in neighbourhoods where residents have not been engaged in fights for greater public investment, such as in higher income neighbourhoods and in neighbourhoods where civic apathy has crowded out a general opposition to municipal intervention. When aiming to co-produce goods and services with community-based organisations and residents in these neighbourhoods, a city may either capitalize on a foundational trust in government's intentions or easily exceed the community's low or absent expectations of government.

When choosing a neighbourhood for its crowdfunding programme, a city should be aware that there may be a negative correlation between previous investment in a neighbourhood and its residents' ability or willingness to fundraise. Even without developing strong, sustainable, and meaningful relationships with government, the groups that participated in the *El-Space Challenge* were largely successful in fundraising for their projects. This may point to a powerful feeling of self-reliance among Brownsville's community members that was necessitated by decades of public and private disinvestment. In wealthier and resource-rich neighbourhoods that are not enduring the same legacies of systemic racism and long-term disinvestment, there may be a higher expectation of government. Unlike Brownsville, where residents do not expect their government to act in their best interest, residents of higher income neighbourhoods may expect their government to assume responsibility for funding and maintaining all civic assets.

To ensure the success of its efforts to co-produce goods and services with community stakeholders who trust the government to be a fair and effective partner, a city should only operate a civic crowdfunding challenge in a neighbourhood where it has made significant and well-received investments in improving the lives of residents. In neighbourhoods where there is a foundation of trust between government and communities, a civic crowdfunding programme may bolster that trust and spur innovation among grass-roots groups. Still, to stem the proliferation of cycles of disinvestment—where wealthy neighbourhoods receive public investment and funding flows more slowly to others because their social, economic, and political problems are too complex to solve easily or quickly—cities should be careful not to focus exclusively on wealthy neighbourhoods. Instead, cities should invest strategically and substantially in these neighbourhoods with histories of disinvestment, demonstrating that the city is willing and ready to begin remedying the damage from inequitable and destructive policies of the past.

Specifically, in neighbourhoods without a foundational and mutual trust between residents and the city, governments may consider any number of investment strategies to build trusting relationships before inviting residents and organisations in those communities to participate in a crowdfunding challenge. These strategies may include redesigning funding structures, funding communities' ideas through a PB process, and beginning to build the capacity of community-based organisations to fundraise for and deliver civic projects:

- *Redesign funding structures*: Cities should prioritize funding for innovative projects in historically disinvested neighbourhoods. By explicitly and meaningfully accounting for policies that have segregated, isolated, and systematically disadvantaged people of colour and low-income households, government may signal to communities that decision-makers are committed to achieving an equitable distribution of resources.
- *Begin funding communities' ideas*: PB may be a useful tool for a government looking to invest in the ideas of residents in disinvested areas. As a supplement to sustained, meaningful investment in these communities, the PB process begins to introduce residents to the notion that

government is willing to enable and equip residents to determine the city's funding priorities. This process is also highly collaborative and creates opportunities for governments to explore communities' priorities and to more deeply understand residents' perceptions and expectations of government.

- *Begin building organisations' capacities to fundraise for and deliver civic projects*: Offering trainings and tools for non-profits seeking to improve their governance structures and fundraising capacities may help local community-based organisations prepare to participate in a partnership with government. At the same time, a government may begin to build trust with neighbourhood groups by making small, unrestricted grants to organisations and tracking their success. Small grant-making without a burdensome compliance requirement demonstrates to the community that the city is willing to place trust in organisations' ability to spend funding on important programmes for their communities. This also gives a city some direct insight into the readiness of each organisation to participate in a future crowdfunding programme, when a greater amount of public funding will be available to each group.

9.2.2. Setting criteria

To solicit the greatest range of project types and sizes possible, while still optimising the benefits it stands to receive from the crowdfunding programme, a city should follow these two guidelines in the design of its eligibility criteria:

- *Build criteria to favour new ideas*: To spur innovation and to maintain residents' trust in government's good intentions, a city should design its civic crowdfunding programme to fund new and experimental projects that the city is unable to fund on its own. These might include pop-up, experimental, or pilot projects.
- *Set eligibility requirements for project leaders*: To increase the likelihood of each project's success, a city should be careful to select projects that are led by residents and community-based organisations with the requisite capacity (i.e. time and resources) to partner with the city on the project's implementation. Consistent with the best practices of

co-production, the government must be able to trust that their contact in the community is available to work in full and equal partnership. Because the government should be concerned with building community partnerships, it should only consider submissions from people with deeply rooted connections to the neighbourhood. This may include residents or community-based organisations.

9.2.3. Choosing a crowdfunding platform

When deciding among crowdfunding platforms with which to partner, a City should be careful to select the platform that adds as much value to a neighbourhood's civic life as possible. By coupling the crowdfunding programme with customized grass-roots fundraising trainings and one-on-one coaching from ioby, both NYC DOT and the Los Angeles Mayor's Office aimed to build local organisations' capacities to fundraise and organize. In both the Los Angeles *Great Streets Challenge* and the NYC DOT *El-Space Challenge*, ioby served as a technical assistance provider to each city and to the groups leading the crowdfunding campaigns. A city should seek the same supplemental capacity-building services from any civic crowdfunding platform with which it chooses to partner.

In addition, a civic crowdfunding platform should be able to carry most or the entire compliance burden for participating groups. Low-capacity and volunteer-led organisations are unduly burdened by agencies' requirements that they submit forms and receipts before receiving their matching funds. Similarly, the delay between purchase and reimbursement may result in a severe restriction of cash flow for groups with small or no budgets. To remedy these problems, the crowdfunding platform should be set up to accept the matching funds on behalf of the groups, submit the required documentation on behalf of the group, and disburse funds to the group. Ideally, the platform would be comfortable accepting the risk of disbursing up front to groups so that they are not forced to make purchases and wait for reimbursement. Given that there is a potential for a substantial amount of funding to flow through the crowdfunding platform, the crowdfunding platform would likely need to enter into contract with the city.

9.2.4. Setting the rules of the programme

As in the Los Angeles *Great Streets Challenge* and the NYC DOT *El-Space Challenge*, a city should commit to matching each contribution coming through the crowdfunding platform. This demonstrates a city's willingness to partner with the community on the development and implementation of the idea and enables the city to limit contributions based on the project's popularity in the community.

If the primary purpose of using matching funds is to correlate city funding to the degree of support that a project receives from affected residents, a city should insist on setting a limit on the amount that it will match per donor. Rather than match the complete amount of every donation, a city should set a cap on the amount that it will match from every donor. For example, a city might settle on a 3:1 match with a $100 match cap per donation. In this case, every donation would be matched up to a maximum of $100. A donation of $90 would receive a match from the city for $90; a donation of $1,000 would receive a match from the city for $100 (the maximum allowable match.) This would encourage leaders of crowdfunding campaigns to seek several smaller donations and help to reduce the influence that a small number of large donations might have on the success of a crowdfunding campaign.

9.2.5. Following-up

A city considering whether it might launch a similar crowdfunding programme should weigh its commitment to working with residents and community-based organisations in the long term. Agencies should only lead the effort if they have time and resources to follow through on all promises made to applicants. To maximize the trust-building effect of the programme, the agency's decision-makers should express a willingness to fundamentally change the ways in which their staff interacts with stakeholders during a project's delivery. When attempting to use civic crowdfunding to build trust with communities, decision-makers should commit to adopting the tenets of co-production and participatory governance (e.g. PB and crowdsourcing) as new standard operating procedure. One-time civic crowdfunding programmes like the *El-Space Challenge* and the

Great Streets Challenge seem to have little effect on the perception of government among communities. As evidenced by the strong, negative reaction elicited by some participating groups in Brownsville, an isolated crowdfunding programme like the *El-Space Challenge* may in fact be damaging to communities' trust in historically disinvested neighbourhoods.

Following the completion of a civic crowdfunding programme, government should meet with residents and the leaders of the campaigns, evaluate all successes and failures, and use these findings to develop a more effective strategy to deliver funds to and co-produce goods and services with the community. Programmes such as the *El-Space Challenge* and the *Great Streets Challenge* offer an array of lessons for governments looking to more effectively serve residents. Governments participating in a similar programme should maintain the new connections with community members and continue to engage them in planning decisions. Especially in neighbourhoods where residents' trust in government has been eroded by betrayed promises by planners and policymakers, it is critical that governments prove their trustworthiness by following through on promises to listen to and meaningfully engage members of the community.

9.3. Implications

By reserving the use of civic crowdfunding to fund new amenities in neighbourhoods where residents already trust the government to work in their best interest, there is a significant risk that disinvested neighbourhoods—where trust in government is more limited—will continue to fall behind neighbourhoods that are rich in resources. To account for this possibility, cities wishing to experiment with civic crowdfunding in higher income neighbourhoods should simultaneously increase their allocation of more traditional investments in lower income neighbourhoods. Governments should not opt out of deep community engagement in low-income neighbourhoods, but should understand how the high-risk and high-touch investment of a civic crowdfunding programme may work against its efforts to build trust in these neighbourhoods.

9.4. On Resource Equity

Because government-run civic crowdfunding programmes are relatively new, their impacts have largely been unstudied. Given its limited application in the United States, it remains unclear whether civic crowdfunding may help to ensure that resources are distributed equitably across neighbourhoods. Although Brent and Lorah's analysis suggests that residents and organisations in lower income neighbourhoods are not at a disadvantage in their crowdfunding efforts, there are insufficient data available to examine the long-term impacts on social and economic outcomes with a concentration of crowdfunded projects in a neighbourhood. A future study might examine the strength of relationships between socioeconomic and demographic distributions across a city, the locations of government-led crowdfunding programmes (e.g. the *El-Space Challenge* and the *Great Streets Challenge*), and the long-term effects on public and private investment across neighbourhoods.

About the Author

David J. Weinberger is a City Partnerships Director at ioby and works with municipalities nationwide to leverage ioby's crowd-resourcing platform and services for inclusive and effective community engagement. He enjoys breaking down barriers to civic engagement and building mechanisms for truly participatory planning. Before joining ioby, David worked with the NYC Department of Transportation to plan and launch WalkNYC, the city's first-ever standardised pedestrian wayfinding system. He has also worked with the Metropolitan Transportation Authority, the Democratic National Committee, the US Department of State, and PBS. David received his M.S. in City and Regional Planning from Pratt Institute, and his B.A. in political science and public policy from the CUNY Macaulay Honors College at Hunter College.

Endnotes

1. Bovaird, Tony. "Beyond Engagement and Participation: User and Community Coproduction of Public Services." *Public Administration Review*, September/October (2007), 858.

2. *Ibid.*, p. 847.

3. *Ibid.*, pp. 856–857.

4. *Ibid.*, p. 856.

5. Brabham, Daren C. "Crowdsourcing the Public Participation Process for Planning Projects." *Planning Theory*, 8(3) (2009), 242–262 (SAGE, doi:10.1177/1473095209104824, p. 243).

6. Saad-Sulonen, Joanna. "The Role of the Creation and Sharing of Digital Media Content in Participatory E-Planning." *International Journal of E-Planning Research*, 1(2) (2012), 1–22 (doi:10.4018/ijepr.2012040101, p. 1).

7. *Ibid.*

8. Brabham (2009), p. 257.

9. Meijer, Albert, Nils Burger and Wolfgang Ebbers. "Citizens4Citizens: Mapping Participatory Practices on the Internet." *Electronic Journal of e-Government*, 7(1) (2009), 99–112 (doc.utwente.nl/94543/1/ejeg-volume7-issue1-article183.pdf, p. 99).

10. *Ibid.*, p. 99.

11. *Ibid.*, p. 111.

12. Meijer, Albert Jacob. "Networked Coproduction of Public Services in Virtual Communities: From a Government-Centric to a Community Approach to Public Service Support." *Public Administration Review*, 71(4) (2011), 598–607 (*JSTOR*, jstor.org/stable/23017469, p. 598).

13. *Ibid.*, p. 606.

14. *Ibid.*, p. 600.

15. *Ibid.*

16. Brabham, Daren C. *Crowdsourcing*. Cambridge, The MIT Press, 2013, p. 3.

17. *Ibid.*

18. *Ibid.*, p. 4.

19. Anttiroiko, Ari-Veikko. "Smart Planning: The Potential of Web 2.0 for Enhancing Collective Intelligence in Urban Planning." In *Emerging Issues, Challenges, and Opportunities in Urban E-Planning*, edited by Silva, Carlos Nunes, IGI Global, 2015, p. 21.

20. *Ibid.*, p. 22.

21. *Ibid.*

22. *Ibid.*

23. Fountain, Jane E. "Connecting Technologies to Citizenship." In *Technology and the Resilience of Metropolitan Regions*, edited by Michael A. Pagano, University of Illinois Press, 2015 (*JSTOR*, jstor.org/stable/10.5406/j.ctt155jmmd.5, p. 47).

24. Fountain, p. 45.

25. *Ibid.*
26. Fountain, p. 46.
27. Bovaird (2007), p. 851.
28. Participatory Budgeting Project. *Participatory Budgeting: Next Generation Democracy*. Brooklyn, NY: Participatory Budgeting Project. participatory-budgeting.org/white-paper, p. 2.
29. Bovaird (2007), p. 14.
30. Participatory Budgeting Project. *Participatory Budgeting: Next Generation Democracy*. Brooklyn, NY: Participatory Budgeting Project. participatory-budgeting.org/white-paper, p. 11.
31. Participatory Budgeting Project, p. 10.
32. Bovaird (2007), 14.
33. Lydon, Mike, and Anthony Garcia. *Tactical Urbanism*, Island Press, Washington, D.C., 2015, p. 3.
34. *Ibid.*, p. 4.
35. *Ibid.*, pp. 21–22.
36. Davies, Rodrigo. "Civic Crowdfunding: Participatory Communities, Entrepreneurs and the Political Economy of Place." Master's thesis, Massachusetts Institute of Technology, 2014 (SSRN, papers.ssrn.com/sol3/papers.cfm?abstract_id=2434615, p.139).
37. *Ibid.*
38. Mayer, Martin. "Civic Crowdfunding and Local Government: An Examination into Projects, Scope, and Implications for Local Government." 2016. Old Dominion University, PhD dissertation (digitalcommons.odu.edu/public service_etds/5, pp. 44–45).
39. *Ibid.*, p. 44.
40. *Ibid.*
41. *Ibid.*
42. Gilman, Hollie Russon. *Democracy Reinvented: Participatory Budgeting and Civic Innovation in America*. Washington, D.C., Brookings Institution Press, 2016 (*JSTOR,* jstor.org/stable/10.7864/j.ctt1c2cr0t, p. 144).
43. Marschall, Melissa J. "Citizen Participation and the Neighborhood Context: A New Look at the Coproduction of Local Public Goods." *Political Research Quarterly*, 57(2) (2004), 231–244 (*JSTOR*, jstor.org/stable/3219867, p. 231).
44. Davies (2014), p. 140.
45. Marschall (2004), p. 231.
46. Clary, Bruce B. "Designing Urban Bureaucracies for Coproduction." *State & Local Government Review*, 17(3) (1985), 265–272 (*JSTOR*, jstor.org/stable/4354856, p. 266).

47. Davies (2014), pp. 135–136.

48. Mayer (2016), pp. 45–46.

49. Brent, Daniel and Lorah, Katie. "The Geography of Civic Crowdfunding: Implications for Social Inequality and Donor-Project Dynamics," Departmental Working Papers 2017-09, Department of Economics, Louisiana State University, p. 4.

50. Lorah, Katie. "Let's Talk About Civic Crowdfunding and Government Responsibility." Planetizen. April 4, 2016. https://www.planetizen.com/node/85437/lets-talk-about-civic-crowdfunding-and-government-responsibility.

51. Spacehive. "The Peckham Coal Line urban park." spacehive.com/peckhamcoalline.

52. Taylor, Marilyn. "Unleashing the Potential: Bringing Residents to the Centre of Regeneration." Joseph Rowntree Foundation, York, UK, 1995.

53. *Ibid.*

54. Brent (2017), p. 28.

55. ioby. "Public Sector Partnerships." ioby.org/gov.

56. Smith, Patrick. Personal interview. October 26, 2017.

57. *Ibid.*

58. *Ibid.*

59. Weinberger, David. "NYC DOT vs. LA Great Streets: The Role of Government Trust." Unpublished Keynote presentation, ioby, 2017, p. 16.

60. Weinberger (2017), p. 26.

61. Weinberger (2017), p. 25.

62. "Timeline." *Not-for-Profit Timeline*, New York City Department of Design and Construction, www1.nyc.gov/site/ddc/contracts/not-for-profit-timeline.page.

63. Smith, Patrick. Personal interview. October 26, 2017.

64. Lee, Layman. Personal interview. November 27, 2017.

65. Smith, Patrick. Personal interview. October 26, 2017.

66. *Ibid.*

67. Mayor Eric Garcetti's Great Streets Initiative. *Great Streets Challenge Application Manual. Great Streets Challenge Application Manual*, ioby, 2015.

68. O'Brien, Lilly. Personal interview. October 24, 2017.

69. *Ibid.*

70. Mayor Eric Garcetti (2015), p. 2.

71. O'Brien, Lilly. Personal interview. October 24, 2017.

72. *Ibid.*

73. Auerswald, Sarah. Personal interview. October 27, 2017.

74. *Ibid.*

75. O'Brien, Lilly. Personal interview. October 24, 2017.

76. *Ibid.*

77. Mayor Eric Garcetti (2015), p. 3.

78. *Ibid.*

79. *Ibid.*

80. Auerswald, Sarah. Personal interview. October 27, 2017.

81. City of Los Angeles. Mayor Eric Garcetti. *Mayor Garcetti Announces Great Streets Challenge Grant. Press Releases.* City of Los Angeles, May 12, 2015. Web. November 1, 2017 <www.lamayor.org>.

82. O'Brien, Lilly. Personal interview. October 24, 2017.

83. Mayor Eric Garcetti (2015), p. 16.

84. Mayor Eric Garcetti (2015), p. 3.

85. Weinberger (2017), p. 7.

86. *Ibid.*

87. *Ibid.*

88. O'Brien, Lilly. Personal interview. October 24, 2017.

89. Auerswald, Sarah. Personal interview. October 27, 2017.

90. Bovaird (2007), p. 856.

91. Auerswald, Sarah. Personal interview. October 27, 2017.

Chapter 4

The Design of Paying Publics

Ann Light and Jo Briggs

Abstract

Crowdfunding offers a different approach to social innovation undertakings at a time of rapidly shrinking state support. Social innovation involves new social practices that aim to better meet the social needs and shape collective futures. The innovation here hinges on the way in which crowdfunding platforms can change the way that society works as well as the financial details of individual campaigns. Working alongside the design features of the platform are the social, economic, and legal aspects of financial systems that evolve over time and shape what platforms can enable. In this chapter, we discuss how publics form around platforms with an interest in what is being supported. We use the term "paying publics" to refer to the way in which the four UK-based platforms we feature are using this relationship with their funders and supporters to change how funding affects communities and environmental behaviour. We can be quite precise about who the individual members of a crowd contributing through a particular platform to a specific campaign might be, whether friends and family or international networks of backers. So, we suggest that we are not served well by the term "crowd." In talking about publics, we refer to the way that particular groups may be brought into being by the actions of the platform. No single platform is redesigning economic life. However, each offers possibilities for linking private, public, and personal money and services in new ways; together, that signals societal as well as financial innovation. New common interests grow around the platform and can take on a life of their own.

1. Introduction

1.1. Crowds vs. Publics

In crowdfunding, multiple individuals co-fund ventures, just as, in crowdsourcing, many individuals co-source answers, solve problems, and offer the wisdom of the crowd.[1,2] In fact, we can be quite precise about who the individual members of a crowd contributing through a particular platform to a specific campaign might be, from friends and family to international networks of funders. So, we suggest that, in looking at activity with and around platforms, we are not served well by the term "crowd." Instead, we argue that participants in a crowdfunding ecosystem for social innovation meet the definition of public,[3] with common interests growing through an investment of time and/or resources in related outcomes.

In talking about publics here, we refer to the way that groups around platforms may be constituted by the actions of the platform. Where people related through platforms are not aware of each other at the outset, the platforms (and campaign organisers) may work to build this awareness. Clearly, it is unnecessary to generate this level of collective sensibility for all crowdfunding, but we are looking at instances when such constellations can be usefully mobilised.

"The public" is identified and constituted by an issue *outside its immediate control*, argues Dewey.[4] We can regard collecting, loaning, and giving money for a particular social end as the construction of community-level publics around an issue. In crowdfunding, possible engagement is determined by design factors in the hands of the platform-makers and in regulatory frameworks such as UK financial legislation, so participants have little control over the form of their engagement. Yet, these local publics can exert influence and play a part in enacting or resisting the goals of those setting the agenda. Murray *et al.*[5] note the impact of/on an early group of funders: "In February 2008, 26,000 people, responding to a web call, each put £35 into a newly formed co-op and bought a football club, Ebbsfleet United. Two months later many of the members—pioneers of a new form of financial collaboration—travelled to Wembley and saw their side win the FA trophy" (p. 112).

2. Methods of Data Collection

We interviewed the people behind several crowdfunding platforms with an interest in social innovation to learn of their motivations and their ensuing design choices. We chose the sample presented here from a wider pool of interviews, some of which span several years, selecting those where the interviewees exhibited a clear intention to change social practice as well as raise money through crowdfunding. We selected the cases, for variety and relevance, from our samples (~15 long-term studies by Light and ~15 more cross-sectional interviews by Briggs), choosing to feature four examples with related themes to make the points in this chapter. We have picked cases of innovators with ambitions for social change that pair well to draw contrasts and find common ground: two attempting to change societal environmental behaviour through providing a new financial mechanism and two that speak to a more general socio-economic trend in reconfiguring local relations (see Table 1). We are not claiming all platforms show these characteristics, but merely claim that attempting to shape social behaviour is a discernible phenomenon. We recorded all interviews and transcribed and analysed them.

Our first interviews specific to digital platforms were conducted in 2009. Those we use come from between 2012 and 2016. Access to interviewees came through personal relationships and intermediary introductions over many years of participating in business, social, and activist networks and government forums. The benefit of having several initiatives right from the start, before launch, is that we have been able to trace the progress and draw on it. In other words, long exposure to the sector determines our conclusions.

Table 1:   *Summary of characteristics of our four case studies.*

Platform	Trillion fund	Patchwork	Newcastle CC (Funding circle)	Crowdfunder
	Driving new societal environmental behaviour		*Changing local socio-economic configurations*	
Overview	Ltd. company—renewable energy investments.	Ltd. company—coordinated gift-giving.	Local authority working with limited company.	Ltd. company—mobilising grass roots.
Motivation beyond profit	Environmental sustainability by moving investment into green fuels.	Minimising waste through coordinated gift-buying.	Fast finance for local firms by networking with local investors.	Support for groups, start-ups, and causes to raise project funds.
Products	P2P lending, community shares, investment fund, bonds, and equity.	Customised donation reward system for aggregating presents.	P2P lending.	Reward, donation, and community shares.

3. Four Examples of Digital Crowdfunding

The descriptions here are drawn from the extended semi-structured interviews with those in charge of (or acting as spokesperson for) a platform, interwoven with observations from watching their progress. Each interview covered aspects of motivation, beliefs, ethical orientation, and ambitions for the platform, but we only draw on a narrow cross-section here to discuss social change elements and how these fit with choices and outcomes. We do not include material from users, either as fundraisers or as funders, limiting this discussion for brevity and focus. We present the studies in two pairs to reflect similarities between goals (see Table 1): two platforms that engage with worldwide environmental sustainability in their choice of services, and another two that seek to make more generic socio-economic change towards regional social and economic sustainability.

4. Case Study: The Trillion Fund

Trillion Fund claimed to be *the UK's largest social crowdfunding platform* in April 2015 (video on platform website), with a turnover in millions and 7,000 members.

The platform is principally a P2P loan, rather than an equity model, enabling the spreading of investments and a range of causes to support. However, there are some opportunities that involve issuing community shares: "By joining, you are becoming part of a community that has a broader objective than just maximising returns for investors."

The team that set up Trillion Fund ("crowdfunding for people, planet and profit") was motivated by the need for large loans and advances to small energy companies wanting to innovate with renewables. They were already successful entrepreneurs and the target (in the name) was to help generate the "$1 trillion of annual global investment required to reach the scale of clean energy generation needed to prevent further global warming."

Motivation was commercial and environmental, a business to make money for the platform owners and divert existing funding to greater global ecological well-being. This mix of missions ran through all aspects, informed by the belief that investors tend to look at returns first: "I don't care immediately why people are investing in renewables. I want them to invest in renewables and then make money. Profit is the motivation that breaks down barriers and makes people aware. We have to preach to the unconverted because that is the majority" (Groves, interview). Investors agree a rate of return on their loan, made directly to any company trying to raise funds, but brokered by the site. The platform acts as a two-sided market, allowing investors to search for and manage opportunities that range from supporting major energy infrastructure projects to social enterprises and providing companies seeking funds with investment. The team promotes their featured energy companies, both on the site and beyond, and campaigns for more funding and subsidy for the renewables sector. This promotion can be seen not only as part of making the site commercially successful but also as a pragmatic strategy to promote renewable investment.

In a shock move, the platform closed to all future business in September 2015. The service was like many P2P lending platforms, but, in specialising in the renewable sector that was made viable by government subsidy, it became vulnerable to changes in the national funding policy. "Most renewable energy projects in the UK are dependent on subsidies to provide returns to lenders. The level of subsidy is dropping rapidly and it is not clear what projects will be seeking funding going forward. As a business, Trillion incurs significant costs in marketing new loans and the Board has decided it is not prudent to continue to offer loans without visibility of future funding and project opportunities" (website). Existing clients, based on more than 120 capital raises in the area of renewable energy, continue to borrow/lend through the site, but there are no new campaigns.

5. Case Study: Patchwork

The second study in this pair involves a gift aggregation platform. Knight left a job at an environmental charity to set up the platform, which she runs from a shop in South London, stressing both local presence and green ambitions: "The principles are about being resourceful, using our money wisely to invest in things that are wanted and needed and not buying each other a ton of crap that we don't need, that ends up in landfill."

Patchwork,[6] with the styling of a rewards-based service but the function of a donation system, supports groups of people buying a single collective present. An item, like a bicycle, or series of elements, like those for a honeymoon, are divided into manageably-priced bundles, shown in a patchwork image on the website, under the slogan: "Get friends and family together to fund one gift that's really wanted—piece by piece." Then each donor pays for a part, directly into the recipient's PayPal account, and the site levies a fraction of the cost. Knight raised the money to fund the project through patchwork principles, involving 25 individual investors who all have 1% of the business. The company raised £250,000 in 20 days.

She says that the idea came to her as she planned her own honeymoon, worrying about the taboo of asking for money. The platform allows for customisation, in choosing images, prices, etc., and in then allowing you to send a personal "thank you" to the individuals who bought each piece. "So you can send pictures of you drinking your beer on the beach to all the people who bought the beer." The effort people put in has impressed Knight. There are ready-made patchworks for use, but people go further, for instance, the couple getting married, who not only made a patchwork themselves, but staged what they wanted to do, took pictures of themselves doing it, and uploaded the images.

The site takes 3% of each contribution. This contrasts with other platforms, such as Kickstarter and JustGiving, which charge more, or TaskRabbit with a 20% service levy. "In terms of competition, our commission is quite low. The sharing economy only really works when you reach scale. So, there won't really be any money to be made until we reach real scale." This figure was chosen from a user's point of view. Knight points out an advertising-supported model would not work, nor would affiliation to retailers, but, even now, it remains early days in terms of financial viability.

6. Case Study: Newcastle City Council

This loan platform and portfolio shows the working, for mutual benefit, of branded white-label platform products (i.e. developed to be branded by others) in a partnership between a council and a technology provider. The account is based on interviews in 2014 with Councillor Michael Johnson, at that time an elected member, who instigated the initiative at this northern English council, and with Sophie Chappelow, at the time of the interview senior marketing and partnership manager at P2P platform Funding Circle.[7] The interviews were conducted a few weeks after the launch of the new council-backed lending scheme.

Newcastle City Council is supporting the supply side of business finance because "traditional methods of raising capital were no longer working for companies" (Johnson, interview). The council believes lending decisions should be based on flexible criteria, with a quicker turnaround than that of traditional banks (which can take months to respond to a loan request, by which time opportunities have often passed). It wanted to address people locally "who didn't have experience of investing in businesses and who didn't have sufficient capital to become Angel investors, but who had enough money to be managing their personal investments." Johnson judged that, being risk averse, these individuals are unlikely to invest in businesses directly, but are "probably socially motivated." These characteristics, and Johnson's own interest in P2P lending prompted him to find a way to "provide them with an investment product that not only presents itself as a lower risk option [but] also has a social benefit of restricted investment criteria—so it has to give back locally."

A policy was introduced enabling the city council to co-design a product with Funding Circle. Previously, "[the council] would have taken on a fund to do that and … it might be the cost of hiring three or four people to manage this pot of money" (Chappelow, interview). The partnership is intended to enable council and lenders to benefit through the low cost of each financial transaction and, overall, a fast, easy, and inexpensive business loan supply service.

Funders provide loans with a duration of 6 months to 5 years. They select companies to invest in based on the credit model given in the proposal (banded A+ to C– depending on risk) and offer loans in return for annual interest of between 6% and 15%. Investors can assess the viability of a loan request by conducting online searches and the platform facilitates queries via direct communication with those seeking funds. Alternatively, the platform's auto-bid function auto-matches an investor to suitable investments, based on their appetite for risk and desired interest-rate reward. Like Trillion Fund, all money is held by clients rather than with the platform; investors make their own decisions. This puts financial risk with investors (including the council which is a lender), rather than with the platform-makers.

Funding Circle views its partnership with a local authority as building "an element of trust. We know we're trustworthy, but conveying that? People will go: 'Oh, well I probably wouldn't have trusted them before but if the council is involved....' Trust is a big element of it" (Chappelow). In this way, the partnership operates as advocacy not only for the platform but also for the P2P sector.

There are, however, tensions specific to the ambitions of the council in using networked finance. The council wants to keep invested money local, whereas the platform auto-bid algorithms factor in diversification (of sector and geography) to spread investment risk for funders. And take-up of council loans on Funding Circle has been slow. Issues concern eligibility, with applicant companies not having a requisite Newcastle postcode and companies finding the credit requirements of the platform as onerous as conventional bank loans, requiring two years' accounts.

7. Case Study: Crowdfunder

The UK's nearest equivalent to Kickstarter is Crowdfunder, Britain's major rewards-based and equity platform, which supports community enterprise, charitable donation, and product generation. Material comes from conversation and collaboration with the platform[8] in 2014–2015, and a 2016 interview with Crowdfunder's fund manager Jason Nuttall.

Crowdfunder focuses on social issues, partly organised by geographical location. This is counter-intuitive given the scaling potential of automated platforms, but works to grow audience for local activity. Crowdfunder Local on the platform website features area-specific campaigns to show off campaigns. The platform is very active in scouting for and helping to encourage both new projects and new partnerships. It fosters small, located enterprises such as Leeds Bread Cooperative (LBC), which used the platform to raise money to buy ovens to set up a business. Their campaign raised £8,690 in 42 days from 114 backers. In the process, the bakery built an online local public that converted to a customer base beyond the campaign. LBC then received grant funding from a cooperative legacy fund and other business loans in what Chappelow (at Funding Circle) refers to as the "multiplier effect ... making that money work a lot harder" (interview).

Funding manager Nuttall visits traditional funders to explain how match-funding funds that communities are managing to collect can improve impact for their budgets and these projects' reach. And much support goes into helping organisers choose a suitable project, plan, and schedule: "to do X by Y with Z" (Nuttall). Would-be founders struggle with this and the necessary practical skills, especially making a video. Crowdfunder provides coaching to people starting out on the platform.

Nuttall is personally interested in developing the platform to gather and redistribute resources beyond money, such as gifts-in-kind and time. These forms of exchange are particularly useful to social enterprises, building up support around them. He points to novel structures that have been enabled by the platform's support, such as a community-owned distillery (which asked for £1.5M, received £2.5M and is now building a visitor centre with the extra money). Such a campaign builds long-term engagement, extending one-off commitments with news, encouragement, and further opportunity. "It can show what that project goes on to do. And, if there's a delay, people are more forgiving as they understand the dynamics and are being kept up to date. It makes for better publicity for any hazards" (Nuttall, interview). He suggests that people are not merely sponsoring, shopping, or even building a brand; they are in the process of becoming the network through which the aims of these enterprises are realised.

8. Discussion

There are features common to all our examples, such as using algorithms to manage transactions and networks to connect up markets. And there are marked differences, such as how campaigns are built and how they anticipate long-term viability. Here, we want to address questions about how paying publics are brought into being.

9. Ambitions for Social Outcomes

Our interviews with platform-makers show that socio-economic intentions informed the design of each, with a goal of assembling people to a specified societal end.

'Trillion Fund is changing money' says the platform's site. Accompanying the owners' aims to capitalise for profit on new forms of financial mechanism are social, political, and environmental ambitions. Their aim was to mobilise the huge sums needed to secure enough renewable energy to impact climate change. Within their strategy, they knew a competitive rate of return on investment was non-negotiable: it would be impossible to woo investors on the strength of a green portfolio alone. Nonetheless, their marketing and careful vetting of portfolio speak to their concern to use the platform to enable green initiatives to compete with the existing forms of energy generation and replace them. Talking to the CEO and other team members reveals a passion to change not financial relations, but directions of funding and a pragmatic goal of using a financial tool to do so. They hoped that people would come for the profits and stay for the politics, intending to form a public that is increasingly interested in renewable energy because they have invested in it. Ironically, their most effective move to generate collective interest may have been to suspend new schemes (after change in subsidy policy), arousing group concern from platform investors about the safety of their investments. But here too they attempted to turn investor and media attention into fuel for a campaign on subsidies and change from traditional energy sources, using the interest to make their point and unite their investors.

Patchwork also aims to have positive environmental impact, changing social practice and enabling personalised services that are more than a sum of its parts. Here, the mechanism for bringing this about is offering a new, easy way for gift-giving, using the power of the network to communicate, connect, broker, and to manage and aggregate money. Knight can point to the elaborate creative work done by users as an endorsement of the effort she put into devising a clear and configurable set of interactive visual elements. The effect of success would be to cut down on poor resource use (unwanted gifts), whether or not her customers use the service because of green principles.

When we turn to Crowdfunder and the white-label Newcastle council platform, we see intended societal remodelling most nakedly. Agnostic about what kind of campaign they support, as long as they are legitimate, both the reward- and loan-based platform stress their potential to move money into enterprises that struggle to find flexible support elsewhere, creating a new wave of entrepreneurship funded collectively by individuals. Both want to work at a local level and Crowdfunder is able to link fundraising with community generation, witnessed here in the new market for artisan bread in Leeds, born of the campaign to launch it.

Although still seen as a novelty by many users, the platforms regard their mission as changing the whole way society distributes and/or makes money (and/or what it lends on). We note that everything from government policy to onerous authentication requirements to lack of advertising opportunity can hamper development. These aspects have a discernible impact on viability, i.e. parts of the constellation create challenges, but it does not negate the potential to meet these challenges and make a big change. All four sites have engaged people in crowdfunding in ways that the owners have chosen in order to make societal change, though only one, Crowdfunder, is visibly successful at this point.

10. Publics: Mutual Awareness and Trust Issues

Publics form around issues and objects representing an issue, and mutual awareness grows through issue formation. Only one of our studies presupposes existing relations. Patchwork Present provides the means to join people in a collaborative act where mutual awareness is already high. It brings together people who wish to do something for someone else and many will be known to each other as the recipient's friends. Most platforms do not assume loyalty from anyone, even if individual campaigns may depend on family and friends.

Nonetheless, other people's behaviour is an influential factor across platforms in attracting funders and giving them confidence to invest. People need to see others taking a risk and making it work. Making enough money to sustain platforms' business may be paramount, but a prerequisite is making crowdfunding socially acceptable.

As noted, the Crowdfunder Local pages make destination campaigns for many cities and regions to promote clusters of people and activities. Local clustering can create what Light and Miskelly call 'a relational asset' in an area[9] (p. 55) where "no one is working in isolation. Borrowing goes on between projects and the effort to find materials, support and funds is often shared." This kind of environment makes a fertile bed for crowdfunding, drawing in new adherents as the idea becomes more integrated and a more *usual* way to raise funds or offer support. We can see a zone of influence constructed beyond immediate actions on the platform.

Relatedly, partnerships are about supplying traffic and signalling trust, shown in the gratitude Funding Circle has for its council partner in Newcastle. The partnership is significant for how the platform is accepted, beyond any individual campaign. It is interesting that a council is acting as the face of the new mechanism, showing an ability to reinvent itself, and also the complexity of the journey for new financial instruments.

There is clearly a difference between supporting a bakery, from which you can buy bread afterwards, and lending only for promised income. Loan models are not intrinsically social and the rewards they promise are profit. They are more likely to have users who remain remote from each other, making it harder to create publics. (That is, until a platform is threatened or legislation changes dramatically, when, like Trillion Fund, there is collective interest in future activities and the safety of lender money.) Yet, Trillion Fund celebrity shareholder Vivienne Westwood held a party to attract investors and raise awareness of renewables, a specifically social occasion for a service that does not need to bring its users into contact. While this is good business and promotion, there is also a rationale for the particular choice of engagement that draws on their social mission to accumulate a constellation around them. This hybridity is visible to some degree across all our examples, with no simple meeting of markets, publics, or both.

11. Designing Paying Publics

We have argued that platforms can be used to configure the constellations of people who are involved in crowdfunding by bringing them into being in particular relations around socio-economic issues. The particular character of these interactions is financial, because all the focus is on the movement and redistribution of funds and what they will buy. This is why we talk about "paying publics," drawing attention to the way that crowdfunding platforms create interests of an economic kind. Yet, we are not just talking about new markets.

Markets are full of crowds, masses of people who happen to be together. We make the distinction between markets, which are purely economic, and publics, which are social and political; between crowds, which are happenstance, and publics, which evolve and shape (as well as being shaped). We can make this distinction since each platform described here has a goal beyond making money through bringing together types of market; it has intended to provide social and ethical interests of the sort that configure publics. There is a dual awareness. The first is of contributing to, and benefitting from, the building of a new type of financial mechanism that could change societal relations. The second is of the impact, on a particular sector, of the specific relations that it has sought to bring into being, be that sector renewable energy or local business. No single platform is redesigning economic life. However, each offers possibilities for linking private, public, and personal money and services in new ways around issues of concern, and, together, that signals societal as well as financial innovation.

In our first pair, the publics constituted are global: united by the use of the site to drive new collective environmental behaviour and, as such, less obviously social in an immediate sense. In the second, concerns are defined as local (as part of a localism agenda) and publics are thus geographical, where new relations among communities are an expected outcome as well as higher level societal structures. This shows that platforms can work in quite different ways to effect social change and the related paying publics look different too.

Platforms can be a combination of off-the-shelf payment systems and a content management system with a fairly standard interface design. This then makes the generosity and size of a particular platform's publics key to its effectiveness and this hinges on its unique design. UK rewards-based platform Crowdfunder in many ways closely resembles the standardised design of world leader Kickstarter (video, text campaign message below, ladder of increasingly expensive rewards in a right-hand column). Yet Crowdfunder's business model and design of services (including coaching to potential founders) is significantly different. As shown, Crowdfunder is actively shaping publics through partnerships and local engagement. Nuttall is teaching would-be founders what a campaign is, securing match-funding for these enterprises, helping them build custom, and thus changing how such organisations operate.

The design *around* the platform, of relations built with each neighbourhood and wider social structures, becomes key to gathering and shaping a paying public. This may also be key to the platform's viability: the ability to attract users. A crowd uses a platform. A paying public emerges as it develops a collective sense as an interest group able to fund particular outcomes through placing its money. As these publics evolve, platforms become more than infrastructural; they are a support, a magnet, and a focus for societal trends and what it is possible to enact with a popular groundswell. We are watching a vanguard, using algorithms to automate interactions. Yet, we may be reminded of the subscription models used in the 18th–19th centuries in Britain and other parts, when people provided and shared activities and resources without state or local government mediation.

We can point to the work of active public-building that our studied platforms engage in: promoting mutual awareness by showing user activities on site, in newsletters, and in blogs; flagging up campaigns (particularly successes) and listing numbers of pledges as well as pledged amounts; running events around campaigns; teaching founders how to organise campaigns and tell a compelling story about their goals; showing how to turn a public into customers and realigning organisations to benefit from match-funding initiatives and crowdfunded outcomes; thinking geographically as well as thematically about who the public might be. We can also note that legal and regulatory matters shape what platforms can do; and changes in financial and political climate have an impact on how publics form and what concerns them. The sector will continue to change fast.

About the Authors

Ann Light is a Professor of Design and Creative Technology at Sussex University and also Professor of Interaction Design at Malmo University in Sweden. Her research concerns the politics of design and how to make culture change towards sustainability, working with scientists, designers and members of the public to explore alternative futures. She writes about the economics of the Sharing Economy, led the project "Design for Sharing" (2013–2014), and has been analysing and supporting grass-roots digital platforms for many years. She is a member of the COST Action "From Sharing to Caring: Examining the Socio-Technical Aspects of the Collaborative Economy" (CA16121, 2017–2021), advising the European Commission on the future of the Sharing Economy.

Jo Briggs is an Associate Professor in the School of Design at Northumbria University, Newcastle-upon-Tyne. She uses design approaches to investigate into and through the collaborative economy and to explore parallel ways of supporting cultural and social operations amid radically shrinking state funding. She led the project "A Taxonomy of UK Crowdfunding" (2014–15) and also "Creative Temporal Costings" (2015), which examined timebanking as a means of collaborative exchange among creative practitioners. Recent publications include a co-authored article on trust and Participatory Budgeting based on interdisciplinary research conducted in the North East.

Endnotes

1. Jeff Howe. *Crowdsourcing: A Definition*. Crowdsourcing Blog, 2006.
2. Ethan Mollick and Ramana Nanda. "Wisdom or Madness? Comparing Crowds with Expert Evaluation in Funding the Arts." *Management Science,* 62(6) (2016), 1533–1553 (doi: 10.1287/mnsc.2015.2207).
3. Compare with: Christopher Le Dantec and Carl DiSalvo. "Infrastructuring and the formation of publics in participatory design." In *Social Studies of Science* 43(2) (2013), 241–264 (doi: 10.1177/0306312712471581).
4. John Dewey.*The Public and its Problems*, Holt, New York, 1927.
5. Robin Murray, Julie Caulier-Grice and Geoff Mulgan. Ways to Design, Develop and Grow Social Innovation: Social Venturing. Nesta and the Young Foundation, 2009. https://youngfoundation.org/wp-content/uploads/2012/10/Social-Venturing.pdf
6. Previously *Patchwork Present* with website at https://patchworkit.com/
7. https://www.fundingcircle.com
8. http://www.crowdfunder.co.uk
9. Ann Light and Clodagh Miskelly. "Sharing Economy vs. Sharing Cultures?" *Designing for Social, Economic and Environmental Good*, IxD&A (2015), 49–62. http://www.mifav.uniroma2.it/inevent/events/idea2010/doc/24_3.pdf

Chapter 5

The Patronicity Crowdgranting Model

Ebrahim Varachia

Abstract

Patronicity is a civic and community-based crowdfunding platform that has partnered with the Michigan Economic Development Corp (MEDC) to develop the first of its kind, innovative crowdgranting model that engages community members to fund place-making and community development projects within their neighbourhoods. The model provides matching grants when funding goals are achieved through crowdfunding. This proves enough support is behind a project while democratising the way funding is raised and granted.

Since then, Patronicity expanded, partnering with MassDevelopment, the Indiana Housing & Community Development Authority, and other granting organisations to manage similar grant-match through crowdfunding programmes targeted at driving economic development and enhancing community engagement.

The chapter will focus on the partnerships with various state entities and use project successes as case studies on how crowdfunding enabled their organisations (municipalities and non-profits) to fund community projects through an increase in civic engagement and participation. We will cover the reasons why the partnerships formed, and the impact it has had which led the partnerships to continuously be renewed.

The crowdgranting model through Patronicity has led to over $8.8 million in grant dollars towards communities and over $10.5 million in crowdfunded dollars (metrics as of October 2018).

Crowdgranting helps to solve problems of blight and access to capital in communities across the entire state while recognising community members are able to make their voices heard by crowdgranting to support projects in their neighbourhoods. For the MEDC, we've developed an innovative model to streamline a grant process from 6 weeks to 72 hours and have received four to five times the publicity and media coverage for the projects they are supporting than traditional methods. Additionally, not only have the grant dollars more than doubled through our crowdgranting model, we've seen an approximate 7.6:1 total private investment leverage on the grant dollars being disbursed.

1. Introduction

Founded in 2013, Patronicity is America's only crowdgranting platform—blending the best aspects of crowdfunding with grants— aimed at serving the needs of grant-giving agencies as well as local governmental units, non-profits, and community groups to engage and revitalise communities large and small. Patronicity currently runs three state-wide sponsor-granted programmes in Michigan, Indiana, and Massachusetts, as well as smaller partnerships with a local foundation and corporation looking to leverage their social responsibility. Since its inception, Patronicity has assisted communities in raising over $10.5 million in crowdfunding, matched by over $8.8 million in funding creating over $54 million in direct economic impact through total leveraged funding connected by our crowdgranting programmes (metrics as of October 2018).

In June 2014, the Michigan Economic Development Corporation (MEDC)—a state-wide economic development agency focused on growing and attracting business, talent retention, and revitalising communities and urban centres—partnered with Patronicity to develop a first of its kind grant-match programme through crowdfunding. This model has been termed as crowdgranting. Through the Public Spaces, Community Places (PSCP) grant programme, qualifying projects receive a matching grant from the MEDC if they can showcase enough community support and civic engagement through crowdfunding. The programme has been renewed for the fifth time for $1.5 million for the 2019 fiscal year (ending October 2019), with plans to be renewed again in 2020. The PSCP programme has inspired other large state governing organisations, cities, corporations, and foundations to develop a similar model framed around its innovation and success.

Patronicity came together with the MEDC to recognise thriving places help define a community's economic vitality. From bike trails, to pocket parks, to public art projects, they contribute to a strong quality of life, help attract and retain talent, and grow stronger local economies. In Michigan in 2014, with declining public revenues and local budgetary concerns on the minds of communities across the State, the MEDC saw a need to empower them to continue these types of improvements during a time of economic distress. Communities tend to stray away from innovative, place-making improvements when limited budgets are focused on hard infrastructure such as waterlines and crumbling roadways. The PSCP programme focuses on creating new or activating distressed public spaces for community use; such as pocket parks, trails, outdoor plazas, public art, farmers markets, art centres, and more.

Matching grant programmes are not new, but they oftentimes create challenges for organisations to come up with the funds. Due to these challenges, a lot of traditional matching grant opportunities go untapped as organisations are unable to find the funding. The Michigan Municipal League, another partner in the PSCP programme, recognised the specific challenge municipalities face in fundraising and proudly supported the programme to advocate for a crowdgranting opportunity to drive municipal projects. The crowdgranting model opens the opportunity for municipalities, non-profits, and community groups to tap into a donor base primed to provide support and funding towards projects. Crowdgranting enables everyday residents, local business owners, and community stakeholders to play a role in supporting the creation of community assets, a space previously reserved for the wealthy elite. The crowdgranting model not only garners deeper community engagement but also reduces the sole burden for organisations to come up with the match alone, while empowering donors with an avenue to fund and support projects meaningful to them.

2. Project Case Study: REACH Art Studio Centre

In August of 2014, the Reach Studio Art Center (REACH), heard about the newly launched Public Spaces, Community Places grant programme and connected with Patronicity to begin the application. REACH is a community art centre located in the REO-Town District of Lansing which aims to make art accessible to community members and at-risk youth. After recently purchasing the neighbouring vacant storefront buildings on their block, they were in need of nearly $100,000 to reno-vate the space, turning it into a new youth art gallery and outdoor court-yard. A nearly impossible amount to raise on their own, REACH turned to the PSCP programme for funding. Being a new applicant and an unknown organisation to the MEDC, the REACH application was almost immediately rejected. Patronicity and the MEDC team took a step back and recognised the value of the project and let the programme work itself out by allowing the community to be the final review committee. REACH went on to raise nearly $50,000 from over 300 donors, thus showcasing engaged community support and demand for the project in the neighbourhood. Other surprising findings showed that 55% of donors never donated to REACH before, and over 38% of donors heard about the campaign through the newspaper. These finding indicated that the crowdgranting campaign was able to target and draw in new donors towards the campaign and that local press generated the most aware-ness on the campaign for neighbourhood residents. An impact video on REACH's success can be found at: https://www.youtube.com/watch?v=jaEroNYvz4U.

The practice of crowdfunding aims to fund projects by raising small amounts of money from a large number of people that invites residents to be engaged in the process from start to finish. The goal of this practice is to have an inclusive platform that allows local residents and stakeholders to play a role in projects that will transform their communities into places where talent wants to live, businesses want to locate, and entrepreneurs want to invest. Due to the innovative crowdgranting component of the programme where qualifying and approved applicants only receive matching funds if they are able to raise their target crowdfunding goal, the programme has enabled the following outcomes:

(a) Democratising the access to capital where non-profits, community groups, and municipalities apply by submitting qualifying projects that fall within the grant parameters and receive funding through community demand, establishing the community as the final review committee.
(b) A deeper sense of civic engagement where the programme enables community members, residents, businesses, and local organisations to vote with their dollars on projects they find valuable within their community.
(c) Increased public awareness and press and media attention towards community projects which enables greater support for public initiatives.
(d) Freed up organisational resources while expanding the number of projects and overall granting budget. This programme builds off of matching grant programmes, however enabling the crowd to be involved through crowdfunding revolutionising the mechanism in which the match is leveraged.

The idea was to flip economic development on its head with the vision of creating vibrant communities where people would love to "Work, Live and Play" in an effort to retain and attract talent. This programme has helped to solve problems of blight and access to capital in communities across the state, from small rural towns to inner urban cities, while recognising community members are able to make their voices heard by crowdgranting to support projects in their neighbourhoods. This method to democratise access to capital enables driven communities to be the final review committee in the grant process.

For the MEDC, Patronicity has developed an innovative model to streamline the grant process from 6 weeks to 72 hours, providing 4.5 full-time employees (FTEs) value at a cost of 1.5 FTE to administer the programme and generates five times the publicity and media coverage for projects they are supporting more than the traditional methods. Additionally, not only have the grant dollars more than doubled through

Patronicity's crowdgranting model, it has led to an approximate 7.6:1 in total private investment leveraged on the grant dollars disbursed.

Since launching this initiative, the MEDC's "Public Spaces, Community Places" programme has been renewed and expanded year after year. To date (November 4, 2018), the programme has granted $5,609,166 while over 33,000 community members have crowdfunded over $6,645,004 towards 184 projects state-wide which leveraged a total of $42,245,420 in private investment. The programme has yielded a 97% funding success rate for projects and has brought the programme international recognition and awards. The programme is an award-winning finalist in the 2017 Harvard-Kennedy School of Government's "Innovations in American Government" awards and winner of 2015 International Economic Development Council GOLD award in Public–Private Partnerships with Patronicity.

Patronicity has built a platform that enables a quick review and approval of projects. Applicants apply directly through Patronicity.com and then the Patronicity team works with those individual projects to evaluate and vet them to qualify for the grant programme. The Patronicity team also provides technical assistance to applicants ensuring they understand the process, fit the grant requirements, and have the knowhow, tools, and readiness to successfully launch and run a community-led crowdfunding campaign. The platform collects any necessary metrics from the projects for the granting agencies review team to mark as "conditionally awarded the grant." Approved projects then launch their crowdfunding campaign, reaching out to the greater community and are only able to achieve the conditionally awarded grant if they are able to reach their funding goal through community support and traction. This ensures the project has enough funding to move forward and can complete the project successfully.

The model continues to evolve, as Patronicity is developing an online reporting tool that allows projects to submit updates during the construction and completion of the project. The reporting mechanism collects other specific metrics reported by the projects to measure success and impact. The crowdgranting model is ready to be adopted by other grant-giving entities who want to engage the community to be champions of their own sustainable development, while employing an innovative tool to fund and generate awareness around impactful projects. Plans to expand ideation opportunities for community members are under testing, with hopes to create an online town hall meeting that can source more ideas for further implementation. Reach out to the Patronicity team to learn how the crowdgranting model can be tailored to fit your granting goals within your government or organisation.

About the Author

Ebrahim Varachia is the President and Co-Founder of Patronicity. Based out of Detroit, he recognized a city filled with people passionate about growth, empowerment and change. He believes the cornerstone between a great idea and it coming to life is often times the funding and started Patronicity as a civic crowdfunding platform to inspire growth and change throughout cities and towns, both urban and rural across Michigan and the country. Growing Patronicity to be more than just a civic crowdfunding platform, he pioneered the crowdgranting model which has led Patronicity to change the way grants are administered and how communities come together to envision, build and create more sustainable and impactful projects meaningful to them. He was selected as NextCity's Vanguard Fellowship 2018 and was awarded the SOCAP18 Entrepreneurship Scholarship.

Chapter 6

Match-Funding Calls for Open Crowdfunding: The Experience from Goteo.Org in New Policies for Crowdvocacy

Enric Senabre Hidalgo, Mayo Fuster Morell,
Cristina Moreno de Alborán and Olivier Schulbaum

Abstract

Since crowdfunding first appeared, and with the proliferation of plat-forms in recent years, various systems and formulas of operation have appeared within the general crowdfunding model. One such system, still in its early days, is match-funding (co-funding between citizens and insti-tutions), which permits public and private organisations to double the financial contributions for projects from individual users. After an analy-sis of the state-of-the-art match-funding practices, this chapter focuses on a case study of the Goteo.org platform, a pioneer in the international development of this model. The advantages and impact of this method of crowdfunding compared to the traditional method are analysed using data collected on the behaviour in 15 match-funding calls for projects from 2013 to 2018. The results show that match-funding campaigns are more likely to be successful, significantly increase average donations, and generate new dynamics of institutional cooperation and proximity in the support for initiatives. A series of challenges and future lines of development for the model are finally discussed, especially how civic match-funding can become a powerful instrument for public participation and policy innovation, with many initiatives crossing boundaries between activism, advocacy, social entrepreneurship, and social innovation. In this sense, the novel concept of "crowdvocacy" could emerge as a distributed but coordinated process among different actors and platforms where civic initiatives increase their influence in public life, from citizens' awareness and engagement to empowerment and wider participation in democratic life.

1. The Emerging Method of "Match-Funding"

The study *Reshaping the crowd's engagement in culture*[1] shows that in the European cultural and creative sector alone individuals and cultural organisations all over Europe have launched some 75,000 crowdfunding campaigns since 2013, collecting a total of €247 million, particularly in the United Kingdom and France. One of the most significant findings from the European data is that only half of crowdfunding campaigns have been successful in meeting their target. Particularly striking is the fact that the €247 million collected in total represents only 7% of the total committed amount (which was €3.4 billion). This means there is a "black hole" of over €3 billion that eventually was not assigned to campaigns, as their minimum funding goals were not reached. Besides suggesting that unsuccessful campaigns are over-ambitious in their demands for money, this also demonstrates one of the most common rules of crowdfunding platforms: when projects fail to reach an established funding target within a given time, all the donations received from users are refunded at no extra cost.

Another significant finding in *Reshaping the crowd's engagement in culture* is the relative "delocalisation" of platforms. Although up to 600 crowdfunding platforms are active at any given time in Europe, almost half the campaigns started by creators of European projects were hosted on US-based platforms, mainly Kickstarter and Indiegogo. These two platforms have a global reach and are, by a large margin, world leaders in hosting campaigns from different countries.[2] In relation to these deficiencies and possibilities in the development of crowdfunding, in parallel with other methods such as equity crowdfunding,[3] in 2013, some platforms started pilot projects in match-funding. The method allows funds successfully obtained in a campaign to be increased with capital from an institution providing additional funds.

Before the emergence of crowdfunding, the concept of match-funding with matching donations or payment of funds was already in use in the contexts of charity, philanthropy, or the public good.[4] For institutions, the most popular model was one whereby a public organisation, sponsoring body, or corporate social responsibility department completed funding in the form of investment (a subsidy or loan) in a project that had already obtained a substantial amount of its target funding from other sources.[5] The first crowdfunding platforms in Europe that tried different forms of match-funding in 2013 were Goteo in Spain, with the support of the International University of Andalusia, and KissKissBankBank in France, with the support of La Poste, while in the US it was Indiegogo, with the support of the Kapor Foundation.[6]

As discussed in the report *Matching the crowd: Combining crowdfunding and institutional funding to get great ideas off the ground,*[7] based on a pilot experience with different campaigns on crowdfunding platforms for the artistic and social sectors, the match-funding method could offer a number of advantages with regard to project funding, as follows:

- helping provide additional funds for project campaigns;
- improving the chances of success for campaigns;
- significantly increasing the average amount of donations;
- diversifying and broadening the profile of campaign donors.

By analysing data on funds raised and behaviours in different campaigns and from interviews with participants, the report shows that the lack of effectiveness in crowdfunding could be overcome by supplementary mechanisms, such as match-funding.

Following other cases and contexts and from the perspective of civic crowdfunding, which has not received much attention in the literature,[8] this chapter aims to provide further evidence and analysis along these same lines, examining similar indicators and offering additional observations. We intend to not only confirm to what extent match-funding represents a more effective formula for crowdfunding beyond the artistic and social sectors but also to identify other key implications in terms of local impact of this type of digital fundraising mechanism as well as recommendations for its development and future study. Our study is based on an analysis of 14 match-funding calls for projects on the Goteo.org platform and interviews with the platform management team.

2. Case Study: Goteo Match-Funding Calls

Goteo was one of the first platforms to start operating locally and internationally from Spain in 2011, covering different sectors and fields within the so-called "civic crowdfunding."[9] Its activity is now undergoing sustained growth, making it especially rich in the types of campaign and diversity of its users and initiatives.[10]

Goteo was one of several pioneering platforms and the first to offer the match-funding method internationally.[11] It now has consolidated experience in this form of funding (between 2013 and 2018, there were a total of 15 match-funding calls for projects varying in size and success). In addition, in line with its philosophy of open knowledge and a commons-based approach,[12] it includes a number of transparency measures that make it particularly suitable for third-party research and analysis and an API that permits the generation of related applications to access open data on its operations (see Figure 1).[13]

Compared to the other match-funding platforms and match-funding calls for projects currently exploring this method, Goteo not only has more experience but also has a number of specific features on how it approaches crowdfunding and its particular version of match-funding. Goteo's calls for projects are coordinated around sponsoring bodies, which are both private and public institutions, who call for projects in specific fields they wish to promote, announcing the total amount of capital provided (the so-called "match-funding pool") to double individual donations, along with the other details and conditions for the call for projects. Managers of projects looking for funding can then offer proposals for specific campaigns within a given period, using the Goteo form. In the final stage, projects are selected that can then access capital from the match-funding pool. This selection phase is normally preceded by a

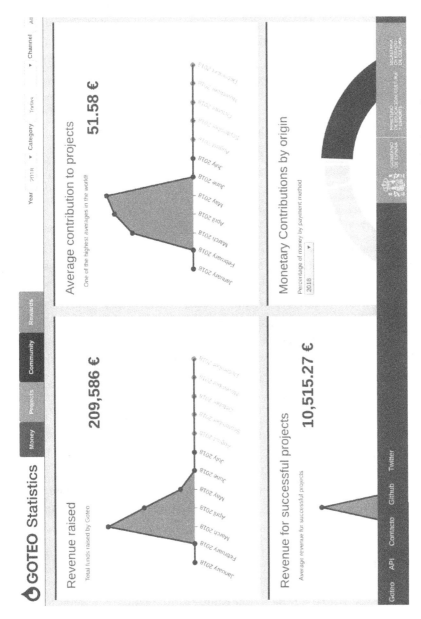

Figure 1: *Website of Goteo open data, with statistics on a specific match-funding call.*

period of training, which involves crowdfunding workshops co-organised by the funding institutions, followed by a final round of dedicated assessment to help project managers design and improve their proposals in line with the most important crowdfunding mechanisms (see Figure 2).

As with all the other campaigns on the platform, each campaign sent for selection by a match-funding call for projects provides a presentation and description, the financial requirements and rewards, the expected social impact, and any other relevant information for its funding if it is eventually accepted and publicised, using the platform application form. This permits a final set of campaigns to be selected, which combine the Goteo experts' viability criteria with the criteria of innovation, originality, or topical importance for the organising institution, which should be closely involved in the selection process.

Since 2013, the platform has worked successfully with universities, regional governments, private foundations, local councils, and local innovation agencies in calls for projects related to sociocultural innovation, educational innovation, childhood and cooperation, culture and public domain assets, entrepreneurial spirit, health, smart city projects, education, cultural heritage, and the arts (see Figure 3).

Once selected and published, the campaigns are given a time limit to reach their established financial target. During this period, each time a user makes a donation, the established match-funding pool directly and simultaneously contributes an equivalent amount to the same campaign. Thus, the contribution instantly doubles the individual donation, displayed on a graphic progress thermometer for each campaign. As agreed beforehand with the organising body, limits are established regarding the extent to which the system can double contributions, so that particularly large contributions (such as hundreds of euros) cannot "drag in" more than 50 euros or other pre-defined maximum amounts. This measure guarantees maximum diversity and a minimum participation level, while preventing fraudulent use of the system. Finally, if campaigns fail to reach their funding target, the same mechanism as used in other campaigns is activated and the displayed donations are refunded, at no extra cost, to both individual users and the organising institutions (whose money pre-assigned to projects with insufficient social crowdfunding support is technically saved) (see Figure 4).

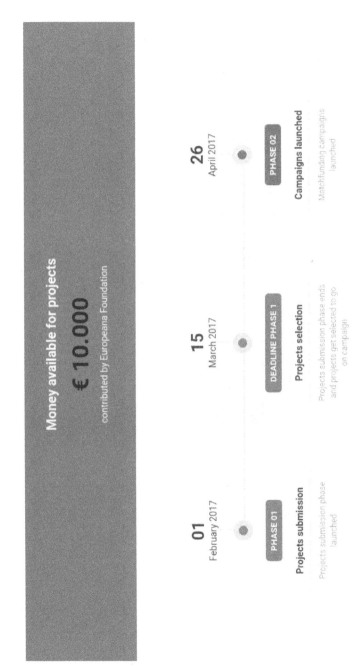

Figure 2: *A match-funding call for projects hosted by the Europeana Foundation.*

Figure 3: *Different active match-funding campaigns within a call for projects.*

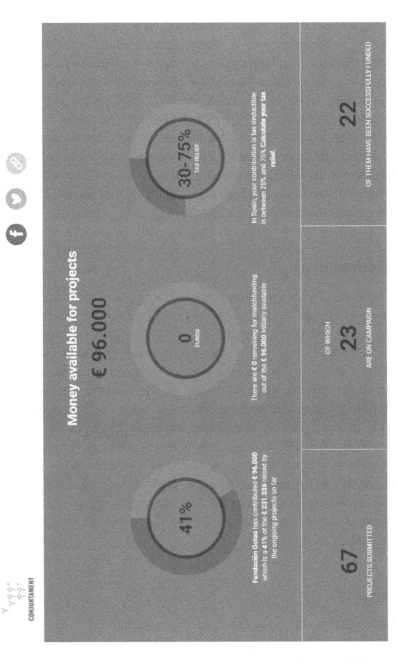

Figure 4: *A match-funding call for projects in the active match-funding campaign phase.*

This system marks one of the main differential features of Goteo, which was designed and developed in 2011 based on a clear need for different forms of crowdfunding, such as digital match-funding, which was non-existent at that time. This is particularly the case with the features of real-time display of how the money is doubled, the call for projects on specific topics, and the creation of *ad hoc* selection committees involving the platform and the organising institutions. It differs from the more common method of match-funding found in other platforms (Indiegogo, KissKissBankBank, Voordekunst, Spacehive, Verkami, and Crowdfunder), which normally establish alliances with trustworthy institutional partners, reaching agreements whereby at the end of the campaigns they add lump sum donations to those that have achieved their target or a minimum agreed percentage.

3. Methodology

Data collection for this analysis was based mainly on access to Goteo's public statistics page,[14] which shows both the aggregate overall behaviour of campaigns and that of specific match-funding calls for projects. As shown in Table 1, the 15 calls for projects since 2013 have involved a variety of different volumes in the selected campaign projects, contributions, and participants, covering a total of 138 initiatives that accessed match-funding, supported by a variety of institutions.

The study is based on an analysis comparing some of these data on the behaviour of match-funding campaigns with averages from the other Goteo campaigns since the platform started in November 2011 (a total of 1,093 projects, at the moment of writing this). In addition, a supplementary part of the analysis is based on data from an experimental website[15] for geolocated viewing of donations and the source of capital in the match-funding pool in Goteo.

4. Results

As Table 1 shows, one of the main issues to consider is how 121 out of the 138 campaigns selected in Goteo match-funding calls for projects since 2013 finally received the previously defined funding. As shown in Table 2,

Table 1: *Statistics on Goteo match-funding campaigns.*

Goteo match-funding calls	Selected projects	Successful projects	Total funding (€)	Feeder capital assigned (€)	Contribution from institution	Capital not mobilised (€)
Sociocultural innovation (2013)	5	4	25,988	10,000	31%	2,000
Innovation in health #1 (2013)	5	3	24,897	12,000	38%	2,663
Innovation in education and open knowledge (2013)	5	5	34,715	10,000	27%	560
Entrepreneurship (2013)	10	5	28,002	20,000	42%	8,322
Innovation in cooperation and childhood care (2014)	5	4	26,882	10,000	28%	2,517
Innovation in health #2 (2014)	5	5	33,949	12,000	35%	0
Cultural heritage and digital remix (2015)	5	4	26,979	10,000	33%	1,064
Crowdfunding ZGZ (2015)	4	3	20,054	12,000	42%	3,485
Gipuzkoa cultural projects (2016)	20	20	151,024	70,000	44%	3,428
Strike a match for education (2016)	3	2	16,572	10,000	44%	3,509
Supporting education (2017)	13	11	62,924	40,000	39%	17,419
Meta Gipuzkoa (2017)	16	15	159,343	70,000	40%	7,369
Crowdfunding ZGZ (2018)	4	4	35,041	14,000	40%	0
Conjuntament BCN (2018)	23	22	231,336	96,000	42%	0
Meta Donostia (2018)	15	14	131,646	70,000	47%	0
Sum/Average of all match-funding calls	**138**	**121**	**1,009,352**	**466,000**	**38%**	**52,336**

Table 2:   *Statistics on Goteo match-funding campaigns.*

Goteo match-funding calls	Average revenue in projects (€)	Average funding goal	Success rate	Average number of donors	Average individual donation (€)
Sociocultural innovation (2013)	6,390	123%	80%	70	31.73
Innovation in health #1 (2013)	7,629	139%	60%	98	40.97
Innovation in education and open knowledge (2013)	6,943	118%	100%	66	36.35
Entrepreneurship (2013)	5,340	135%	50%	44	52.2
Innovation in cooperation and childhood care (2014)	6,549	119%	80%	68	42.02
Innovation in health #2 (2014)	6,790	117%	100%	66	49.87
Cultural heritage and digital remix (2015)	6,715	116%	80%	67	44.76
Crowdfunding ZGZ (2015)	6,548	109%	75%	54	58.02
Gipuzkoa cultural projects (2016)	7,551	166%	100%	70	54.42
Strike a match for education (2016)	7,551	118%	67%	60	55.54
Supporting education (2017)	5,276	137%	85%	51	48.82
Meta Gipuzkoa (2017)	10,447	171%	94%	81	64.74
Crowdfunding ZGZ (2018)	8,760	125%	100%	63	47.42
Conjuntament BCN (2018)	10,518	131%	96%	104	51.58
Meta Donostia (2018)	10,554	136%	93%	84	55.06
Sum/Average of all match-funding calls	**7,571**	**131%**	**84%**	**69**	**49.00**
Average of the rest of the crowdfunding campaigns (no match-funding)	**5,288**	**121%**	**65%**	**123**	**41.8**

this 88% success rate is significantly higher than the normal 75% rate for the platform (already among the highest success rates in crowdfunding platforms on the market[16]).

This indicates that the match-funding method can indeed achieve a higher rate in channelling individual and common funds to projects, an important factor for projects and initiatives that choose crowdfunding at a specific moment in their development.

This is even more relevant when one considers that Goteo campaigns normally require an average of 123 individual donors to successfully obtain funding. By contrast, in the match-funding campaigns analysed since this method was first offered, the number drops by half, with an average of 69 users required. Another observation from the overall analysis of match-funding campaign behaviour compared to that of traditional crowd-funding campaigns lacking this additional multiplying component is that the average contribution from users is higher in the case of match-funding (average €49) than in other Goteo campaigns (approximately €42).

Another relevant aspect with regard to general match-funding behaviour on Goteo over time is that only 38% of the €1,009,352 collected up to end of 2018 by this system came from the institutional match-funding pool (therefore, 62% came from individual donors). Other similar large-scale initiatives, such as the recent Arts Council England and Heritage Lottery Fund match-funding case study, carried out by NESTA in the United Kingdom, indicate similar behaviour, although based on less developed match-funding formulas (where the contribution is provided at the end): from a starting multipliable capital of £251,500 an additional £405,941 was eventually raised (i.e. twice the capital initially committed for multiplying).

5. The Local Dimension

Another key aspect, noted in the introductory section of this study on international "delocalisation" dynamics of crowdfunding using large American platforms such as Kickstarter and Indiegogo as the technological solutions, suggests a need to establish mechanisms for local participation in this sector. Only 50% of crowdfunding campaigns for European organisations and projects use platforms whose headquarters and main business activity are in EU member states or the European continent, while the rest mostly choose the two large global platforms.

This is a problem, given that match-funding requires agreements and alliances with public and private bodies, often aiming to have an impact on a local area or a specific field. Thus, mechanisms that strengthen proximity between crowdfunding platforms and organisations that have traditionally funded innovative, risky, or minority projects need to be explored. An example of this behaviour from match-funding, favouring dynamics of local support among institutions and local communities, can be found in the Goteo geolocation tool, which provides a clear display of how funds activated by the match-funding usually flow from the area of influence of the "match-funder" institutions (see Figure 5).

6. Implications for Public Policy

Our aim was to determine if match-funding represents a viable alternative to crowdfunding and to what extent it can represent a set of improvements in the usual mechanisms and rules behind the usual crowdfunding platform mechanisms. The data analysed from Goteo support the idea that the match-funding model of crowdfunding can help provide additional funds to projects, increase the chances of a campaign's success, significantly increase the average amounts donated, and generate new and agile dynamics of institutional transparency, cooperation, and proximity in the support for initiatives.

From a public policy perspective, this suggests that match-funding offers a motivation and incentive for the public to finance projects and is thus a formula that supports the assignment of funds and validates social interest in new initiatives, without covering all or even half of the funds needed for their implementation. In other words, the way in which match-funding works on Goteo could provide appropriate dynamics for channelling significant amounts from public or private funds through a donation system that facilitates public participation above and beyond what the funds themselves could achieve (see Figure 6).

In line with the report *Reshaping the crowd's engagement in culture*, in both culture and other frequently related sectors (education, technology, and social sectors, among others), the growing use of crowdfunding by

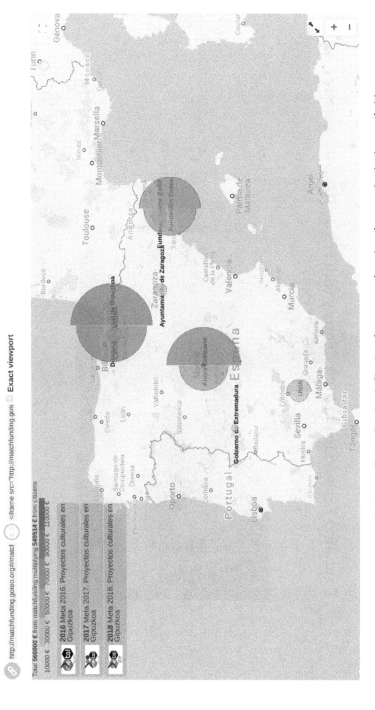

Figure 5: *A match-funding call visualisation indicating the average donations between institutions and citizens.*

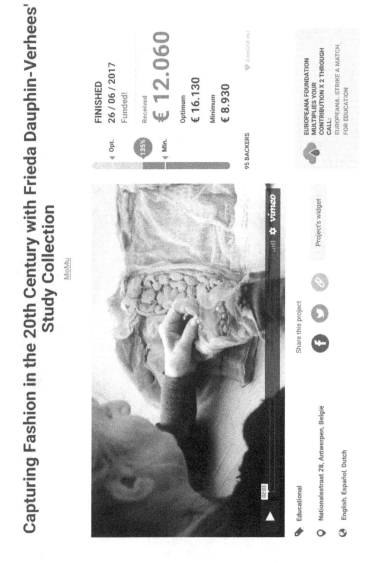

Figure 6: *Screenshot of a Goteo campaign within a specific match-funding call.*

communities of creators in Europe and the world will increase further in the forthcoming years. Therefore, options such as match-funding and its tendency to associate trans-institutional funding strategies, where initiatives with a social impact are required, could represent an improvement not just in effectiveness but also in the local visibility and impact of these still new hybrid forms of funding.

In the case of the match-funding model analysed here, based on the specific characteristics of the Goteo platform, a number of additional values are involved which seem to make this specific form of crowdfunding more effective in connecting with the commons-based model (with additional training so that projects are selected for general publication, but particularly based on a subset of indicators such as social impact, replicability, and open knowledge). Furthermore, mechanisms for viewing data help track and analyse the behaviour in the call for projects.

This tracking and analysis also permit comparisons of campaign behaviour (as in this case) or interpretations by public and private institutions of their expanded area of influence. This can help provide proactive mapping of areas of needs and initiatives to which funding with institutional cooperation should be targeted. When one also considers that another specific feature of Goteo campaigns is that they indicate the non-monetary resources that might be needed for their implementation (materials, infrastructure, donations of time, etc.), match-funding thus becomes a tool that, if broadened, could facilitate coordinated collective and institutional action in the participatory assignment of resources, with an open data mechanism and transparent management. It is also significant that the match-funding training actions provided with calls for projects are co-organised. These tools involve various levels of learning and familiarisation with the specific Goteo model, while creating a symbiotic relationship between institutions and the common goal.

7. Opportunities for Further Development

Based on the behaviour of match-funding campaigns described above, a wider analysis would need to extend the comparison to new match-funding campaigns and also permit comparisons with other platforms that begin to apply the model. This is particularly important because the application of the model to other types of institutions is steadily growing in Goteo, as is the total volume of capital in match-funding pools for multiplying funds. As an example, in 2018 Goteo has launched a new a model of match-funding, slightly different from the previously described "open calls." The main difference is that in this model (a) the call for projects is open until the available funds are fully distributed; (b) the "matching" institution has more flexibility to decide which projects to support; and (c) a specific algorithm is programmed according to the needs of the "matcher." As of November 2018, the described model has been implemented with an institution (a local political party from Madrid) which has allocated €100,000 to support civic projects that contribute to improve the city, resulting in this new modality of match-funding covering a total of 22 projects and matching an additional €148,755 in donations from individual users.

On the contrary, further research would also need to carry out new analyses of data comparing match-funding platform users as opposed to other collaborative funding mechanisms, such as municipal participative budgets, social currencies, and digital time banks. Advancing in this direction could provide more in-depth knowledge of viable alternatives for social initiatives that otherwise lack resources in a number of areas. Advances are also needed in the development of the Goteo data viewing mechanisms, both graphs and maps. It is hoped that other platforms will start applying match-funding mechanisms that provide similar open data, as this would permit even broader and richer comparisons with regard to their scope, thereby advancing this significant phenomenon in online collaboration dynamics.

Another element to mention is the dynamic of continuous improvement and development of the platform, which, as well as being an open-code repository,[17] continues to incorporate different functions in a modular format. Many of these have an impact on the match-funding model analysed here. Specifically, in the opinion of its developers, results from match-funding so far as described in this study would lead to a number of specific actions as follows:

- more and better tools for viewing all local needs in a given call for projects, thus making it a more effective tool for the organising institutions in general;

- a user profile that permits capital contributions to the match-funding pool by individuals and groups of users, not just institutions, and the activation of similar mechanisms to attract third-party donations;
- activating data entry forms for projects and the organising institutions, which permit continuous match-funding actions (not just for specific topic-based calls for projects) that dynamically match "supply and demand" for capital in match-funding pools;
- advancing in the development of more versatile configurations to view and adapt match-funding to different contexts and types of user, following the example of new algorithms that permit the display of the minimum number of micro-sponsors required to reach the minimum funding level.

In our opinion, these potential improvements for an open-source crowd-funding platform, confirmed and validated by the type of action-research that makes the basis of this chapter, as well as the results we have discussed here, represent an opportunity for both open innovation and civic crowdfunding, which have in match-funding practices a growing and promising field.

8. From Crowdfunding to Crowdvocacy

As indicated above, an important part of the Goteo match-funding model has to do with the need to innovate in current crowdfunding models and adapt funding mechanisms between public organisations and other key actors in social innovation, in order to provide policymakers and institutions with effective mechanisms for generating impact at different levels.

Coining the term *crowdvocacy*, Goteo has witnessed how civic match-funding can become a powerful instrument for public participation and policy innovation, with many initiatives crossing boundaries between activism, advocacy, social entrepreneurship, and social innovation. Promoting the link between political participation platforms, digital campaigning tools, civic crowdfunding, and civic technology could expand the voice and enlarge the influence of socially innovative initiatives. The match-funding model described here could play an important part in that approach, among other open-source tools and initiatives currently under development. Goteo envisions crowdvocacy as a distributed but coordinated process among different platforms where civic initiatives increase their influence in public life, from citizens' awareness and engagement to empowerment, achieving, finally, full participation in democratic life.

The concept is already in its early development but as a working prototype that is being tested in parallel to other digital infrastructures for participatory democracy developed in Spain, like Decide Madrid (https://decide.madrid.es/) and Decidim Barcelona (https://decidim.org/), where the Goteo Foundation is collaborating to identify and improve the scalability of participation patterns (see Figure 7).

The following questions represent a series of challenges and opportunities identified, based on lessons learned since 2013, for expanding the Goteo match-funding model to other domains, institutions, and even platforms:

- The impact derived from match-funding represents a huge opportunity for all sorts of organisations that can publicly and transparently demonstrate their commitment with the civil society for activating resources in collaborative solutions for civic initiatives around the issues of interest.

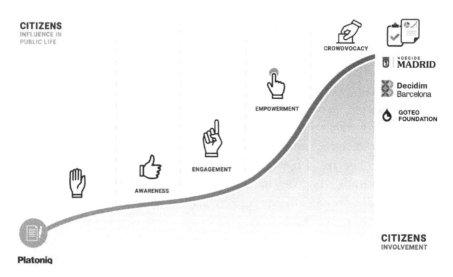

Figure 7: *Visual representation of the concept of crowdvocacy as being developed with Decide Madrid and Decidim Barcelona.*

- The potential of match-funding mechanisms like the one described here needs to focus on projects which have a clear social goal or at least initiatives that can provide in parallel to their development some level of engagement with open outputs.
- The most important implementation issues and barriers have to do with rigidity in public institutions and lack of expertise in collaborating with non-profit organisations, since the channelling of match-funding pools is a model not foreseen in the usual collaboration agreements.
- The impact derived from match-funding represents a huge opportunity for all sorts of organisations that can publicly and transparently demonstrate their commitment with the civil society for activating resources in collaborative solutions for issues of interest.
- The match-funding model has the demonstrated capacity to impact in terms of both funding and increased capacities. Further work needs to be done to reach potential funders interested in multiplying their impact, especially those of the private sector.
- To develop different models of engagement between the matchers and the matched initiatives once the funding is distributed. How can we support mutually beneficial long-standing relations?

About the Authors

Enric Senabre Hidalgo works at the Dimmons.net Research Group (Internet Interdisciplinary Institute—UOC) on co-design methodologies and agile frameworks for research processes and for the development of digital commons and public policies for the platform economy. He is a research fellow at the CECAN Research Centre (University of Surrey) and guest researcher at the Austrian Centre for Digital Humanities (Austrian Academy of Sciences). He has been Vice President of the Observatory for CyberSociety and Associate Professor of Software Studies at the Open University of Catalonia, where he holds a Master's degree in the Information and Knowledge Society. Enric has also worked as a coordinator of the City and Collaboration Department at Citilab-Cornellà and as online content coordinator of the Infonomía network as well as implemented eLearning projects and open educational resources for various institutions, like the Government of Catalonia and the Mozilla Foundation.

Mayo Fuster Morell is the Dimmons Director of Research at the Internet Interdisciplinary Institute of the Open University of Catalonia. She is faculty affiliated at the Berkman Center for Internet and Society at Harvard University and at the Institute of Govern and Public Policies at Autonomous University of Barcelona (IGOPnet). In 2010, obtained her PhD from the European University Institute in Florence on the governance of common-based peer production and has contributed numerous publications in the field. She is the Principal Investigator (PI) at UOC for the European project DECODE—Building the next generation of cooperative data platforms for digital sovereignty. She was PI for IGOPNet of the European project P2Pvalue: Techno-social platform for sustainable models and value generation in commons-based peer production. She is also responsible for the experts group BarCola on collaborative economy and commons production at the Barcelona City Council.

Cristina Moreno de Alborán is Co-Director of Fundación Goteo. She holds an MSc. in Human Geography by the University of Amsterdam and a BA in Political Science and Public Administration by the Universidad Complutense de Madrid. She developed her career in the third sector working with local and international organisations in diverse countries such as Guatemala, Honduras, Tanzania or Indonesia and has managed programs related to the strengthening of civil society, justice reform, good governance, and access to public information. Back in Europe since 2013, she has been working with two of the most promising Spanish civil society organisations that use technology to empower citizens, first in Fundación Civio as Business Developer Manager and now with Fundación Goteo.

Olivier Schulbaum is a Social Entrepreneur, Goteo.org co-founder, President, and Managing Director of Platoniq. He is currently researching on how to amplify direct democracy by bridging political participation, digital campaigning platforms, civic crowdfunding, and technology. He is also working closely with major digital platforms for citizen participation launched by the municipalities of Madrid and Barcelona—Decide Madrid (Consul) and Decidim. He is also a workshop facilitator focusing on cooperation and distributed social innovation processes via dynamics and methodologies introducing cultural shifts within civic organisations such as NGOs, public institutions, cooperatives, and social businesses.

Endnotes

1. IDEA Consult (2017). *Reshaping the crowd's engagement in culture.* Publications Office of the European Union. Retrieved March 14, 2018 from https://publications.europa.eu/en/publication-detail/-/publication/7e10916d-677c-11e7-b2f2-01aa75ed71a1/language-en

2. Zohrabyan, S., Shavarsh, Z., Paula, O. F., Rui, P. L. and José, Á. G. Connecting funding to entrepreneurs: a profile of the main crowdfunding platforms. In: *Cooperative and Networking Strategies in Small Business.* Springer, Cham (2017), 97–129.

3. Nevin, S., Sean, N., Rob, G., Philip, O'R., Joseph, F., Shanping L. and Jerry, C. Social Identity and Social Media Activities in Equity Crowdfunding. In: *Proceedings of the 13th International Symposium on Open Collaboration.* ACM (2017).

4. Görsch, M. *Komplementäre Kulturfinanzierung: das Zusammenwirken von staatlichen und privaten Zuwendungen bei der Finanzierung von Kunst und Kultur.* Dissertation.De (2001).

5. Walker, C. *A Great Match.* Charities Trust and RBS (2017). Retrieved November 14, 2018 from https://cdn.thebiggive.org.uk/static/docs/A-Great+Match-EVersion.pdf

6. Davies, R. Three provocations for civic crowdfunding. *Information, Communication & Society* 18.3 (2015): 342–355.

7. Baeck, P., Jonathan B., and Mitchell, S. Matching the crowd. (2017). Retrieved November 14, 2018 from https://www.nesta.org.uk/publications/matching-crowd-combining-crowdfunding-and-institutional-funding-get-great-ideas-ground

8. Stiver, A., Alexandra, S., Leonor, B., Shailey, M., Mike, R. and Dave, R. Civic crowdfunding research: challenges, opportunities, and future agenda. *New Media & Society* 17.2 (2015): 249–271.

9. Davies, Rodrigo. Civic Crowdfunding: Participatory Communities, Entrepreneurs and the Political Economy of Place. Massachusetts Institute of Technology (2014).

10. Gascón, G., Joan-Francesc, F.-G., Josep, R.-R., Judit, M.-M., Eva, S.-L. and Pere, M.-M. Crowdfunding as a formula for the financing of projects: An empirical analysis. *Revista Científica Hermes* 14 (2015).

11. Senabre, E. Goteo: Crowdfunding to build new Commons. In: D. Bollier, S. Helfrich, (eds.), *Patterns of Commoning.* Commons Strategy Group and Off the Common Press (2015).

12. Fuster Morell, M., Joan, Subirats, H., Marco, B., Rubén, M., Jorge, S., *Procomún digital y cultura libre: "Hacia un cambio de época."* Icaria, Barcelona (2015).

13. Vergés Pascual, I. Una API per a la plataforma de crowdfunding Goteo. (2016).

14. Goteo.org statistic page. Retrieved November 14, 2018 from https://stats. goteo.org/home/es

15. Goteo Match-funding Visualizations. Retrieved November 14, 2018 from https://matchfunding.goteo.org/#/home

16. Navarro, F. S., Fernández-Delgado, F. C. Crowdfunding para la producción cultural basada en el procomún: el caso de Goteo (2011–2014)/Crowdfunding for commons-based cultural production: the case of Goteo (2011–2014). *Historia y Comunicación Social* 20.2 (2015): 447.

17. Goteo Version 3, the Open Source Crowdfunding Platform. Retrieved November 14, 2018 from https://github.com/GoteoFoundation/goteo

Part 3.2

Civic and City Level

Chapter 7

Co-Creating Cities—Practical Experiences from Crowdsourcing and Crowdfunding Urban Areas in Turin, Brussels, and London

Conny Weber and Reinhard Willfort

Abstract

The aim of this chapter is to share experiences gained within the Incubators project (2014–2017), where a prototype for an online platform allowing citizens to co-create their local neighbourhoods in Brussels, London, and Turin has been developed. One of the main learning is that citizens who are actively looking to co-create public space are willing to invest time and energy. Therefore, we recommend to public authorities to consider active participation in the creation of public spaces by inviting citizens through open calls and defined challenges to provide their ideas. Crowd-based technologies such as crowdsourcing and crowdfunding are a means for smart and user-oriented involvement and can be used by local governments or other stakeholders without much background knowledge or time to set everything up.

155

1. Introduction

Co-creating cities is not new as such; however, user-centred urban development is not a common practice, not yet. Classic examples for people co-creating landscapes are paths, which have been created by people stomping along the same trail. Usually, landscape architects design their trails according to their own vision, without considering that people might create their own paths. But there are also exemptions. A nice example for user-oriented landscape design, or co-creating urban areas, is the landscape architect Günther Grzimek (1915–1996). He was famous for his user-centred approach not including pathways or trails in his green areas but waiting until pathways were well established by the people in order to create a trail. This is a great example of urban co-creation, and also more importantly this approach turned out to be very cost-efficient as expensive mistakes of planning could be avoided.

In a management context, the phenomenon of involving people in the innovation process is coined by the term *open innovation* and refers to including external resources, i.e. stakeholders, end users, or communities in the innovation process. According to Chesbrough,[1] open innovation "is a paradigm that assumes that firms can and should use external ideas as well as internal ideas, and internal and external paths to market, as the firms look to advance their technology." The concept of open innovation can be very well applied to urban planning—involving several stakeholders, such as local public authorities, citizens, local businesses, and other experts in the planning process which allows involving their ideas and expectations. Even more precise for describing the phenomenon of involving inhabitants and stakeholders in urban planning processes might be the term "crowdsourcing," which was coined in 2006 by Jeff Howe[2] and describes an organisation leveraging the power of crowds for generating and assessing new ideas. With the common uptake of collaborative technologies, these approaches have become increasingly computer based while enabling access to large user communities. Co-creative processes are the outcome of a shift in urban design; they move from experts towards giving actors the capability to directly contribute their experience.[3]

Allowing people to co-create their cities, e.g. through crowdsourcing, is a cost-efficient methodology and avoids expensive misplannings.

Co-financing urban development by making arrangements between two or more public and private sectors is also not a new phenomenon, e.g. through public–private partnerships (PPPs).[4] However, the last two decades demonstrate a clear trend towards public authorities making greater use of various co-financing arrangements, such as PPP[5] or civic crowdfunding.[6] Crowdfunding has become a promising tool for leveraging funds. Having started with private projects mostly from the creative scene, public organisations and local authorities[7] have also start experimenting with this kind of alternative finance.

Besides the acquisition of financial resources, crowd-related activities offer several added values regarding innovation aspects and risk management, as the financial involvement of private persons as well as their ideas provide a proof of concept and important feedback.

In this section, we report on some key findings gained in the Incubators for Public Places[8] research project, a JPI Urban Europe-funded project (2014–2017), which involves developing and applying an online platform for public participation in the design and redevelopment of public spaces in local neighbourhoods in Brussels, London, and Turin. In these neighbourhoods, three living labs[9] have been established.

2. Co-Creating Cities with Crowdsourcing and Crowdfunding

Incubators aim to support the co-creation of urban areas, enhancing the factors that motivate, encourage, and enable the urban actors to reach a common understanding and coordinate actions by consensus and cooperation rather than following strict development plans.[10] The means to this goal are crowd-based information and communication technologies to advance the co-creation capabilities of urban areas. Within the project, we developed (1) a methodology and (2) a software platform and tested our concepts in (3) three Urban Living Labs in cities across Europe (from London, over Brussels to Turin).

3. The Methodology

The theoretical framework supporting the Incubators project describes a participatory innovation management process combining crowdsourcing

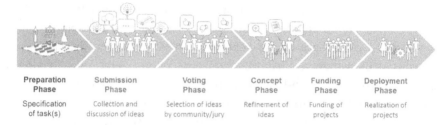

Preparation Phase	Submission Phase	Voting Phase	Concept Phase	Funding Phase	Deployment Phase
Specification of task(s)	Collection and discussion of ideas	Selection of ideas by community/jury	Refinement of ideas	Funding of projects	Realization of projects

Figure 1:   *Overview of the crowdsourcing and crowdfunding process.*

and crowdfunding. Independent of the urban area or the project type, the citizens and stakeholders, i.e. the crowd, can co-create their urban areas. In the early stages of the innovation process, such as the idea-finding phase, the methodology allows, for example, the management of crowd-sourcing campaigns by inviting a wide target group to submit their ideas in response to a defined challenge, to provide feedback on project ideas, or to vote for the best ones. Ideas that have been evaluated successfully in the first phase are further supported for gathering further funds through crowdfunding.

The Incubators methodology (Figure 1) builds on a traditional innovation management process, including the "wisdom of crowds"[11] in all stages. It starts with defining a challenge (preparation), the creative finding of ideas (Submission), evaluating these ideas (Voting), realising an idea (Concept), crowd-financing an idea (Funding), and finally deploying (Deployment) the idea.

4. Crowdsourcing

In the crowdsourcing phase (Figure 2), local authorities, citizens, or other interested stakeholders can describe a challenge and start an open idea contest for gathering ideas from the crowd. In the next step, these ideas are evaluated and commented on by the community. This community is open to anyone but consists ideally of heterogeneous individuals, including creatives, potential stakeholders, and experts from the urban area to be developed. The aim of the crowdsourcing phase is to take advantage of a network consisting of *co-thinkers*, creatives, entrepreneurs, and organisations who collaboratively submit ideas, evaluate the ideas, and shape them

Figure 2: *The crowdsourcing process.*

into a final concept for urban development. Thus, architects, authorities, and citizens can collect feedback and reduce the innovation risk at a very early stage, and at the same time leverage the chance for a successful crowdfunding campaign.

5. Crowdfunding

The crowdfunding phase (Figure 3) aims to support the realisation of a promising project idea by providing both the know-how of experienced innovation experts, investors and multipliers, and financial support. Within this phase, a jury, e.g. consisting of financial experts, architects, local authorities, and other stakeholders, selects the most promising ideas from the crowdsourcing phase and prepares a crowdfunding campaign. This so-called civic crowdfunding provides a valuable means for financing or co-financing urban development projects together with citizens. However, factors associated with success and failures among crowd-funded projects are very diverse and may depend not only on the project's quality but also, for example, on the number of friends in social networks, geographical aspects, the duration, domain, or the funding goal.[12]

6. The Platform

The overall aim of the project was to develop and implement ICT tools to empower citizens in urban planning, usually marginalised groups, to co-create, evaluate, and co-fund scenarios and discover new ways to realise

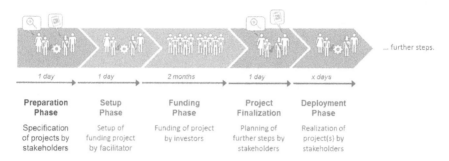

Preparation Phase	Setup Phase	Funding Phase	Project Finalization	Deployment Phase
Specification of projects by stakeholders	Setup of funding project by facilitator	Funding of project by investors	Planning of further steps by stakeholders	Realization of project(s) by stakeholders

Figure 3: *The crowdfunding process.*

and support these. For this purpose, we followed a user-centred approach, involving the users in the design and development phase.

As an essential aspect of the co-creative model, it is crucial to deploy the web-based tools appropriate to and attractive for the different types of actors contributing to the Incubators. Based on interviews and focus groups with envisioned users from the living labs, we started defining the user types and created the so-called personas.

In the prototyping process, these four types of users have been involved:

(1) people (i.e. citizens who want to solve their real-life problems),
(2) utilisers (enterprises that want to develop their businesses in the area),
(3) enablers (public-sector actors, developers), and
(4) providers (domain experts, e.g. universities, consultants, and technicians).

A persona[13] (also known as an archetypal user) is an invented person that is used to represent a type of user of a use case. A persona can be based on a real person or on a set of real people. The idea is to make up personas to enable talking concretely about them as users of the Incubators system. After collecting the users' knowledge, we continue translating it into design concepts. From these, the design prototypes of the additional services required, the crowdsourcing, and the crowdfunding activities could be derived.

Based on the interviews with relevant stakeholders of the Incubators project, some of the personas were developed as shown in Figure 4.

Figure 4:   *Overview of the incubators personas.*

Having the so-called personas in mind, we further developed use cases and scenarios, from which we derived the systems requirements describing the scope and boundaries of the platforms. A first mock-up was evaluated in small settings always including people from the three living labs. Step-by-step, the paper mock-ups were refined and finally the web-based prototype was created.

In order to allow many users to contribute their ideas and funds, the Incubators platform prototype was accessible via any browser and mobile phone (Figure 5). People had to login and could then navigate via the menu to current challenges and vote for already existing ideas, prepare their ideas, or access open crowdfunding campaigns (Figure 6).

7. The Urban Living Labs

Within their specific socio-spatial and political contexts, the three Urban Living Labs, London, Brussels, and Turin, explored the processes of co-creation on urban spaces. It was the intention of each case to unfold its own particular and context-based configuration and to look for its potential to develop through a self-organised participatory process.

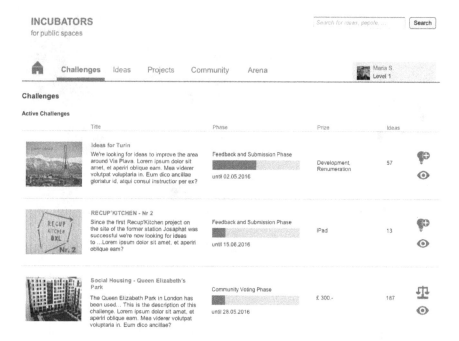

Figure 5:   Incubators platform—screenshot crowdsourcing.

7.1. Brussels

The objective of the Brussels[14] use case was to develop an Urban Living Lab structure for the revitalisation of Josaphat Oud Station, Schaarbeek in the north of the Brussels, where a zone of regional interest is planned to be developed in a new sustainable neighbourhood. The Incubator was set and operated in collaboration with BRAL, which is one of the four recognised Belgian environmental federations founded in 1973. In discussion with BRAL and the SAU-MSI (public owner of the site), it was agreed to test the possibilities of the Incubators platform to facilitate citizen initiatives in the transitory use of Josaphat or other to-be-developed areas in Brussels. The tool was tested in a small scale in relation to the already existing uses of the site (see Figure 7).

The challenge for the Brussels case was to provide ideas for the transitory use of Josaphat within the given area. For this, both the crowdsourcing and crowdfunding modules as well as a 3D environment have been

Figure 6: *Incubators platform—screenshot crowdfunding.*

Figure 7:   Brussels Urban Living Lab.

integrated. The users were asked to relate their ideas to the five values that were collectively discussed and agreed upon among the current users of the site in earlier workshops as follows:

Josaphat as a (1) natural environment, (2) a common and integrated place, (3) a laboratory, a workshop space, (4) a circular and transitory use, and (5) a serene and convivial atmosphere.

7.2. London

The Pollards Hill housing estate is in south London, at the eastern extremity of the borough of Merton, and bordering on the borough of Croydon. There are some local shops and community facilities; the area is served by a number of local buses. The estate generally comprises high-density, low-rise housing completed in the early 1970s.

The challenge for the London case[15] was to provide ideas for redesigning the courtyards of the Pollards Hill housing estate. Within the framework of the "Pollards Hill Regeneration" project—which covers the extensive refurbishment of over 800 homes—the Incubators system was used to collect design ideas about open spaces. In discussion with Moat Homes Ltd. (manager of the Pollards Hill housing estate) and Potter Raper Partnership (landscape architects), it has been agreed that the Incubators system will be deployed in parallel with a more traditional

Figure 8: *London Urban Living Lab.*

consultation process. The traditional consultation covered all nine court-yards over a period of 1 year (July 2017–June 2018), while the Incubators system was used for a trial period of about 2 months (August–September 2017). This trial was based on a single courtyard as an example, and all residents have been invited to submit their design ideas through the online platform (see Figure 8).

In the London case, the Incubators system has been adapted in order to fit within the more general "Pollards Hill Regeneration" project. This project already foresees a traditional consultation process staffed and facilitated by Moat which, for the first period, was supplemented by the Incubators' online consultation. All residents have been provided with a personal username and password so that they could access the platform anytime and anywhere, while Moat was able to keep track of their identities. Upon request, logins were made available to residents at the Community Centre/Consultation House.

Once logged in, residents were able to design their own courtyard drawing inspiration from a set of interventions devised in collaboration with Moat and Potter Raper. A total of 27 Living Labbing sessions were held periodically at the Community Centre/Consultation House to assist non-expert users in the idea submission process. At the end of this period, the outcomes of the online consultation were merged with the outcomes of the workshops. It was then the landscape architect's task to interpret residents' ideas into blueprints, for the Municipal Offices in charge of the regeneration process.

7.3. Turin

The aim of the Turin use case was to develop an Urban Living Lab structure for urban, social, and architectural regeneration that was to be pursued in a public housing neighbourhood located in the outskirts of south of Turin. Mirafiori Sud, close to the Fiat Mirafiori factory, was built in the mid-1960s to provide housing for the new population arriving from other Italian regions due to the automotive industry. Today, the area presents some critical issues: degraded public spaces, lack of commercial activities, and the aging of the community. For the support and structuring of public spaces, local actors contributed to the definition of places, starting from the planning of infrastructures in public spaces: renovation of the central "square" and the covered market. The project has a strong multi-scalar governance dimension (see Figure 9).

In close collaboration with Città di Torino, Circoscrizione 10, and Fondazione della Comunità di Mirafiori, it has been agreed that the

Figure 9:   *Turin Urban Living Lab.*

Incubators project will involve the inhabitants of Mirafiori Sud in a number of participatory workshops. During these workshops, participants have contributed to Living Lab workshops, arranged as serious games, and were invited to elaborate design scenarios. Furthermore, in-depth interviews were carried out with the local stakeholders (e.g. decision-makers, opinion leaders, representatives of local organisations) (for more details, refer to Caneparo *et al.*[16]).

With the aim of gaining some preliminary knowledge for the implementation of the Incubators system, the Turin case managed Living Lab workshops. These early workshops gave insights for the design of the Incubators online system. The participants went through all the phases (proposing, voting, and funding) over a period of 3 hours in the Living Lab workshop. The facilitators managed the workshops and the discussion in a three-tier structure. During the first round, the participants convened around a model of the neighbourhood: in the early experimentations, the model was physical, while later it was digital. The participants were invited to place pictures of potential interventions on the area of interest in the model. During the second round, they were asked to comment on other participants' ideas. In the third and final round, they had prioritised and assigned funds to the interventions, managing a given budget. The aim of the Living Lab was twofold. On the one hand, the workshops were to verify and to assess the basics of the Living Labbing process, in order to get co-creative contributions for the Incubators online platform. On the other hand, starting from the citizens' wishes and integrating the workshop outcomes, new design concepts were developed, to advance alternative design concepts to the Municipal Offices.

8. Conclusions and Outlook

The aim of this chapter was to share some experiences gained within the Incubators project (2014–2017), where a prototype for an online platform allowing citizens to co-create urban areas in their local neighbourhoods in Brussels, London, and Turin has been developed.

One of the main learning is that citizens who are actively looking to co-create public space are willing to invest time and energy. Therefore, for those who are responsible, it can be an interesting first step to allow active participation in the creation of public spaces by inviting citizens through open calls and defined challenges to provide their ideas. Crowd-based technologies such as crowdsourcing and crowdfunding are a means for smart and user-oriented involvement and can be used by local governments or other stakeholders without much background knowledge or time to set everything up.

The main challenge to overcome when using digital technology for co-creating urban areas is that in spite of communication efforts, citizen participation was not high. However, those who participate in the process, do so enthusiastically. A further challenge especially in Turin and London was that the skills of many people involved in co-creation were not sufficient to deal with crowd-based technology. For this reason, within the project we supported the test phase with a mixed approach, based on both real and virtual participatory instruments. A further learning is that the number and scope of defined challenges should be clear in order to improve the effectiveness of the participatory process (see Figure 10).

Crowd-based technology has opened up great potential for local authorities, architects, and citizens to play a more active role in interpreting and proactively shaping their urban environments. These developments not only pose technological challenges—with regard to design of human–computer interactions—but also raise questions on how those challenges are bound up with the aptitudes and inclinations of different kinds of users. Hence, they are raising questions about who is able to

Figure 10:   *Co-creating cities with crowdsourcing and crowdfunding.*

make the most use of these technological processes and how best they may be embedded in specific participatory planning processes. In this project, we figured out that intermediating services or facilitators interacting between the citizens, architects, and public authorities are important to make the most out of the co-creative process. But further learning from current-use cases and pilots, e.g. in the field of civic crowdfunding,[17] will provide more answers.

Acknowledgements

This work was supported by the Joint Programming Initiative Urban Europe under grant 414896 and by ERDF under grant 186 245 C.

About the Authors

Conny Weber is working for the ISN (Innovation Service Network) group since 2006. She has an extensive background in the field of innovation management and experience with EC projects' management. Her work and research topics are mainly related to innovative business models and alternative financing instruments, enhancing access to finance for entrepreneurs and SMEs. She holds a degree in information science from Saarland University (Germany). And a PhD in business studies from Karl-Franzens University Graz (Austria).

Reinhard Willfort is a consultant, innovation manager, trainer and university lecturer and deals with fostering and supporting open innovation, especially crowdsourcing and crowdfunding processes. Since 2001 he is founder and CEO of ISN (Innovation Service Network), Austrian's leading innovation company. As entrepreneur and founder of five further start-ups, he supervises the innovation and knowledge management processes of several leading Austrian companies. In 2012 he established Austria's first crowdinvesting platform 1000×1000.at.

Endnotes

1. Chesbrough, H.-W. *Open Innovation: The New Imperative for Creating and Profiting from Technology*, Harvard Business Review Press, 2003.

2. Howe, J. "The Rise of Crowdsourcing." *Wired* 14 (2006). http://www.wired.com/wired/archive/14.06/crowds.html

3. Sanders, E.B.-N. and Stappers, P.J. "Co-creation and the New Landscapes of Design." *CoDesign*, 4(1) (2008), 5–18.

4. Wettenhall, R. Mixes and Partnerships Through Time. In Hodge, G.A., Greve, C. and Boardman, A. (eds.), *International Handbook in Public–Private Partnerships*, Edward Elgar, Cheltenham UK, 2010.

5. Hodge, G. A. and Greve, C. "Public–Private Partnerships: An International Performance Review." *Public Administration Review*, 67(3) (2007), 545–558.

6. Passeri, F. The European Dimension of Civic Crowdfunding—The Potential of Crowdfunding for Boosting the Economic and Social Effectiveness of European Structural and Investment Funds. (Ed) European Crowdfunding Network, www.eurocrowd.org. 2017. https://eurocrowd.org/wp-content/blogs.dir/sites/85/2017/11/Civic-Crowdfunding-and-ESF_Final.pdf

7. Two European Lighthouse examples are the initiatives from London and Spacehive (https://www.london.gov.uk/what-we-do/regeneration/funding-opportunities/crowdfund-london) and Barcelona and Goteo (http://fundacion.goteo.org/blog/una-ola-de-proyectos-sociales-llega-a-barcelona-comienza-el-viaje-de-los-24-de-conjuntament).

8. For more information, visit https://jpi-urbaneurope.eu/project/incubators/

9. Veeckman, C., Schuurman, D., Leminen, S. and Westerlund, M. "Linking Living Lab Characteristics and Their Outcomes: Towards a Conceptual Framework." *Technology Innovation Management Review*, December (2013), 6–1.

10. Caneparo, L., Rolfo, D., Bonavero, F., Van Reusel, H., Verbeke, J., Marshall, S., Hudson-Smith, A. and Karadimitriou, N. "Semantic Analysis of Public Spaces in Brussels, London and Turin living labs: A Taxonomy of the Interventions." In: *Incubators Conference – Urban Living Labs for Public Space. A New Generation of Planning?* Brussels, Belgium, April 10–11, 2017.

11. Surowiecki, J. *The Wisdom of Crowds*, Anchor Books, New York, 2005.

12. Mollick, E. "The Dynamics of Crowdfunding: An Exploratory Study." *Journal of Business Venturing*, 29(1) (2014), 1–16.

13. Cooper, A. "The Inmates are Running the Asylum." *SAMS* (1999).

14. Van Reusel, H., Verbeke, J., Bonavero, F., Caneparo, L. and Rolfo, D. "Incubators of Public Spaces: A Digital Agora to Support and Empower Self-organised Participatory Processes in Urban (re)development." In: *Critical*

Alternatives 2015. Unfolding Participation, Aahrus, Denmark, August 17–21, 2015, 1–4.

15. Caneparo, L. and Bonavero, F. "Neighbourhood Regeneration at the Grassroots Participation: Incubators' Co-Creative Process and System." *ArchNet International Journal of Architectural Research*, 10(2) (2016), 204–218.

16. Caneparo, L., Bonavero, F. and Melis, B. "Incubators of Public Spaces (2): Tools for Self-organisation in Urban Regeneration. In: *31st International PLEA Conference*, Bologna, Italy, September 9–11, 2015, 1–8.

17. Passeri, F. (2017).

Chapter 8

Crowdfunding Social or Community-Focused Projects

Jonathan Bone

Abstract

Over the last few years, crowdfunding has exploded in popularity and is now a commonly used tool for start-ups and SMEs looking to raise finance. However, the potential of crowdfunding remains relatively untapped by more socially focused organisations such as charities, community groups, and social enterprises. In this chapter, based on previous primary research, we explore the opportunities and challenges in crowdfunding for social or community-focused projects.

In addition to funding projects that would struggle to access finance elsewhere, opportunities include many potential non-financial benefits such as boosting volunteering, increasing transparency, allowing more experimentation with innovative ideas, and providing a new way to raise awareness for social causes. However, crowdfunding for social purpose projects does not come without its challenges. Here, we discuss potential issues around having a potential negative impact on equality and participation in projects, too much focus on short-term initiatives rather than long-term projects, and concerns that it may encourage the withdrawal of public funding.

Finally, we explore how institutional funders can support socially focused projects to unlock the potential benefits of crowdfunding, through

*matched crowdfunding schemes by combining their grant funding with
the money projects raise through crowdfunding. Such schemes can also
help institutions to find new social purpose projects to support and
make their own grant funding go further, allowing them to support more
projects.*

1. Introduction

Crowdfunding has exploded in popularity around the world, from a
marginal fundraising tool just a few years ago to, in some countries, now
being a mainstream form of finance.[1,2,3] However, the vast majority of this
market growth has stemmed from equity finance for start-ups or loans for
businesses and individuals, with relatively little being raised for social or
community-focused projects. This is despite there clearly being potential
for the sector to use this new tool, not only for fundraising but also for
increasing engagement with projects and causes.

I will discuss insights from a study, published in 2016, which
explored the opportunities and challenges in crowdfunding for social or
community-focused projects.[4] Here, I will also explore how institutional
funders can help support smaller organisations and projects through
matched crowdfunding schemes by combining their grant funding with
the money projects raise through crowdfunding. Matched crowdfunding
will be discussed within the context of what was learnt from a matched
crowdfunding pilot fund for the arts and heritage sectors.[5]

Throughout, I use the term "social or community-focused projects" in
a broad sense to refer to organisations that act in the public's interest and
are not primarily motivated by profit, including charities, NGOs, community groups, arts and heritage organisations, and social enterprises.

2. Opportunities and Challenges in Crowdfunding for Socially Focused Projects

Public donations have been used to fund community-based projects for
hundreds of years. Among other things, this has included funding the
building of the base of the Statue of Liberty,[6] mosques on the Indian
subcontinent,[7] and public parks in Victorian England.[8] The advent of

digital technology now means that this can take place online, making it quicker and easier for fundraisers to reach potential backers. Crowdfunding, in this modern sense, is being used to raise money for a broad range of socially focused projects including running events, restoring buildings, building community parks and gardens, and buying essential equipment.

In comparison to other sectors, crowdfunding for socially focused projects has been relatively slow to take off. For example, in the UK—the largest crowdfunding market in Europe—crowdfunding makes up around 17% of seed and venture-stage equity investment and 15% of new loans to small businesses,[9] but less than 0.5% of charitable giving.[10] This lack of traction was of interest, as many of the key features of crowdfunding seem to align closely with what is known about people's social motivations for giving. For instance, crowdfunding campaigns raise money for specific projects or activities rather than organisations more generally, so they typically tell a personal story and convey the direct impact backers will have on other people's lives, something which has been shown to increase people's tendency to give.[11] Further to this, crowdfunding platforms typically tap into peer effects by making the names of those who have contributed visible[12] or encourage reciprocity by offering small rewards such as a postcard or a badge in return for contributing.[13] Combined, these characteristics would seem to present a strong case that crowdfunding could be a key fundraising tool in this field.

The mismatch between the huge potential and lack of use of crowdfunding by socially focused projects motivated me and my colleagues to conduct research into the opportunities and challenges in crowdfunding for this sector. Crowdfunding these types of projects typically takes place through donation-based, rewards-based, or community shares models of crowdfunding,[14] so when I mention crowdfunding in this chapter, I will be referring specifically to those models, rather than the equity-based or lending-based models.

3. Opportunities

Our research has uncovered several potential benefits, both financial and non-financial, associated with crowdfunding for socially focused projects. Most significantly, crowdfunding can help fund projects that would otherwise struggle to raise money elsewhere. In one survey conducted by NESTA and the University of Cambridge, 64% of respondents who had raised funds via donation-based crowdfunding reported that they were unlikely or very unlikely to have received finance elsewhere. Crucially, the same study found evidence to suggest that this is the result of additional money being raised for social projects rather than just money being given via different methods or to different types of projects. When surveyed, 77% of individuals who donated money to support a social project said that the money they gave was in addition to what they what would normally give to charity.

Alongside directing more money to social causes in general, crowdfunding also appears to help get projects off the ground that are overlooked by institutional funders (e.g. local and national governments, trusts and foundations, and other conventional grant makers) but that the public sees a significant need for or is otherwise interested in seeing happen. In this way, it could be argued to represent the democratisation of funding for social and community-focused projects. At the very least, it demonstrates some tangible interest in the project taking place from a community or set of individuals willing to back it.

Crowdfunding can also help those with more experimental projects to test whether there is a demand for their idea with little financial loss, because if there isn't support from the crowd, they will not reach their crowdfunding target and the project won't go ahead. In this way crowdfunding may help drive innovation in solving social issues by getting more experimental ideas off the ground. In a survey of more than 450 socially focused organisations, 66% of those that had used crowdfunding said that it is better than other fundraising methods for its possibility to fund innovative projects.[15]

The "proof of concept" provided by a successful campaign can also subsequently help convince larger, more traditional funders to get behind a project and scale it up. Support for this claim comes from a survey of participants of a matched crowdfunding pilot, which will be discussed in detail later, over a third of which reported that they had used their crowdfunding experience to leverage some form of further funding.[16]

Looking beyond the money it can help projects raise, running a crowdfunding campaign also brings with it non-financial benefits, such as helping projects to promote their cause and building communities of supporters and volunteers. Evidence for this comes from a study in which it was found that 90% of people who gave money to support a social project subsequently promoted it to their social network (typically via social media), 29% gave feedback or advice to those running the campaign, 27% offered to volunteer directly with the project, and 27% made introductions on behalf of the project.[17]

Further to this, in the survey of matched crowdfunding pilot participants, 82% of fundraisers agreed or strongly agreed that their organisation had a stronger public profile and over three-quarters reported that they had built new audiences that they will continue to engage with as a result of the campaign.[18] The same survey showed that crowdfunding can be instrumental in developing skills within organisations, and two-thirds of fundraisers reported that running a crowdfunding campaign significantly improved their pitching and fundraising skills. Specific skills that they reported to have improved included film creation, image creation, and social media skills.

It is this mixture of financial and non-financial benefits that can help build resilience within organisations working on social or community-focused projects and reduce the risks associated with innovative new projects. Building resilience is particularly crucial for socially focused organisations which are typically small, and thus, single funding awards can be the difference between survival and closure.

4. Challenges

It has been argued that crowdfunding may have a potentially negative effect on diversity, equality, and participation. This concern is centred on the idea that people and communities that are wealthier, more "tech-savvy," and have larger social networks may be in a better position to run successful crowdfunding campaigns. This means that crowdfunding successes may become concentrated in better-off areas of a city or country or on issues which are more relevant or interesting to wealthier people.[19]

There has been very little research into this question.[20] However, in one study, it was found that crowdfunding for social projects involved equal numbers of men and women in both funding and fundraising and includes people from a diverse mix of geographical and economic backgrounds.[21] This finding was echoed in our own matched crowdfunding pilot.[22] However, further research is needed to fully understand who is (and who isn't) benefiting from crowdfunding both in terms of those running socially focused projects and those using them.

Crowdfunding focuses on raising money for one-off projects. For projects that require maintenance (for example, a community playground or garden), this raises a challenge around who will pay for this maintenance after the money raised from the crowdfunding has all been spent. Linked to this, there is a question about how the organisations that crowdfund can become sustainable so their projects can not only be maintained in the future but also they can scale their positive social impact. I believe that future work aiming to understand this question would be valuable. One possible avenue of research is into the role that new models for community investment, governance, and ownership—such as community shares— may produce more sustainable and resilient community projects.

It is often argued that while it is dressed up as democratisation, crowdfunding is in fact a Trojan horse for privatisation.[23,24,25] This argument centres on the concern that crowdfunding risks encouraging public sector funders to withdraw from funding services or facilities that should be paid for by the taxpayer. It relates to ideological debates about the role of the state versus civil society and is not specific to crowdfunded projects but is applicable to any project supported by public donations or volunteering. This argument should be taken seriously; however, even with its growing popularity, it is unlikely that crowdfunding itself is the force behind cuts to state funding, which is being driven by numerous complex economic and political factors. Nevertheless, it is important to be clear that as crowdfunding focuses on funding one-off or short-term, and often fragmented, projects, it cannot substitute broad-based and long-term funding structures.

Crowdfunding puts the decision of who receives funding and therefore which projects can go ahead in the hands of the crowd, who often have little or no specific expertise of the social issues which those projects are trying to address. This raises potential conflicts between what the crowd wants to fund and what institutional funders, who do have expertise in these areas, deem to be priority causes. For example, in interviews with crowdfunding platforms, we heard anecdotal evidence that crowdfunding projects aimed at helping the elderly, homeless people or people with addictions can be difficult because of the stigma attached to these issues or simply because they not seen to be as "sexy" as, for example, projects to build a new community theatre.

Here, I have explored a handful of the potential challenges that have been raised around crowdfunding projects with a social purpose. Though I believe that the opportunities do outweigh the potential challenges, it is important for fundraisers and policymakers to be aware of these challenges so they can make decisions about where crowdfunding should and shouldn't be used and act to mitigate any potential negative effects.

5. Matched Crowdfunding

A growing number of institutional funders—including governments, trusts, and foundations, businesses with a focus on CSR and universities—have been exploring how they can harness the potential of crowdfunding by setting up matched crowdfunding schemes whereby they combine their grants with money raised through crowdfunding. For example, if a project can raise the first 75% of its fundraising target from the crowd, an institutional funder might offer them a grant to top up the remaining 25%. Matched crowdfunding schemes are being set up by a diverse range of funders around Europe. Their reasons for doing so are myriad but typically centre on either finding new projects or ideas to support, particularly projects that the public have already shown that they are interested in by backing with their own money, or making their money go further by mixing it with crowd donations.

6. A Matched Crowdfunding Pilot for Arts and Heritage

At NESTA, we wanted to understand how matched crowdfunding could be used to help more social purpose projects to unlock the potential financial and non-financial benefits associated with crowdfunding as well as help institutional funders to identify new projects to support and make their money go further. In order to explore these questions, we set up a matched crowdfunding pilot in collaboration with two UK-based funders, Arts Council England and Heritage Lottery Fund,[26] each of which provided approximately GBP 125,000 (€135,363)[27] of grant funding for arts and heritage projects raising between GBP 4,000 (€4,332) and GBP 40,000 (€43,316) through Crowdfunder,[28] a crowdfunding platform selected from a competitive tender process. To be eligible for the match, projects were required to meet a set of criteria that conformed to the funders' usual guidelines, such as not being commercial or income-generating projects, being aligned with a specific art form and outlining who the beneficiaries of the activity would be.

As part of the pilot, we tested the effect of different matching ratios on backers' contributions, with some campaigns receiving 25% of their target as a match and others receiving 50% of their target. We also experimented with different matching methods. Crowdfunding projects were randomly assigned to either receive, a "top up match" where the project was required to reach a specified percentage of its target (in this case 75% or 50%) before the institutional funder provided the remaining amount required for the project to reach its target or a "bridge match" where the match happened when a certain goal has been met (in this case 25% of the target) in order to "bridge" the middle part of the campaign, which is typically slower in terms of crowd donations, The fundraiser was then required to raise the remaining amount from the crowd (see Figure 1).

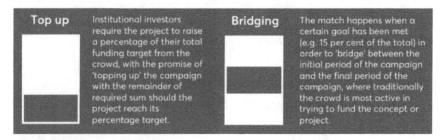

Figure 1: *Matching methods explored in pilot.*

Source: Adapted from Baeck *et al.* (2017).

In total, 59 projects were funded through the pilot. This included a range of project types including art installations, plays, restorations, and performances. Arguably the most important insight from the pilot was that matched crowdfunding can help drive significant amounts of funding towards socially focused projects. The GBP 251,500 (€272,349) provided by the funders leveraged an additional GBP 405,941 (€439,593) from a crowd of 4,970 backers. This means that every GBP 1 invested by the funders leveraged a further GBP 1.61 from the crowd.

We found that crowdfunding campaigns that had the offer of a match were significantly more likely to reach their target compared to a control group of similar campaigns that did not have the chance for a match. Further analysis indicated that the increase in success rates for those offered a match could be attributed to larger average contributions to these campaigns, rather than an increase in the total number of people backing them. The pilot found that a 50% match is more effective at increasing the average contributions of backers than a 25% match, but the effect of the matching method used (i.e. top-up or bridging) was inconclusive.

Alongside this, we found that the pilot was able to reach arts and heritage organisations that were not otherwise known to the institutional funders. For example, 42% of the match-funded projects report that they had never applied for funding from the funders before. This suggests that matched crowdfunding pilots can, as we previously hypothesised, help institutional funders to identify new project to support.

The pilot, however, did not come without its own challenges. From the funders' point of view, although subjective, it was indicated that the quality of the projects funded was lower in quality to projects funded through conventional grant-funding streams. This relates to the challenge raised earlier around potential conflicts between what the crowd wants to fund and what institutional funders see as priority causes.

It was also felt that information gathered on projects was not always strong enough to make detailed value judgements on their work—for example, the lack of transparent budgeting or precise descriptions of activity presented difficulties throughout the pilot. The functionality to collect this information is something that we believe would be reasonably simple for crowdfunding platforms to integrate into their current offering.

On the fundraisers side, running a successful crowdfunding campaign requires a lot of hard work, often with careful planning months in advance of going live and regular updates to keep up the momentum throughout the campaign. Although a relatively small number, around one in 10 fundraisers reported that they thought that the resulting impacts they experienced were not worth the effort involved in running a crowdfunding campaign.

7. Conclusions and Recommendations for Policymakers, Funders, and Other Supporters of Social Purpose Projects

Crowdfunding presents a significant untapped potential for social purpose projects to raise money, awareness for their cause, and get more people involved in campaigning and volunteering. Given the potential benefits associated with crowdfunding, it may be in the interest of policymakers, funders, and other supporters of social purpose initiatives to do what they can to support the use of this new tool.

Our matched crowdfunding pilot for arts and heritage projects showed that this type of scheme can help institutional funders find new projects to support and make their money go further, while helping socially focused projects to unlock the financial and non-financial benefits associated with crowdfunding.

This is not the only way in which policymakers and other institutional funders can help social purpose organisations make the most of crowdfunding. A lack of skills and knowledge has been shown to be the biggest barriers to using crowdfunding for social purpose organisations.[29] Funders and policymakers can help bridge this gap by investing in crowdfunding training, as well as more general digital training, with a focus on the specific needs of social purpose organisations and projects.

Further to this, funders can support socially focused projects that they are not able to fund themselves by setting up referral schemes, where unsuccessful applicants are directed to a crowdfunding platform or a third-party directory of platforms. This will help raise awareness on crowdfunding as a potential alternative route to funding.

Crowdfunding for socially focused projects is still relatively new and there is still a lot to learn about how we can make the most of this tool.

It is, therefore, important that policymakers and institutional funders rigorously measure the impact of any interventions undertaken and share any evidence produced so that both themselves and the wider sector can continue to learn what works and what doesn't.

About the Author

Jonathan Bone is a Senior Researcher in NESTA's Policy, Analysis and Research team. His research at NESTA has focused on two key areas: (1) exploring how crowdfunding and other forms of alternative finance can be used as an innovative way of financing businesses, charities and community-led initiatives and (2) understanding how digital social innovation and entrepreneurship can be supported across Europe. Before joining NESTA Jonathan completed a PhD in Evolutionary Biology at University College London, where he used experimental economics methods to investigate the evolution of human cooperation.

Endnotes

1. Ziegler, Tania, Reedy, E.J., Le, Annie, Zhang, Bryan, Kroszner, Randall S. and Garvey, Kieran. *Hitting Stride: The 2nd Americas Alternative Finance Industry Report*, The University of Cambridge, Cambridge, 2017, https://www.jbs.cam.ac.uk/fileadmin/user_upload/research/centres/alternative-finance/downloads/2017-06-americas-alternative-finance-industry-report.pdf
2. Garvey, Kieran, Chen, Hung-Yi, Zhang, Bryan, Buckingham, Edward, Ralston, Deborah, Katiforis, Yianni, Deer, Luke, Ziegler, Tania, Ying, Kong, Maddock, Rodney, Shenglin, Ben, Xinwei, Zheng, Jenweeranon, Pawee, Li, Wenwei, Hao, Rui, Huang, Eva and Zhang, Jingxuan. *Cultivating Grow: The 2nd Asia Pacific Region Alternative Finance Industry Report*, The University of Cambridge, Cambridge, 2017, https://www.jbs.cam.ac.uk/fileadmin/user_upload/research/centres/alternative-finance/downloads/2017-12-cultivating-growth.pdf
3. Ziegler, Tania, Shneor, Rotem, Garvey, Kieran, Wenzlaff, Karsten, Wenzlaff, Karsten, Yerolemou, Nikos, Hao, Rui and Zhang, Bryan. *Expanding Horizons: The 3rd European Alternative Finance Industry Report*, The University of Cambridge, Cambridge, 2017, https://www.jbs.cam.ac.uk/fileadmin/user_upload/research/centres/alternative-finance/downloads/2018-ccaf-exp-horizons.pdf

4. Bone, J. and Baeck, P. *Crowdfunding Good Causes*, NESTA, London, 2016, http://www.nesta.org.uk/publications/crowdfunding-good-causes

5. Mitchell, S., Bone, J. and Baeck, P. *Matching the Crowd*, NESTA, London, 2017, https://media.nesta.org.uk/documents/matching_the_crowd_main_report_0.pdf

6. The Statue of Liberty and America's Crowdfunding Pioneer." *BBC Magazine* (2013), http://www.bbc.co.uk/news/magazine-21932675

7. M, Mines. *Public Faces, Private Lives: Community and Individuality in South India*, University of California Press. Berkeley and Los Angeles, California, 1994.

8. Jordan, H. "Public Parks, 1885–1914." *Garden History*, 22(1) (1994), 85–113, https://www.jstor.org/stable/1587004?seq=1#page_scan_tab_contents

9. Zhang, Bryan, Ziegler, Tania, Garvey, Kieran, Ridler, Samantha, Yerolemou, Nikos and Hai, Rui. *Entrenching Innovation: The 4th UK Alternative Finance Industry Report*, The University of Cambridge, Cambridge, 2018, https://www.jbs.cam.ac.uk/fileadmin/user_upload/research/centres/alternative-finance/downloads/2017-12-ccaf-entrenching-innov.pdf

10. Estimate is based on NCVO (2018; see end of this footnote) estimate of individual giving in the UK being £22.3 billion in 2015/2016 and crowdfunding for good causes being estimated at £93.6 million by taking the sum of all donation-based crowdfunding, all community shares and 20 per cent of rewards-based crowdfunding based on data from Zhang *et al.* (2018; see endnote 1). We estimated that 20 per cent of rewards-based crowdfunding went to good causes based on a search of keywords associated with charity, social enterprise and community group projects using the crowdfunding analytics company TAB (https://www.insidetab.io). NCVO. The UK Civil Society Almanac, NCVO, London 2018, https://data.ncvo.org.uk/a/almanac18/income-from-individuals-2015-16/

11. Jenni, K. and Loewenstein, G. "Explaining the Identifiable Victim Effect." *Journal of Risk and Uncertainty*, 14 (1997), 235–257, https://doi.org/10.1023/A:1007740225484

12. Cotterill, Sarah, John, Peter and Richardson, Elizabeth. "The Impacts of a Pledge Campaign and the Promise of Publicity: A Randomized Controlled Trial of Charitable Donations." *SSRN* (2011), http://dx.doi.org/10.2139/ssrn.1833487

13. Alpizar, Carlsson and Johansson, Stenman. "Anonymity, reciprocity, and conformity: Evidence from voluntary contributions to a national park in Costa Rica." *Journal of Public Economics*, 92(5) (2008), 1047–1060.

14. In the UK, community shares are a unique form of share capital that can only be issued by co-operative societies, community benefit societies and charitable community benefit societies. Investors in community shares have a democratic

say in the project's social aims and can be paid interest on their shares if the society's trading performance allows it. Investors may also cash-in their shares with the society at some point in the future. However, unlike company shares, they cannot go up in value—though they can go down in value.

15. Bone, J. and Baeck, P. *Crowdfunding Good Causes*, NESTA, London 2016, http://www.nesta.org.uk/publications/crowdfunding-good-causes

16. Mitchell, S., Bone, J. and Baeck, P. *Matching the Crowd*, NESTA, London 2017, https://media.nesta.org.uk/documents/matching_the_crowd_main_report_0.pdf

17. Baeck, P., Collins, L. and Zhang, B. *Understanding Alternative Finance: The UK Alternative Finance Industry Report*, NESTA, London 2014, https://www.nesta.org.uk/report/understanding-alternative-finance-the-uk-alternative-finance-industry-report-2014/

18. Mitchell, S., Bone, J. and Baeck, P. *Matching the Crowd*, NESTA, London 2017, https://media.nesta.org.uk/documents/matching_the_crowd_main_report_0.pdf

19. Sullivan, P. "Raising Money for Civic Projects Raises Questions about Democracy." *The Washington Post*, 2016. See: https://www. washingtonpost.com/local/raising-money-for-civic-projects-raises-questions-about-democracy/2016/04/24/9dd7cb0c-058e-11e6-b283-e79d81c63c1b_story.htm

20. Evidence on the effect of equity crowdfunding on reducing inequality is positive concerning gender, mixed with respect to geography, and less encouraging in terms of race. Greenberg, J. Inequality and Crowdfunding. SSRN, 2017 http://dx.doi.org/10.2139/ssrn.1833487

21. Baeck, P., Collins, L. and Zhang, B. *Understanding Alternative Finance: The UK Alternative Finance Industry Report*, NESTA, London 2014, https://www.nesta.org.uk/report/understanding-alternative-finance-the-uk-alternative-finance-industry-report-2014/

22. Mitchell, S., Bone, J. and Baeck, P. *Matching the Crowd*, NESTA, London 2017, https://media.nesta.org.uk/documents/matching_the_crowd_main_report_0.pdf

23. Moskowitz, P. Crowdfunding for the Public Good Is Evil. Wired. March 2016, http://www.wired.com/2016/03/crowdfunding-is-evil/

24. Minton, A. "Civic Crowdfunding is Privatisation Masquerading as Democracy." *The Guardian*, October 2017. https://www.theguardian.com/cities/2017/oct/24/civic-crowdfunding-erode-democracy-local-authority

25. Lange, A. "Making something big happen at an urban scale is more than a popularity contest." Dezeen. June 2014, https://www.dezeen.com/2014/06/19/alexandra-lange-opinion-crowdfunding/

26. With funding from DCMS; The pilot ran from September 2016 until October 2017.

27. At September 2017 exchange rates (GBP 1 = Eur 1.0829); Although the match fund was originally £250,000. an additional £1,500 was provided by Heritage Lottery Fund to ensure a 50% match for the last project through the pilot.

28. https://www.crowdfunder.co.uk/

29. Bone, J. and Baeck, P. *Crowdfunding Good Causes*, NESTA, London, 2016.

Chapter 9

Crowdfund London: Civic Crowdfunding as a Tool for Collaborative Urban Regeneration in London

James Parkinson

Abstract

Through Crowdfund London, City Hall is exploring how many small projects, delivered by local people, can have a big social impact and how the city's government can catalyse a powerful mix of public, private, and local collaboration.

Crowdfund London is a capital investment programme that uses civic crowdfunding to source grass-roots innovation, promote local engagement, and empower communities to collaborate in regeneration. It has grown into an internationally recognised and pioneering tool for city government to work more directly with citizens and align local interventions with strategic investments.

The programme has supported over 100 successful campaigns to date. The Mayor has pledged over £1.8 million matched by more than £2.2 million from a crowd of more than 14,000 backers.

By running campaigns, communities are demonstrating local buy-in, developing new networks of support, accessing a mix of funds and funders (who often contribute expertise or resource to make things happen), and developing skills and experience for long-term resilience. They are also learning about the conflicts at the heart of regeneration and gaining an understanding of the planning system. These skills are helping people in their personal and professional lives.

City Hall is learning how public funds can be used responsibly in this way; from governance and due diligence, to innovation in monitoring and evaluating success. There are key challenges for public agencies around the democracy of campaigns, ensuring we motivate all Londoners to participate. But we have evidence of many successes.

1. What is Crowdfund London?

Crowdfund London is a capital investment programme that uses civic crowdfunding to source grass-roots innovation, promote local engagement, and empower communities to collaborate in regeneration. It has grown into an internationally recognised and pioneering tool for the city's government to work directly with citizens and align local interventions with strategic investments.

Through Crowdfund London, City Hall is exploring how small projects delivered by communities can have a big social impact and how city government can catalyse a powerful mix of public, private, and local collaboration to improve places across the city.

The programme has supported over 100 successful crowdfunding campaigns to date. The Mayor of London has pledged over GBP 1.8 million matched by over GBP 1.5 million from a crowd of more than 14,000 people, businesses, and institutions.

By running campaigns, communities are demonstrating local buy-in, developing new networks of support, accessing a mix of funds and funders (who often contribute expertise or resource to make things happen) and developing skills, confidence, and experience for long-term resilience. They are also learning about the conflicts at the heart of regeneration and gaining a better understanding of the planning system. City Hall is learning how public funds can be used responsibly in this way; from governance and due diligence, to innovation in monitoring and evaluating success. There are key challenges for public agencies around the democracy of campaigns, ensuring we motivate all Londoners to participate. We have evidence of many successes and believe there is real value in this approach.

2. How Did Crowdfund London Begin?

The Mayor of London's Regeneration Team delivers strategic capital investment programmes and projects. These aim to improve the prosperity of places undergoing change or accommodating growth across the city. With London's population growing rapidly, the Mayor's vision for "good growth" is one where development works for all Londoners to create successful, inclusive, and sustainable places.

Building strong and resilient communities takes well-designed interventions that are sensitive to a range of local needs. These must increase opportunities for local people while respecting the unique character of London's diverse neighbourhoods and town centres. A blend of professional expertise is often required to get this balance right, but we also need to create more opportunities for communities to be involved in this process, as active citizens, to help find solutions that work for them.

Crowdfund London began in 2014 as a response to this challenge. We know that communities understand their area best and are well placed to contribute innovative ideas or solutions to address local issues. We also know that involving communities in a meaningful way is easier said than done. There are many examples of good practice in co-design and participatory engagement in regeneration, but this takes time, skill, resources, and commitment. It is often an imposed, top-down process.

We wanted to explore a way to give communities collective agency to bring forward ideas themselves, from the bottom-up, knowing that the city will listen and try to help. An online platform for Londoners to propose ideas directly to City Hall was initially considered, thereby utilising modern technology to engage citizens in civic change by crowdsourcing innovation. But we also wanted a solution that would result in positive action, rather than the negative reaction typical to most citizen engagement with the planning and development process in the UK. The process had to promote local collaboration and address the deficit of direct control when local communities are involved in regeneration; to genuinely acknowledge grass-roots voices in a meaningful way and instil local ownership.

Ideas owned, part-funded, and delivered by the community led us to civic crowdfunding. This would take the principle of initiatives like Seattle's Neighborhood Matching Fund[1] and blend this with the dynamic fundraising potential of modern crowdfunding platforms. The crowdfunding—or public campaigning—element would incentivise wider local engagement and debate, with ideas presented and developed in public rather than a "closed" grant fund application. Would this new approach mean that we move away from the "usual suspects" applying for community-based funding?

There were emerging examples worldwide of crowdfunding platforms supporting civic improvement, many of these in America, but only a few programmes led by city governments or other public funders at any kind of scale with similar ambitions to promote local collaboration. We were going to have to experiment. We were particularly interested in exploring some of the ideas proposed by Dan Hill and others in Brickstarter[2]—a primer for people working on problems at the intersection of crowdfunding and crowdsourcing, social media, urban planning, and decision-making.

Civic crowdfunding was an emerging marketplace at the time; something else that was important to us. The sharing economy was developing rapidly with new digital tools helping people to organise themselves and their resources in different ways. This activity was considered innovative and liberating, but we were equally aware of the limitations. There would almost certainly be unforeseen consequences when these platforms start to disrupt existing and complex systems—such as a city—very quickly, especially when these systems are often governed by slow-moving policy frameworks.

Airbnb, Uber, and other peer-to-peer online services that have grown in popularity and success since the late 1990s see themselves as facilitators of an individual's freedom and enablers of more flexible and efficient organisation. This approach often means they take little strategic responsibility for the effect that their natural disruption might have; there is a disconnect between the art of the possible and ensuring the fine balance of a city that works for everyone. If policy can't keep up, how else could we contribute to the development of this activity? Could we help shape good practice and minimise negative consequences when it came to civic crowdfunding?

We'd seen an example from 2011 in America, when Twitter user @ MT used the online news and social networking service to directly engage with the Mayor of Detroit to ask for a statue of Robocop.[3] He got a direct, public response. Not the one he wanted (the Mayor had no plans for such a statue, thank you) but a genuine dialogue, all the same, with the Governor of the city facilitated by social media. This was an interesting development for democracy, but still quite superficial as a response and arguably frivolous as an idea.

However, this was not the end of the story. The tweet went viral inspiring an online crowdfunding campaign[4] to see if anyone else would be willing to pay for it. The campaign was a success and raised almost US$70,000. So, you can have an idea, pitch it to the Mayor, get some support, and fund it. All from a laptop (or smartphone) in your bedroom. You could argue that this empowers people, which can strengthen civic society. However, there is a risk that it will happen in a vacuum, where communities still aren't really engaged in the collective production of their local neighbourhood, individuals are just throwing out ideas to see what will stick. Ideas that could be anywhere. The platforms involved— consistent with the sharing economy attitude—focus on the positive of this exchange; they have enabled something that somebody wanted to happen, to be realised. More often than not, they don't seem to be taking responsibility themselves for whether the ideas are meaningful in the first place, whether they are deliverable nor who will benefit and who won't. It's all about the campaign. If the campaign is a success, then there is enough legitimacy to the idea and it's now a collective endeavour that should be realised.

The parts of the Robocop statue in Detroit were manufactured but remained sitting in a warehouse on the edge of the city, years later. The permissions were not in place to actually build it, from copyright negotiations to agreeing on an appropriate location in the city. In that time, costs have almost certainly risen and, at the time of writing, it remains unbuilt. The platforms utilised have surely moved on from any involvement, so even if it is built, is that what Detroit needs, or even wants? It's hard to be sure, because we know very little about the crowd of backers in the first place. They could be a global community of Robocop fans—a crowd with little interest in the needs or desires of the people of Detroit.

Just because you can raise the money for something, from a "crowd," doesn't mean you can build something, nor that it's the right thing to do nor the best thing to do in the first place.

This shows all the potential and all the problems with crowdfunding civic projects and applying the sharing economy to real-life city making. In such a complex cause-and-effect system, a more sophisticated framework is required to achieve genuinely positive and representative action.

We wanted to explore the role that a city government could play in promoting a serious and effective process, especially in an emerging (and rapidly expanding) marketplace. This raised important questions. What would good look like? How could we take all this positive energy and channel it in a useful way? By getting involved could we leverage our position as a city authority, key funder, and experienced delivery agent for this type of a project to achieve better outcomes? That is, help to shape these platforms, from the outset, into sophisticated tools for people to use in a more meaningful way? Can our action incentivise new and exciting collaboration around projects that can have genuine social impact?

Civic projects—typically more complex and costly compared to your average book publication or one-off event—will inevitably require public or institutional investment and/or engagement at some point (from permissions and licences to long-term maintenance). We felt that we could contribute to thought leadership in this growing space to ensure the sector finds a clear and coherent way to engage public institutions and understand our constraints. This could help unlock a pathway for London's local authorities to experiment with these platforms using a clear reference point, which would promote consistency at scale across the city.

We did not envisage any of these activities to replace the need for strategic regeneration or well-delivered co-design processes, but the potential to deliver many small projects quickly is a powerful concept; regeneration often takes time and this activity could support and test long-term goals at a different scale with a new dynamism and more diverse actors.

3. How Does it Work? Create, Fund, Launch

Working with the UK-based civic crowdfunding platform Spacehive, we set up a pilot fund in 2014 to directly pledge public money to citizen-led initiatives pitched on the site as local campaigns. Spacehive allows for ideas to be shared publicly, before campaigns begin—a powerful design tool when utilised appropriately to engage local people in the development of the project. Budgets can also be broken down into distinct components. These two features were important to us from the beginning.

Our aim was to design the programme in an agile way, based on the principles of testing and iterative development. We knew we wouldn't have all the answers (or even an understanding of all the questions) at the outset and some process innovation would almost certainly be required for us to scale things up. Our initial proposal was to demonstrate proof of concept and digital innovation in the programme management procedure.

To be considered for a pledge from the Mayor, communities need to organise themselves to create, fund, and launch projects themselves.

3.1. Create

We set the brief as a call for ideas. To access public funds, you must respond to public need. We look for creative and positive ideas that celebrate local character, respond to a clear challenge or opportunity, and have wide support.

To help communities come together and propose fundable propositions, we've developed a support programme that precedes each funding round. We run events across the city to introduce City Hall's offer and explain how to get involved. At these sessions, groups learn from previous, successful project creators and gain insight about the journey they are beginning. They get a takeaway pack of resources,[5] we've developed to encourage them to think like a designer, critique their proposal, and plan ahead. With this platform, we then offer follow-up sessions to refine things through specialist design, planning advice, and a "crowdfunding campaign" masterclass.

Any constituted (although this can be quite informal) local group, organised around a shared objective can participate in the programme. This represents a lower barrier to entry when compared to many public-sector grant funding programmes.

3.2. Fund

To secure funding, communities need to run a public crowdfunding campaign, demonstrating support through donations and other engagement with their Spacehive[6] webpage. This is their pitch to both us and the wider crowd. We make it clear that we'll appraise this effort, which helps to incentivise local engagement. In addition, we appraise projects based on their approach to our brief and make pledges in a bridging model; so, groups begin their campaigns and demonstrate support, then we pledge to drive momentum to campaigns, leveraging our reputation to catalyse success. They then go on to crowdfund the rest and hit their target. City Hall's role in de-risking and legitimising projects that often require substantial amounts of cash is vital to unlocking other institutional support for the most impactful ideas.

Initially, we could pledge up to GBP 20,000 per campaign or up to 50% of a project value but this has since been increased to GBP 50,000 and 75%, respectively, to allow the programme to support more ambitious ideas. Typically, we won't match more than 50% of the budget, but the 75% threshold gives us the option to invest more when we can clearly see a lot of local support from a community without the financial means to contribute greatly. This allows a mechanism to offset the natural tendency for wealthier and well-networked communities to succeed in reaching their target more easily.

Campaigns typically run for 3 months. Our pledge is made 6–8 weeks into the campaign to ensure we not only have time to appraise and conduct our decision-making process but also have a meaningful catalytic impact. It is important to align our decision-making to the live campaign process and we've had to develop a tight but realistic process that runs to strict deadlines.

If campaigns that are awarded City Hall's pledge go on to reach their target, the project goes ahead, it's an all-or-nothing model.

3.3. Launch

Crucially, once funded, the projects move forward with the support of the GLA Regeneration Team to help make sure that things happen. We can advise on technical issues, broker key connections with relevant (often hard to access) stakeholders, and align activity with strategic investments to maximise impact. This is important to ensure ideas actually come to fruition, especially when working with inexperienced delivery partners. Fundraising alone will not guarantee success.

4. What Have We Achieved So Far?

The programme has supported over 100[7] successful campaigns to date, with more than GBP 1.8 million pledged by the Mayor matched by over GBP 2.2 million in pledges from a crowd of more than 14,000 people, businesses, and institutions.

We've seen a strong demand for this kind of opportunity across London with over 300 local ideas submitted to date; they represent a project pipeline worth GBP 15 million in potential small-scale interventions that could be realised quickly.

Campaigns represent an incredible range of ideas; from new public green spaces and the re-use of vacant buildings to a market supporting jobless people back into work; from local neighbourhood plans to ambitious feasibility studies and from new shared resources—a library of things—to local training opportunities. Ideas have come from all over London, from a range of different communities, all adding something new to an area while aspiring to have a wider social benefit. Projects are not just coming from well-off communities either, we've seen just as much commitment and desire from more disadvantaged communities and this is where we feel we can have real impact in focusing the attention and resources of other investors. In many cases, these projects have given us a new awareness of—and engagement with—local organisations valuable to our wider work.

Initially, we saw groups we had worked with before—or even helped to set up—coming forward and there was some reluctance to crowdfund with experienced and resourceful organisations often finding the match funds from traditional sources, and failing to take advantage of the opportunity presented by the platform to build consensus, awareness, and engagement around a project. This was largely due to lack of clear communication of the benefits in utilising the platform and the perception that it was too much (unnecessary) work.

However, there was still value in these projects and many built on something we had initially helped to establish, like the refurbishment a shopping arcade in North London.[8] A local group were able to take over one of the units, which allowed them to establish a legacy and sense of ownership after our original investment had ended.

It was also a powerful way for local groups to meaningfully inform emerging or ongoing work. Queen's Park Community Council came together to develop a local plan for the area, on the border of two London boroughs.[9] We could then embed this thinking into a wider strategic project we were developing alongside the local authority.

As things progressed, we put more emphasis on the need to run a public-facing crowdfunding campaign; it wasn't about the money, but we wanted to see support and local enthusiasm alongside commitment to the process. This prompted more variety in the types of organisations (or partnerships) coming forward and more emphasis on community-oriented proposals.

Lydia Gardener and the team behind TWIST on Station Rise[10] delivered a series of monthly afternoon market events with unemployed or underemployed people from the local Tulse Hill estate and surrounding area. Those involved were given the opportunity to trade on the market to test new enterprises and retail ideas at a low cost with a low risk in this busy commuter location. For many of the traders, the positive environment on Station Rise and the relationships and connections that grew out of it led to business incubation and career development. Devised and coordinated by Tree Shepherd (a social enterprise focused on promoting and supporting employment and business growth in the UK's most marginalised communities), these events were made possible by local volunteers at each event. The community crowdfunding campaign proved enormously helpful in this respect; as well as securing over 75 funders from the local community, it helped the project tap into voluntary support in areas as varied as social media, marketing, film, and photography. This undoubtedly contributed to the success of the project and demonstrates that a blend of funders or supporters can be a much more attractive proposition than one institutional funder.

In Surbiton, South West London, a community organisation—The Community Brain[11]—has now been successful in three successive rounds of the programme, each time proposing a smart, modest project that builds momentum and consolidates a wider vision.

Firstly, they proposed a new community space in a vacant shop unit on the high street.[12] Not only did this bring community presence back to the heart of the town centre but also it served as a local resource and meeting point for wider engagement on the needs and desires of local people and businesses. Through events and discussion, their second campaign (more proactively championed) aimed to add a professional kitchen facility to the space, operating on a pay-as-you-feel model.[13] The community had identified this as a valuable shared resource to support local businesses (and incubate new start-ups), community members, and local school children alike. Finally, a campaign to rejuvenate a nearby allotment as a suburban farm[14] to grow food and run educational events linked back to the community kitchen, completed a series of linked interventions that tell a story of local sustainability and demonstrate a sophisticated example of community-led regeneration. By the third campaign, the group were convinced of the benefits of the crowdfunding model and their increasing commitment to the process is clear in the way these campaigns demonstrate an increasing level of support and engagement.

Completing these projects gave the group momentum and a platform of confidence and proven track record to better engage with the local authority on more ambitious, strategic initiatives linked to local planning. They have gone on to secure more substantial funding from the GLA's own Good Growth Fund,[15] have won national awards, and most recently had an installation in the courtyard of the V&A Museum in London. This "heritage shed" later won an RHS silver medal at the Hampton

Court Flower Show. As a consequence of their approach to community engagement in the planning process, they were a winner at the 2019 London Planning Awards. This demonstrates how Crowdfund London is beginning to deliver tangible impact against our own objectives with groups becoming more ambitious in how they help direct meaningful and inclusive participatory processes at a grass-roots level.

The Peckham Coal Line campaign remains one of the best examples to demonstrate the real potential of this approach for strategic, community-led influence.

Essentially a feasibility study for an urban park on a disused rail line, linking two high streets in South London, the idea had been kicking around for a year or two with the local authority initially hesitant to support something so ambitious and risky. Typical of many locally led initiatives, it was proving difficult for the group to move forward.

The Crowdfund London programme convinced this group that now was the time to try again and provided the structure (and, crucially, deadlines) necessary to focus minds and rekindle enthusiasm. They used their campaign[16] to build a strong local base, running a series of engagement events and workshops to invite the wider community to be part of the proposal. This drove momentum, with residents and businesses feeling a sense of local pride and ownership in the idea. As an example, the group had created a map of local independent businesses—cafes, restaurants, and shops—to hand out at the end of the tours and events they were running around the site. This was a real spotlight on the local economy and those businesses got behind the campaign; a local microbrewery even developed a special "coal line porter" beer, with proceeds from sales going towards the fundraising effort. Building relationships in this way coupled with the inclusive and open approach from the project group is exactly what we'd hoped to see.

However, it was still quite an ambitious idea—something that would be difficult to pull off, beyond any feasibility work. This would always deter backers to a campaign. They had approximately GBP 10,000 towards their target of GBP 60,000 when we pledged a further GBP 10,000, and over the next 6 weeks, they surpassed their target with nearly 1,000 backers.

It was our role in de-risking and legitimising the idea that really changed the game as people began to believe something was possible. Our pledge paved the way for other large pledges from charities, businesses, and the public sector; the council went on to match our pledge with GBP 10,000 of their own, embedding the idea into their local plan and becoming a strategic delivery partner.

Once the campaign was a success, we helped to coordinate this crucial early access to other essential stakeholders—such as Transport for London and landowner Network Rail, something that would have been very difficult for the group to do alone a few months earlier. We were able to advise and support the group through the procurement of a design team for the feasibility work and, with our help, they have achieved another ambitious goal to have a genuine and representative community client for the work. There's a long way to go to fund and deliver an eventual park and green link on the old coal line, but the group are now well placed and well networked to build the momentum and funding required via phased delivery. With a viable proposal, they have been able to influence the strategic thinking of other local development projects to ensure the coal line was considered, which is a significant achievement for a relatively new local organisation. This shows that local people can have real strategic influence when a framework is established for them to build capacity and credibility as a meaningful contributor.

5. What Have We Learnt?

Through these pioneering projects, we've been able to reflect on our own impact and the strengths and weaknesses of the programme when considering positive regeneration outcomes.

5.1. Local Ownership over Small Projects, Facilitated by a Crowdfunding Process, Can Empower Communities

The GLA's Opinion Research and Statistics team conducted a qualitative analysis of some of our projects and the people behind them to understand how their experience has shaped their perceptions of regeneration. They concluded[17] that:

- Community groups do feel an increased sense of civic pride and ownership with an improved perception that they can influence change in a positive way. Many are very positive that we offer this opportunity in the first place and value the crowdfunding process, even though they didn't expect to at the outset.
- Capacity and confidence is being built within groups with people learning new and valuable transferable skills which they can then share or use in their personal and professional lives.
- The process contributes to a greater understanding of the pressures and conflicts at the heart of our regeneration work and increased awareness of the planning system, so more engaged and knowledgeable citizens with a platform of experience contribute to a range of local issues on urban development.
- Projects have helped to consolidate robust local networks for support and resource.

In addition, the process has encouraged new organisations to apply for public funding; 23% of the campaigns we've backed have come from grass-roots local community organisations. A further 25% of the campaigns we've funded to date have come from social enterprises or community interest companies. The rest tend to be more established local charities or business improvement districts. In 2018, 53% of groups that successfully secured a Mayoral pledge were accessing public funds for the first time with 65% accessing City Hall funds for the first time.

This empowerment is building local resilience and active citizenship.

5.2. City Authorities Have a Key Role to Play to Promote Socially Inclusive Civic Crowdfunding Campaigns, Ensure their Success, and Maximise the Social Impact of the Outcomes

Around 95% of the campaigns backed by us have gone on to hit their crowdfunding targets. Compare this to about 50% average on Spacehive and closer to 25% for crowdfunding campaigns generally, around the world. This is significant in two ways:

(1) City government or other public institutions clearly have an important effect of de-risking or legitimising projects of this nature giving confidence to others, so it's not just our cash that is useful at the campaign stage. In 2018, 92% of pledges by businesses or institutions to campaigns supported by the Mayor came after we pledged. Cities can help communities do more, by unlocking this latent power. With this comes increased responsibility in that we accelerate certain campaigns over others, so it is important to ensure that our decision-making is fair and transparent. But it means we can channel a lot of resources into important city priorities. To date, 24 campaigns pitching to Crowdfund London have been successful despite not receiving a pledge from us. The number has grown, year-on-year, leading us to conclude a knock-on benefit to an association with a promoted City Hall initiative which again shows the programme has influence beyond the scope of City Hall funds.

(2) A high success rate greatly reduces the risk of running a campaign in the first place, provided groups think carefully to ensure proposals meet our social impact objectives. This incentivises projects that focus attention on worthy but less glamorous ideas and this helps mitigate the "popularity" contest or "media-friendly" element that typifies success in models of online/viral campaign activity. It also opens up this type of opportunity to a more diverse range of community groups, including those with limited resources who must make a significant commitment to participate.

Post-campaign, we have a 96% success rate in the project delivery phase, which shows that our hands-on advice and support plays an important role in getting things done. This again breeds confidence in the model which is important in building momentum. Our work up front to help groups to get their project proposals right (from reliable budgets with contingency factored in, to an understanding of permissions and project management expertise that they will need later on) builds local capacity and helps to de-risk the delivery phase later. Being coached to think like a designer means we can encourage communities to propose increasingly sophisticated or ambitious ideas that would have even greater impact, deliver faster, and experience fewer issues. In delivery, being able to broker

crucial conversations with stakeholders, including local authority partners, is pivotal to successful civic improvement and a sustained legacy for
activities.

The investment in these aspects secures not only the positive impact
of our funds but also the additional investment now focused on city objectives. Through this leadership, we're raising the bar for civic crowdfunding activity across the city and setting standards for others to follow; this
might include other funds led by local authorities or philanthropic institutions or emerging platforms looking to access public funding.

5.3. The Public Sector Should Not be Afraid to Experiment with Civic Crowdfunding to Explore Innovation in Citizen Engagement and Distribution of Public Funds

Re-positioning ourselves more as "collaborators" rather than the "authority" has benefits not only for both the perception of City Hall and our
regeneration work but also a more even platform for citizen–City Hall
engagement. We are used to being the sole or majority funder, dictating
the terms—a consequence of shouldering the main burden of risk. In this
model, we are one of a broader crowd alongside citizens, business, and
foundations. This mix of funders each brings a different value to support
the project objectives or solve problems in delivery, and this spreads the
risk and allows us to support innovation by funding more experimental
proposals. It also means we can support more groups with limited experience, which opens up access to public funds and strengthens
democracy.

This shifts the mentality away from public grantmaking and more
towards local collaboration and development of resilient networks.

When it comes to the development of funding programmes that
engage with new technology, starting small and having an agile and iterative approach was key to success. At the outset, it was difficult to make a
convincing case for the potential impact of this activity, while managing
risk in public expenditure. A pilot programme with a ring-fenced budget
focusing on the way the process contributed to a broader priority (active
citizenship and participatory regeneration) allowed the flexibility to test
and reflect.

Our existing governance and finance procedures do not lend themselves to this type of activity—pledging public money to live campaigns via a third-party online platform—but involving colleagues in a collaborative way to solve problems was crucial to process innovation and building internal support for what we were doing. This has gone on to influence the way City Hall thinks about working more directly with citizens on a number of delivery-focused initiatives.

5.4. We Need to Better Understand the Crowd and Push Civic Crowdfunding Platforms to Take this Seriously

We're aware that the most able and engaged groups or communities will always have the quickest route to success in this kind of model. In the short term, this is fine, we need them as trailblazers, to inspire others and build momentum around this kind of activity. The key challenge for us is working to build awareness and continue to lower the barriers to entry to inspire every Londoner to want to take part in some way to ensure that the crowd is representative of the local feeling.

Communicating the programme is key and has been a challenge to get right. We want to take potential applicants on a journey through the create, fund, and launch, narrative where they feel that they can build the capacity and confidence to succeed over time, no matter what their background or experience. However, we also need them to understand what to expect and the commitment required at the outset. It's a delicate balance, but too little or overly simplistic information at the outset has led to more issues later down the line. We also want to raise general awareness of both the initiative and then the campaigns we're supporting.

We've developed a Mayoral marketing campaign to promote the programme and raise awareness, and it has been important for the Mayor to front the initiative and personally invite communities to contribute their ideas to improve our city. When it comes to raising awareness of specific campaigns, it's the projects themselves that are having the best impact in drumming up local press and engagement. Revealing successes through case studies that people can relate to have helped encourage more people to have the confidence to participate; previous project creators present their experience at our outreach events across the city and present the initiative in a way that feels achievable for local groups.

There is a wider challenge to build capacity to increase engagement in more marginalised communities that may not have access to skills, experience, or resource to participate on their own. This is a long-term challenge that requires a strategic approach. In the short term, greater engagement or participation with existing campaigns can help build awareness and aspiration while also ensuring these ideas become more socially inclusive. We've been promoting campaigns that the Mayor has backed in the communities they will impact to raise that awareness.

We've learnt that the key challenge for the future of civic crowdfunding is about developing a much greater understanding of how democratic and inclusive the process really is. Who is proposing and backing ideas and what are their motives? Who benefits the most and who is excluded? As a public funder, we need to weigh up the opportunity cost of proposals—popular ideas might not always be the best thing long term in any given location or may have a detrimental impact on others living in the area who didn't exercise their voice. You really need a certain scale of local involvement to justify any democratic credentials in this model and much better collection of data. Our experience has been that the platforms are reluctant to collect meaningful data on backers—it's another series of barriers to making a pledge—and, in the current climate around data protection, citizens will be reluctant to give too much up about who they are and where they live. We need to find a way to solve this, to fully realise the potential of this model to contribute to community-led decision-making.

5.5. There is Work to Do to Develop the Wider Framework of Support for Crowdfunded Civic Improvement

While crowdfunding platforms present a powerful potential tool for citizens, there are still limitations as their business models develop. Our experience in developing this programme and speaking to others involved in civic crowdfunding across the world is that platforms have much to do to understand civic projects, city making, and collaboration with public institutions. Things are improving, but platforms must move beyond the idea that the campaign is king. Fundraising is one part of the picture, but it cannot happen in a vacuum when it comes to civic improvement. Platforms are good at communication—the presentation of a difficult or complex journey as something accessible and achievable. Their tools are potentially empowering and powerful. They can offer the kind of one-to-one support that large institutions would struggle to match, providing emotional support to citizens as much as professional advice. They understand what sells and how to pitch ideas with broad appeal. This is all important and helps encourage more people to get involved. However, we cannot scale meaningful change through civic crowdfunding unless platforms gain a better understanding the way the projects are generated, who benefits from them, how they will be delivered, and what their impact is. Understanding the project life cycle is key to success and, therefore, credibility in this approach. If large pledges or institutional funding are to be harnessed (as is often required), then it does matter that things happen as intended and platforms need to take some responsibility to minimise risks by understanding exactly what groups will need to do to get things done. Due diligence must go beyond fraud checks to interrogate budgets and their contingency, permissions and consents, resourcing, and experience within the project team, all proportionate to the scale of the challenge they are setting themselves. Design is important, as is understanding the opportunity cost in doing one thing over another and this should be questioned by all to assess the value of the approach. Campaigns need local backers at scale for legitimacy and platforms need to do more to understand the motivations of backers and whether broad demographics see consensus in the benefits. Post-campaign, things need to happen as described and platforms need to validate this and be able to understand the impact consistently. If platforms see themselves as collaborators in creating better neighbourhoods or cities, then we'll get good results, at scale, more quickly. If they see themselves only as facilitators of fundraising campaigns, there is little to separate civic crowdfunding platforms from any others and the burden of responsibility will fall on those with the most to lose, which will prohibit the public sector from truly embracing this approach.

Beyond the platform and city authority (or public sector) agents, there could be much more of a supportive ecosystem around activity to help ensure quality, deliverability, and democracy of ideas. This could come through innovation in the wider sharing economy as civic crowdfunding movements grow, with new platforms developed to support peer-to-peer exchange for resource sharing, sourcing skills, and accessing expert advice. This could be best implemented via innovation within the platforms themselves, through an expanded business model to better support all phases of community organising, ideation, crowdfunding, and delivery of proposals. Alternatively, it could be public-sector led if we use our position to convene supportive partners and coordinate the strategic infrastructure around this activity to focus a wide range of resources on shared objectives to improve the city.

6. Concluding Remarks

We believe that regeneration is something that more Londoners should feel they are part of. Everyone should have opportunities to contribute to making and remaking the place where they live by coming together to develop common spaces and shared resources and this will strengthen society and democracy in an increasingly uncertain world. Our experience has shown that civic crowdfunding can be used as a tool to engage community and civil society groups to harness local innovation through new and sustainable solutions to local challenges or opportunities. And the process can contribute to local resilience.

We can use civic crowdfunding to challenge the way we think about public space, ownership, and civic responsibility to find new models to create commons spaces and shared resources that work for local places. It can promote exciting new finance and partnership structures to support Mayoral objectives by aligning the public, private, and third-sector organisations alongside citizens as powerful new local delivery agents.

There is a lot of potential in the tools offered by existing platforms and much promise in the value of crowdfunding campaigns as indicators of public support. However, much more needs to be done before campaigns of this sort can be understood as participatory democracy and public institutions have a vital role to play to ensure that increasing activity in this space reaches its potential.

We want to see more projects that build community cohesion, develop a shared sense of civic pride, and involve local people in a meaningful and collaborative regeneration effort in long-term strategic change and we believe that civic crowdfunding has the potential to become an important and established part of this vision.

About the Author

James Parkinson has a background in architecture and urban design and has worked for ten years in regeneration and urban policy in both Yorkshire and London. Currently he is a Programme Manager in the Greater London Authority Regeneration Team, working to deliver place-based investment projects, develop strategic regeneration programmes and support good design in the capital through research to inform policy. Recently he has led the development of Crowdfund London; an internationally recognised, £5 million initiative to support citizen-led regeneration and explore how crowdfunding can foster civic participation and democracy in urban development.

Endnotes

1. In 1988, Seattle's Neighborhood Matching Fund (NMF) was created to provide matching funding for neighbourhood improvement, organize, and develop and implement projects put forth by community members. Central to NMF is the community match which requires awardees to match their award with contributions from the community whether as volunteer time, donated materials, donated professional services, or cash.

    Since NMF's creation, more than 5,000 community projects throughout Seattle have received over $49 million in funding and generated an additional $72 million from community match. http://www.seattle.gov/neighborhoods/programs-and-services/neighborhood-matching-fund
2. Boyer, Bryan and Dan Hill, (eds.), Brickstarter. Sitra. Helsinki, 2013 http://brickstarter.org/
3. https://twitter.com/MT/status/34702937087418368
4. https://www.kickstarter.com/projects/imaginationstation/detroit-needs-a-statue-of-robocop
5. Our Project Planning Worksheets and a Community Projects Handbook are available to download and print at home, alongside tips for developing successful crowdfunding campaigns: https://www.london.gov.uk/what-we-do/regeneration/funding-opportunities/crowdfund-london/resources

6. Spacehive was launched in 2012 and is the world's first crowdfunding platform for civic projects. A London based start-up, they have been our platform partner since the inception of the programme. https://www.spacehive.com/

7. We have pledged to 82 campaigns in total with five campaigns failing to reach their target and three projects failing in delivery. There is always a risk with these types of projects that not everything will happen as planned, but GLA Regeneration Team involvement helps to secure success in the delivery phase.

8. Ten Grand Arcade (Barnet)—A community managed and curated space for events, workshops, and collaboration. The Mayor pledged £16,000. https://www.spacehive.com/ten-grand-arcade

9. Queen's Park High Street: The Harrow Road (Westminster)—Carrying out a study of the High Street and building up a set of recommendations and design guidance to inform a neighbourhood plan. The Mayor pledged £20,000. https://www.spacehive.com/queensparkhighstreettheharrowroad

10. TWIST pop-up on Station Rise (Lambeth)—A monthly pop-up market that provides a much-needed economic boost for the area and personal development opportunities to the underemployed residents. The Mayor pledged: £14,000. https://www.spacehive.com/twist-pop-up-on-station-rise-tulse-hill

11. The Community Brain. https://www.thecommunitybrain.org/

12. The Museum of Futures (Kingston)—A new community space in a vacant shop, to encourage engagement and to develop a vision for the kind of place people will celebrate with pride. The Mayor pledged £20,000 https://www.spacehive.com/themuseumoffutures

13. The Community Kitchen (Kingston)—Pay-as-you feel community kitchen, supporting local start-up or home-based food businesses and the wider community. The Mayor pledged £17,000 https://www.spacehive.com/the-community-kitchen

14. Create a Suburban Farm for Tolworth (Kingston)—Reclaiming the abandoned Tolworth Allotments Society building and surroundings with the aim of introducing the community to suburban farming possibilities. The Mayor pledged £5,000. https://www.spacehive.com/suburban-farming

15. The Good Growth Fund is our flagship £70 million regeneration programme to support growth and community development in London. https://www.london.gov.uk/what-we-do/regeneration/funding-opportunities/good-growth-fund-supporting-regeneration-london

16. The Peckham Coal Line Urban Park (Southwark)—Feasibility study for a community-led elevated park on a disused railway line linking Queens Road and Rye Lane. The Mayor pledged £10,000. https://www.spacehive.com/peckhamcoalline

17. Warbis, M. (2016). 'Understanding the social impacts of the Mayor's Crowdfunding Programme: a qualitative retrospective on rounds 1 and 2.' London: Greater London Authority. https://data.london.gov.uk/dataset/mayor-s-crowdfunding-programme-social-impacts

Chapter 10

Lessons Learned from Civic Crowdfunding and Match-Funding Schemes in Europe[1]

Francesca Passeri

Abstract

Cities such as London, Barcelona, Utrecht, Milan, and many more have set in place match-funding schemes through which citizens have the opportunity of staying actively involved throughout the decision-making process of regional development. In this context, crowdfunding—as per its own open, inclusive, and participatory nature—stands out as a reliable ally for public authorities, significantly increasing the opportunities for wider participation, stimulating citizens' ownership and sense of belonging to a community, and offering concrete validation of the need for specific local actions as they are perceived by citizens. In this chapter, we collect and present seven existing best practices in match-funding in the EU, highlighting the different roles that public authorities can play in supporting and amplifying the impact of crowdfunding campaigns for territorial and social development in their jurisdictions. The presented best practices should also be considered as benchmarks, in terms of procedures, indicators, and outputs, for the replication of match-funding schemes in other European cities and regions.

1. Introduction

The peculiarity of civic crowdfunding is that, by leveraging on the close ties that crowdfunding platforms enjoy with local communities, it can promote a sense of engagement and belonging among citizens by enabling them to contribute to specific projects that will generate common good in their territories. Building on this concept, the peculiarity of *civic match-funding* arises from the participation in the funding scheme of a public authority—generally regional or local—which provides co-funding to successful crowdfunding campaigns in specific sectors.

Cities and regions all over Europe have found different roles within the civic crowdfunding paradigm, ranging from simple sponsors to co-funders. Davies[2] suggests four models to get cities involved in civic crowdfunding:

- **Sponsor:** The public authority runs its own campaign for a specific project on an existing civic crowdfunding platform.
- **Manager**[3]**:** A subnational government creates its own crowdfunding platform to foster the development of its territory. Subnational governments usually create generic platforms which promote both entrepreneurial for-profit projects and non-profit civic initiatives.
- **Curator:** The local authority selects a list of projects that reflect their agenda from an existing crowdfunding platform.
- **Facilitator:** Subnational governments can play an important role in facilitating citizen and civil society empowerment. This includes planning permission, financial as well as technical expertise support, co-screening and/or co-designing projects, etc. Basically, it represents a new type of public–non-profit private partnership with citizens/inhabitants and civic crowdfunding platforms. Some subnational governments have already set up partnerships with civic crowdfunding platforms, which are mainly based on co-financing projects alongside the rest of the crowd.

Although there are a significant number of initiatives where public authorities have been involved as sponsors,[4] best practices presented in this chapter fall within the scope of the last three categories identified by Davies. In this perspective, the roles of Manager, Curator, and Facilitator deserve to receive not only greater visibility but also a closer description of their functioning mechanisms and designs, to inspire and provide guidance to a wider audience of policymakers and crowdfunding platforms on how to best replicate these schemes.

Through analysis of the case studies presented here, we can deduct key findings which we believe will help policymakers to further their agenda in regional development by facilitating match-funding for the following:

- Considering the geographical coverage of this report, it is safe to assume that civic crowdfunding and match-funding practices have multiplied in the European Union over the past 5 years. The intrinsic flexibility of crowdfunding models and platforms has allowed for new partnerships with public administrations in a range of different organisational, political, social, and economic contexts.
- The flexibility of crowdfunding is however not sufficient to overcome obstacles posed by the limited adaptability of public administrations. In the current state-of-the-art, governments at all levels must be remarkably committed and creative in finding suitable forms of cooperation with platforms, so as to implement new models of distribution in of public funding.
- Specific sectors in which crowdfunding and match-funding schemes have been activated by public authorities cover a wide spectrum of initiatives. Taking into consideration solely the cases presented in this report, combined funding has been awarded to projects in the area of environment, social inclusion, entrepreneurship, agriculture, cultural and creative industries, and health research. Looking at the European level, it is also possible to identify potential synergies with ESIF thematic objectives 3 (enhancing the competitiveness of SMEs), 8 (promoting sustainable and quality employment and supporting labour mobility), and 9 (promoting social inclusion and combating poverty and any discrimination).
- Availability of additional resources is only one of the many reasons that drive public administrations in designing a crowdfunding or match-funding scheme. As reported in the case studies, increased citizen participation and sense of ownership, a return on the investment in terms of visibility and accountability in their jurisdiction, better communication

with regards to perceived needs and priorities, and increased likelihood of funded initiatives' success are as strong as the economic factor when deciding on public budget allocation.

- The development of a crowdfunding campaign allows project owners and public officials to acquire a whole new set of skills, through a "learning-by-doing" process. In the case of project owners—i.e. organisations or individuals who decide to seek funding for their projects by activating a crowdfunding campaign—the set of newly acquired skills is a combination of communication, planning, and management competences that can then become valuable assets in the broader labour market, as well as the expansion of their existing networks of contacts. From the public officials' perspective, the design of a match-funding scheme allows them to enlarge the set of options that they can activate in relation to economic development and social inclusion, as well as to strengthen the relationship with different territorial stakeholders by renovating consolidated partnerships and engaging in new opportunities.

- Match-funding schemes and crowdfunding campaigns that enjoy the participation of a public authority have significantly higher chances of achieving their funding goal. When a public authority partners with a crowdfunding platform and commits to supporting a match-funding or crowdfunding scheme, the overall success rate of projects funded under that programme increases from an average of 60% up to an average of 80–90% in most of the presented case studies. The increase in the success rate can be primarily attributed to the combination of training and support actions that all platforms offer to project owners, together with the presence of a trustworthy partner and co-funder such as a regional or local government.

- The major obstacle to the diffusion of match-funding schemes remains the lack of in-depth knowledge about the functioning mechanisms of crowdfunding and the underestimation of the advantages that it can produce by empowering and connecting different groups of stakeholders with common aims.

In what follows, we will present different case studies in which crowdfunding platforms have been partnering—in different forms and to various extents—with public authorities, in order to improve the ongoing dialogue with their citizens and to more efficiently and transparently allocate funds for territorial and social development. Each case study will detail why and how crowdfunding and match-funding have come to be selected as funding schemes, in which areas public calls and project ideas have been mainly issued in order to develop a crowdfunding campaign, which have been the main obstacles that parties involved have encountered in the design of the match-funding scheme, and which results have the different match-funding and crowdfunding programmes produced. In summary, the cases covered are as follows:

- 5 countries
- 6 crowdfunding platforms
- 7 public authorities

Total funds collected through the platforms:

- €69.829 million

Number of funded projects:

- 104

Number of contributors involved:

- 4007

Funded projects' areas of activity:

- Agriculture
- Cultural and creative industries
- Entrepreneurship
- Environment
- Health research
- Social inclusion
- Technology

Average duration of projects:

- 50 days

Average amount collected per project:

- €14.818

2. Case Studies

2.1. Public Authority as Manager

A subnational government creates its own crowdfunding platform to foster the development of its territory. Subnational governments usually create generic platforms that promote both entrepreneurial for-profit projects and non-profit civic initiatives.

2.1.1. Crowdfunding Südtirol

Crowdfunding Südtirol is the first regional crowdfunding platform for South Tyrol. The platform was founded in June 2016 with the aim to sustain local companies to bring innovative ideas to the market. The platform owner and provider is lvh Wirtschaftsverband Handwerk und Dienstleister who offers a 360° consulting in launching and processing a crowdfunding campaign. The platform only hosts projects from South Tyrolean companies coming from different sectors, mainly from craft, agriculture, and film and culture. However, the backers come from the entire German- and Italian-speaking countries. Until today, 14 projects have been presented, with 12 of them being successful. In total, the project owners collected more than €125,000 thanks to Crowdfunding Südtirol.

Name of platform: Crowdfunding Südtirol – Alto Adige	Year of beginning of operations: June 2016	Crowdfunding model adopted: Rewards	Country/ies of operation: Italy	Total funds collected on the platform since launch of operations: €127.771

(*Continued*)

Services offered:	Total backers on platform:	Outreach:	Areas of activities:	Number of applicants to the match-funding call for projects:
360° consulting; preparation of campaign. Setting up of the payment system account; realization of the campaign. Post-management of the campaign.	513	Facebook: 4.061 Instagram: 340 Twitter: 94 Newsletter: 650 Website access: 100.778 sessions.	craft, agriculture, and film and culture.	30/Year **Number of selected projects:** 8/year **Number of successful projects:** 12 successful projects; 80% success rate.

Reasons for setting up the partnership:	Public resources committed:	Impact of the initiative:
To have one place where South Tyrolean projects can be promoted. To sustain and strengthen regional economic circles. To motivate local people to sustain local projects.	Since March 2017, there is an European Regional Development Fund (ERDF) project ongoing with the aim to inform companies about the potential of crowdfunding and to offer them consultancy in realising a campaign. Budget: €150.200	Increased knowledge of crowdfunding in the local community through the organisation of events. Interviews with about 100 companies/persons interested in launching a campaign. Promotion of the topic on the political level: in future, there will be public match-funding to realise a crowdfunding campaign.

Costs borne by public authorities in addition to the pledged resources—Thanks to the ERDF funding, we can cover personnel costs in accompanying a campaign, and offline events and sensitisation activities might also be funded. Before receiving funding from ERDF, lvh invested its own resources. Crowdfunding Südtirol differs from other platforms because project owners are not required to pay any fee for uploading their campaign on the platform: a specific contract is set up with each project owner, depending on their needs. The only cost that they must cover is represented by a fee to the payment system provider.

- Project: LAAB from Jasmin Castagnaro
- Location: Bressanone

- Sector: Craft
- Duration of the campaign: 16.10.17 to 17.11.17
- Backers: 30
- Funding: €10.790 collected

Description of the project

Jasmin Castagnaro makes lamps from the leaves that fall in autumn. She recycles this resource and creates wonderful lamps by hand.

See campaign: https://bit.ly/2ysY77K[5]

Outputs and impact

Jasmin Castagnaro could bring her first series of lamps to the market. Today she sells them on her online shop: https://www.miyuca.it/laab.[6] She was able to perform a market test and see if the price was accepted by a community of potential customers.

Barriers and bottlenecks

In this specific case, the media were fascinated by the product, so we did not have big problems in the campaign.

Problems arise when a project is too complex, too many actors are involved, and it is a project idea rather than a product idea that is funded—since projects are much more difficult to explain in a short time. Also, it can be difficult to be successful when trying to collect funding for an innovative technology, which is too complex to explain, or when the project owner is not fully dedicated to communicate his/her own campaign.

Success factors

- an innovative idea that amazes people;
- there must be a good reason for people to participate in a campaign: e.g. there is a limitation in the availability of the product as in the case of Jasmin. She also offered the product for a special price;
- the idea must be understandable in 30 seconds for common people;
- a good story is needed to promote the idea;
- a good network is essential to promote the campaign;
- to have some backers like companies that book a big perk;

- facebook is a great vehicle to communicate the campaign;
- the platform itself has to guarantee a good mix: In a short time, we had four film projects, so the last project was much more difficult to promote because the people didn't show that interest anymore because it was film.

Do's

- examine if crowdfunding really is appropriate for the project;
- try to involve 2–3 companies in supporting the campaign, and they can back the campaign with a higher amount of money;
- invest a lot of energy and time in the campaign.

Don'ts

- crowdfunding doesn't work well for B2B ideas;
- don't expect that once the project is published the money comes without effort.

Lessons learned

- people like to pay with a bank transfer, and we have this possibility and about 50% select this payment method;
- it is very hard to find the right payment system provider; we had many talks with the local banks, but they did not have the right tools;
- a lot of time and energy goes into the communication work for the campaign;
- it is necessary to select the projects that are presented to the platform: otherwise, the risk in investing time and money for non-appropriate projects would be high;
- project owners are mostly interested in having a market test and gain publicity rather than on the collected money itself.

2.1.2. Koalect and Streekmotor

Koalect is a white-label provider of crowdfunding and crowdsourcing platforms. We build tailor-made platforms for non-profit and social profit organisations according to their strengths, objectives, and context.

This way, we increase engagement between an organisation and its crowd to tackle societal issues and opportunities. Hereunder, Koalect discusses a crowdfunding platform that was built for the regional fund "Streekmotor23." Further, we also discuss a specific project "Buitengewoon Fruitig" that was launched on this platform.

Name of platform: Koalect	Year of beginning of operations: 2014	Crowdfunding model adopted: Rewards, donations, and crowdsourcing	Country/ies of operation: EU, Belgium, France, UK, and the Netherlands	Total funds collected on the platform since launch of operations: €3.2 million
Services offered: Training to organisations crowdfunding platform management; workshops; online knowledge centre with articles and videos; tailor-made user guides; and live-chat support on our platforms to assist users.	**Total backers on platform:** 50,000	**Outreach:** Organisations will communicate about their platform through their own channels, as to be perceived as valuable and trustworthy by potential backers.	**Areas of activities:** Healthcare, education, social entrepreneurship, local development, citizenship, journalism, environment, culture, humanitarian aid, corporate social responsibility.	**Number of applicants to the match-funding call for projects:** 5 so far (pilot phase) **Number of selected projects:** 5 **Number of successful projects:** 5/5; 100% success rate.

Reasons for setting up the partnership: Offer local initiatives the opportunity to set up a crowdfunding campaign; position Streekmotor23 as innovator and facilitator for local initiatives; stimulate local engagement and provide funding.	Public resources committed: Every project has a specific goal between €3,500 and €6,000—when the goal is achieved, Streekmotor23 doubles the amount.	Impact of the initiative: Projects reached more than 300 backers, and besides the financial injection, they were also able to recruit new volunteers for their projects.

		The crowdfunding mind-set allows projects to tell their story through a strong digital tool that gives them visibility and a new digital identity. It also allows them to grow and strengthen their existing services.

Costs borne by public authorities in addition to the pledged resources—
Streekmotor23 covers expenses related to the set-up and licensing cost of the platform. Budget is also allocated to communication and workshops to guide the local initiatives towards successful crowdfunding campaigns (strong train-the-trainer approach).

Project: Buitengewoon Fruitig by Leefboerderij De Kanteling (non-profit)
Location: Southeast Flanders
Sector: Ecology for people with disability
Duration of the campaign: 16 weeks
Number of backers: 51
Funding: €5000 (doubled by Streekmotor23 resources)
Project link: https://streekmotor23.be/project/6659[7]

Aim

Funding for a green meeting area for visitors.

Description of the project

"Buitengewoon Fruitig" is a project from a small non-profit organisation called "Leefboerderij De Kanteling" located in Southeast Flanders. They offer services and day-time accommodation with a strong focus on ecology for people with a disability.

Leefboerderij De Kanteling applied for the open call for projects from Streekmotor23 and was selected as a pilot project to start a crowdfunding campaign on the platform. Leefboerderij De Kanteling easily raised €5,000 in 6 weeks by organising offline and online campaigns that contributed to the success of the crowdfunding campaign. Through this crowdfunding, they were not only able to make new financial investments but also able to engage a new group of people around their organisation.

Outputs and impact

- New financial means for "Leefboerderij De Kanteling;"
- local engagement from citizens and small local communities resulting in new volunteers and recurrent donors on the long term;
- the crowdfunding campaign gave a boost for the digitalisation of the "Leefboerderij De Kanteling;"
- the success of the campaign showed "Leefboerderij De Kanteling" and other small organisations in this region the potential of crowdfunding.

Barriers and bottlenecks

- Lack of experience in crowdfunding—we did a workshop with them on how to set up a successful crowdfunding campaign;
- user support: the Streekmotor23 platform has a user-support live chat that is managed by Koalect;
- as Streekmotor23 has its own dedicated platform, it also has the possibility to offer administrative and procedural support through the platform; the necessary administrative documents and procedures can be directly found on the platform.

Success factors

- Having a tailor-made platform; this way, all projects from Streekmotor23 are gathered on one central point in a Streekmotor23-branded environment;
- live support on the platform;
- crowdfunding workshops for the local initiatives;
- train-the-trainer approach;
- strong project management by Streekmotor23;
- communication skills from Streekmotor23;
- motivated and well-prepared local initiatives.

Do's

- Guide projects through successful campaigns;
- train-the-trainer approach;

- start with pilot projects;
- set realistic objectives;
- follow-up of projects during their campaigns.

Lessons learned

Crowdfunding can be a very strong and effective tool for local engagement and economic development. Trustworthy tools and strategic guidance are the key success factors for crowdfunding to fulfil this role.

2.2. Public Authority as Curator

The local authority selects a list of projects that reflect their agenda from an existing crowdfunding platform.

2.2.1. KICK-ER and Emilia-Romagna region

KICK-ER (www.kick-er.it[8]) is a guidance and first-support service provider for the realization of crowdfunding campaigns. In 2015, ASTER launched this service within its Finance for Innovation Unit, connecting this service with Investor Readiness Analysis & IPR Helpdesk. It is addressed to companies, start-ups, research laboratories, and public institutions based in Emilia-Romagna—or with an impact project on this territory—which would carry out a crowdfunding campaign to launch their own innovative products and/or services. KICK-ER is complementarily integrated with ASTER services dedicated to support business creation and finance innovations and responds to the needs more and is more clearly perceived by start-ups and entrepreneurs: how to launch an innovative product on the market. To do business, innovate, and give shape to your idea, it is necessary—on one hand—to collect the funds needed to make the initial investments and—on the other—to plan an effective marketing strategy: improve your idea, experiment with the working team, test the market and identify its positioning, expand the stakeholders' network, explore the willingness to pay of potential customers, and much more.

Targets of KICK-ER:

- Companies and start-ups that get to know the potential of their innovation;
- researchers who need fundraising for specific research projects, for example, to overcome bottlenecks linked to scale-up;
- public institutions that can foster their accountability and develop community-chosen projects;
- civil society that gets closer to innovation and to policymaking.

Potential beneficiaries go online, register, and submit a short version of their business plan. If it is compliant to crowdfunding criteria and there are the conditions to start a crowdfunding campaign, KICK-ER staff meets the "wannabe" campaigners. During the meeting, the staff works on creating the strategy for a fruitful campaign, helping the beneficiary in the implementation of it, giving promotional support, communication strategy, and post-campaign analysis. The second aspect of the service is linked to training and dissemination activities across the region in order to spread the crowdfunding culture and get potential targets closer to it through witnesses of successful campaigns

Name of Service: KICK-ER	Year of beginning of operations: November 2015	Crowdfunding model adopted: Rewards	Country/ies of operation: Emilia-Romagna Region, Italy
Services offered: Guidance service for the development of the crowdfunding campaign; communication and PR.	Areas of activities: Social entrepreneurship; cancer research; educational activities; performing arts; innovative technologies and cultural projects.	Number of applicants: 93 Number of selected projects: 11 Number of successful projects: 11; 100% success rate.	

Reasons for setting up the partnership:	Public resources committed:	Impact of the initiative:
Provide the regional ecosystem with a toolkit in order to investigate new forms of financing and expand the pool of financial instruments; support the entry into the market of companies with innovative marketing strategies; experiment new forms of support that are complementary to research funds; develop the entrepreneurial potential of research world, bring the public closer to research world and vice versa; support and promote local innovation; raise awareness around crowdfunding in the world of start-ups and innovative SMEs; provide an operational toolkit, to encourage the launch of crowdfunding campaigns; and spread and promote start-ups that use crowdfunding in Emilia-Romagna and discouraging the incorrect use of it.	One man dedicated within the Regional Agency for Innovation and Technology. **Specific Partnership with:** Emilia-Romagna Region.	A total of 93 requests of contact/filled-in forms; 50 meetings; 25 idea pre-screenings support and reports; 35 in-depth calls; and 22 customised *vade mecum*.

Costs borne by public authorities in addition to the pledged resources—None.

Project: TRACe by Istituto Scientifico Romagnolo per lo Studio e la Cura
dei Tumori (I.R.S.T. IRCCS)
Location: Meldola (Emilia-Romagna, Italy)
Sector: Cancer Research
Duration of the campaign: 3 months
Number of backers: 200
Funding: €25.095
Project link: https://www.ideaginger.it/progetti/trace.html[9]

Aim

Support an oncological research project to study circulating tumour cells
in cancer patients.

Description of the project

TRACe is the first crowdfunding campaign to fund a translational
research project of IRCCS (Scientific Institute of medical Research)
IRST in Meldola. Launched in March 2017 and active for 3 months,
"TRACe" crowdfunding campaign had the primary objective of raising
€20,000 in order to support an oncological research project to study circulating tumour cells in cancer patients. The IRST research team aimed
at virtually opening the doors of the IRST bioscience laboratory to the
citizens, not only to guarantee them on how money will be used but
above all to try to break the wall between research and society, especially
in Italy. In fact, TRACe's goal was not only to raise funds to advance in
cancer research but also to link cancer research and citizenship, laboratory work and daily life, and the contribution of everyone and the end
result.

Outputs and impact

- TRACe team was able to involve all stakeholders in every step of the
  research, both during and after the campaign, updating the supporters
  on the goals achieved through fundraising and laboratory results;
- establish a new relationship between the "in the lab" world and the
  one "outside the lab," making the results of research accessible and

communicable to as many people as possible, setting up a new mode of interaction between the research community and society;
- transparent way of reporting where, to whom, and for what the funds are used.

Barriers and bottlenecks

- How researchers could be communicative to the "outside the lab"—researchers learned how to translate the research project to a description which was informative, interesting, and engaging;
- how to engage and reward the backers—successful storytelling activity and gaming strategy to involve as much crowd as possible. Researchers also identified the right social media to spread around the right key message to each target group. They also designed a symbolic set of rewards, which leveraged on the engagement of people.

This service was developed within this region to foster the innovation of the local ecosystem, giving a toolkit to explore new ways to raise fund and realise a communication and marketing strategy, not only for start-ups but also for research labs, entrepreneurs from civil society, and public institutions.

The players involved in the development of KICK-ER were as follows:

- the innovation agency of the region (ASTER, re-branded as ART-ER in 2019): ASTER is in charge of managing the networks of start-ups, incubators, and research labs of the region and can reach the targeted audience of the service (spread contents and disseminate info);
- crowdfunding experts: can co-design the service and be involved in the guidance activity in order to transfer the know-how;
- start-up training and consulting staff: helping to analyse start-up business plans;
- regional government: to enable the networking among all the players and to address policies towards a more integrated development of crowdfunding.

The service was developed within Emilia-Romagna Region—General Directorate for Knowledge, Labour and Enterprise Economy and it is consistent with the policy instrument 2014–2020 ERDF ROP Emilia-Romagna Region—Axis 1 Research and Innovation.

2.2.2. Growfunding and VGC

growfunding / bxl
build your city

Growfunding is a non-profit, Brussels-based, socio-cultural organisation. By using civic crowdfunding as an instrument, we support local, social impact projects not only in financing their project but also in creating a communication strategy, enforcing and enlarging their network, and building a participatory community. Growfunding doesn't focus on one specific sector, but rather takes a geographical scope, i.e. the social tissue of Brussels, as the starting point. Therefore, they support all types of projects (profit and non-profit) active in very different fields (like circular economy, culture, food, etc.). Growfunding was launched in 2014 as a research project, funded by the European Structural Fund, at the University College in Brussels. Since 2016, it was launched as an independent spin-off. Over the past 5 years, Growfunding launched a bit more than 100 projects. Last year, the average success rate was 86%. The community of Growfunding counts 8.600 Growfunders who raised in total an amount of €900,000.

Since last year, Growfunding partners up with local authorities in matching funds. For now, two types of projects are qualified: projects carried out by youngsters (<26 years) and projects that involve 'community development work.'

The case study presents the project of Cyrille, a young man with Togolese roots who arrived some years ago in Belgium. He gained experience as a chef in the socio-cultural organisation called Cultureghem. They organise several activities on the site of the slaughterhouses in Anderlecht. One of them is called KOOKMET, for which they also did a

Growfunding campaign some years ago. Cyrille decided to launch its own project: a mobile kitchen on a cargo bike in order to cook in public spaces.

Name of platform: Growfunding	Year of beginning of operations: December 2013– February 2014: *European social fund* (ESF) funding as research project from Odisee University College.	Crowdfunding model adopted: Rewards	Country/ies of operation: Brussels (Belgium)	Total funds collected on the platform since launch of operations: €902.000
Services offered: Training for project owners; public workshops; tailor-made follow-up; co-organisation of offline events; translation of the project page in ENG-FR-NL; and social media support	Total backers on platform: 8.900	Outreach: Facebook: 5,192 Instagram: 155 Twitter: 489 Newsletter: 1,900 Website access: 25,000 unique visitors.	Areas of activities: All areas, as long as projects are linked to Brussels.	Number of applicants to the match-funding call for projects: No application required, match-funding only accessible over €3,000 Number of selected projects: 3 Number of successful projects: 3 (pilot phase), 100% success rate

(Continued)

(Continued)

Reasons for setting up the partnership:	Public resources committed:	Impact of the initiative:
Both VGC (the local authority who gives subsidies for youth projects and community development projects) and Growfunding try to support small-scale experimental projects with social added value for the city. The shared goal and shared focus on youth/community development resulted in a more structural cooperation: projects that need more than €3,000 can benefit from the matched funds. By linking the projects to participative rewards, through a crowdfunding campaign, the social impact is expected to be bigger.	Maximum €3,000/project	Expectation is to substantially increase the success rate of projects. Most of the youngsters who want to launch a project don't have a strong network. A successful campaign also influences the confidence of the youngsters and stimulates their ambitions for the future. Launching a crowdfunding campaign brings them several skills (planning, communication, financial literacy, etc.) that will also benefit them in their future (professional) life. A matching fund will strengthen the network, and thus, the social impact of projects, as both networks of Growfunding and the local authority will function as a lever.

Costs borne by public authorities in addition to the pledged resources—The public authority (VGC) paid for the development of the software behind the matching-fund. Growfunding, as non-profit organisation itself, also receives a subsidy for community development work.

KooKVéLo is a project run by Cyrille and Rinke who need €7,500 to fund their mobile kitchen on a cargo bike. The KooKVéLo can move through Brussels and can basically be present on every public space.

Location: Brussels
Sector: Food, social inclusion, and mobility
Project holders: Cyrille and Rinke, two inhabitants
Duration of campaign: 62 days
Collected funds: €8,080
Number of backers: 46 growfunders

De KooKVéLo is a mobile kitchen, open for everyone to cook together. This project was created by lovers of the kitchen who only have one wish: sharing their love of cooking.

The idea of KooKVéLo is not only to inform different kinds of people about healthy food but also to let them see the fun of cooking and eating together. The project wants to connect different types of people who are present in the public space at various events in Brussels. Cooking and eating are two basic activities that are existing in every culture. We want to mix all the cultural richness of Brussels by making a unique recipe.

KooKVéLo will be a collection kitchen through the recycling of food surpluses. In this way, we want to show that food waste is not necessary and that we can re-use food. It is an opportunity to recycle in a positive and creative way.

De KooKVéLo also gives an opportunity to cook together and above all an opportunity to meet different cultures. It wants to create a link in the different neighbourhoods of Brussels through different events and work-shops where cooking together is central.

Outputs and impact

- The purchase of a cargo bike for Cyrille and Rinke;
- participation in some events (with Growfunding);
- creation of a network;
- some organisations and companies supported the campaign with a bigger donation.

This could help KooKVéLo to set up a sustainable network and bring in new clients.

Success factors

- Local authorities get a lot of project proposals for project subsidies, and for them, it is sometimes hard to distinguish strong projects from less qualitative proposals. By combining this public funding to an instrument like crowdfunding, you give 'the crowd' the last word to say whether a project is valuable enough or not. In this way, a top-down instrument like subsidies is combined to bottom-up voices and you give the final word to the citizens. In our case, we already had the luck to work closely together with our local authority. Since they know which methodology we apply and 'trust' us, it helped them to convince to step into the matching fund.

- By working together with a credible actor like a local authority, projects that participate in the MF get more credibility. Some enterprises backed the case study with a bigger donation, which made it easier to reach the target goal. Though convincing enterprises and companies to back a crowdfunding campaign is for many projects a time- and energy-intensive job, a good preparation to find out which enterprises or organisations can help financially support a campaign can influence the success of a campaign substantially.

Do's

- Invest enough time not only in the preparation of a crowdfunding campaign (make a network analysis, a timeline with a communication plan, offline events) but also in the realisation of the rewards. In the case of our case study, Growfunding also helped in the realisation of the rewards by co-organising an event. The post-campaign support for some projects is also necessary, especially for more precarious groups (like youngsters).
- Combine the online campaign with offline events, since for some people doing an online transaction is still a big step, offline events also give great visibility.
- The project idea needs to be concrete, tangible and unique, and easily understandable for the wider crowd.
- Transparent communication during and after the campaign (e.g. how to get your reward) is important for the credibility of not only the platform and local authority but also the project itself on the long term.

Don'ts

* It's suggested not to set the matching fund percentage at too high a rate (maximum 30%) to make sure projects still create a social impact (and need the crowd).

Lessons learned

In this case, the matching fund is applied to support additional precarious youth projects. If the local authority gives their support, by approving the match-fund, this already gives the project owners a big boost to start their campaign and the trust that their project is valuable for the community. This positive energy is important, because a crowdfunding campaign is time and energy demanding and can be stressful. The visibility of the matching fund on the website is also important.

2.3. Public Authority as Facilitator

Facilitator—Subnational governments can play an important role in facilitating citizen and civil society empowerment. This includes planning permission, financial as well as technical expertise support, co-screening and/or co-designing projects, etc. Basically, it represents a new type of public–non-profit private partnership with citizens/inhabitants and civic crowdfunding platforms. Some subnational governments have already set up partnerships with civic crowdfunding platforms, which are mainly based on co-financing projects alongside the rest of the crowd.

2.3.1. Goteo and Ajuntament de Barcelona

Goteo is a platform for civic crowdfunding and collaboration on citizens' initiatives, including social, cultural, technological, and educational projects. Goteo has replicas and alliances in several countries, thanks to its open-source code. It is a tool for generating resources "drop by drop" for a community of communities consisting of over 123,000 people, with a funding success rate over 75%.

There is a non-profit foundation (with consequent tax advantages for donors) behind the platform and a multidisciplinary team developing tools and services for co-creation and collective funding. Our commons mission is tightly linked to principles of transparency, progress, and societal improvement.

Goteo supports social impact and commons project to succeed and improve their communities, be it physical or digital. To make it a reality, we created an operating system of open tools and methodologies.

One of these tools developed by us is the match-funding calls, involving software, methodologies, policy recommendations, and protocols. Such as the match-funding Conjuntament.

The initiative presented as a case study is a platform coop in ecomobility called Som Mobilitat. This project participated in a programme, called "La Comunificadora," organized to promote collaborative economy initiatives in Barcelona. Thereafter, and with the aim of creating their technological platform, expand its member's base and make itself known, the coop decided to launch a crowdfunding campaign in Goteo. This was really successful and had a huge impact that resulted in new cooperatives created in the Spanish State replicating their model, developing a software platform that is nowadays being used by several coops in Europe, a large increase of cooperative members (from 300 to +700 during the campaign) and the beginning of an European joint cooperative formed by regional cooperatives.

Section A—General Overview of Platform and Partnership

Name of platform: Goteo	Year of beginning of operations: 2011	Crowdfunding model adopted: Rewards and donations	Country/ies of operation: World Spain	Total funds collected on the platform since launch of operations: €6,000,000

Services offered:	Total backers on platform:	Areas of activities:	Number of applicants to the match-funding call for projects:	
Coaching and tutoring of project owners; specific workshops on crowdfunding, co-creation, and communication; adaptive offer and customised configuration of match-funding models, including technological development when needed; a collaborative crowdfunding platform under FLOSS code; communication support for campaigns and match-funding calls.	+84.500 **Outreach:** Facebook: 12.200 Instagram: 960 Twitter: 23.600 Newsletter: 123.000 Website access: +200.000/ month.	Social including social economy; Education; Environment; Communication; Culture; Technological; Entrepreneurship; Scientific; Design	67 **Number of selected projects:** 24 **Number of successful projects:** 23 95.65% success rate vs. average 73.44% on Goteo	

Reasons for setting up the partnership:	Public resources committed:	Impact of the initiative:
Promote projects that come from organised civil society and that have a special potential to change, in many aspects, the dynamics of the city. Citizens' participation is a political priority and one of the fields where the City Council is investing more effort and energy. Know the reality of the civic ecosystem as well	€96.000	At the end of the match-funding call, there has been an injection of +€230.000 into the sharing and social economy of Barcelona resulting in several jobs created, an increase in the scalability and sustainability of new projects and others already established such as cooperative cinemas, autonomous initiatives

(*Continued*)

(Continued)

as the needs and dreams of the inhabitants of the city. Invest public funds in a more just and efficient way in those projects that are more interesting for the citizens, Legitimise the city budget. Generate data and have a map of information that can be useful to design public policies. Create empowered citizens who are more aware of the city as space and common resource.		of occupational inclusion, circular economy and sharing economy cooperatives, platform coops, etc. Impact on media: We appeared on several newspapers, we participated in national level and thematic radio programmes, and we were featured in the blog of P2P Foundation. We observed a triplication of the projects that start a campaign in our platform since the announcement of the match-funding Conjuntament.

Costs borne by public authorities in addition to the pledged resources—The services (training via workshops, coaching 24 projects during all the process, design & IT, communication, etc.) related to this match-funding call amount to €54.000. From this total, the Barcelona City Council covers €39.960 (74%) and the Goteo Foundation the remaining 26%.

Project: Posem en marxa Som Mobilitat a Barcelona (We launch Som Mobilitat in Barcelona) by Som Mobilitat Coop.

Location: Barcelona (Spain)
Sector: Social entrepreneurship and environment, cooperative platform.
Duration of the campaign: 3 months
Number of backers: 206 + 82
Funding: €22.825 + 9.150
Link: http://goteo.cc/sommobbcn[10] (previously, they had another successful campaign in Goteo: https://www.goteo.org/project/compartim-vehicles-electrics[11])

Aim

To create local groups of the sharing economy platform coop Som Mobilitat, to improve the sustainable mobility in Barcelona.

Som Mobilitat is a non-profit consumer cooperative (using platformcoop model) around car-sharing and the ecomobility that wants cities with fewer cars, less noise, and cleaner air. To this end, they are committed to develop a renewable, shared electric mobility service whereby the citizens can rent and share electric vehicles from the cooperative or partners (p2p) through their mobile phones. One of their main goals is to spread the model and create a big European network of ecomobility based on the cooperative values and structure.

The specific goals of the first crowdfunding campaign were as follows:

- To contribute to the development of the shared electric mobility (Som mobilitat) platform.
- Integrate the tech platform into a network of European cooperatives of electric mobility.
- Accelerate the arrival of 100% shared electric mobility throughout Spain.
- Create a social and cooperative alternative by way of shared electric mobility.

Outputs and impact

- Before the campaign, the team participated in "La Comunificadora," a support programme for sharing economy initiatives in which Goteo is one of the co-organizers. The participation in this project gave them tools, resources, and the needed guidance to improve and boost their project, as well as a crowdfunding workshop that helped them reach more efficiently their campaign goals.
- During the first campaign, Som Mobilitat duplicated their cooperative members, from ~300 to ~700.
- Using the funds of their first campaign (€22.285), they developed an integral tech platform to satisfy their internal needs as a platform coop and to manage the car rental. Thanks to the fact that the objective of the

campaign was to use this technology openly and offer it to the world, today there are already projects that use the platform.

- A group of people from País Valencià chose the reward "workshop & test of Som Mobilitat" and, due to this, they were able to replicate the project (http://alternacoop.com).
- A lot of projects from all of Europe and the Spanish State contacted them using Goteo to create synergies, to replicate the project in other regions, and to federate with other projects; for example, with REScoop, a European federation for renewable energy cooperatives.
- It was a huge communicative tool that lit the fuse of a European cooperative of ecomobility cooperatives (http://www.themobilityfactory.eu[12]).
- Project owners had an incredible opportunity to spread their project among social economy networks, and they are still known for their successful crowdfunding campaign.
- Project owner used Goteo as a validation tool to check their communication and organisational abilities, which have then been employed in the cooperative itself.
- There has been a significant impact in press and media, thanks to the crowdfunding campaign.
- Based on the communication needs of the crowdfunding campaign, a network of ambassadors was created and maintained afterwards, so as to develop the community and support the creation of local groups all around Catalonia.
- Barriers and bottlenecks (communication, administrative, procedural, etc.), and how they have been addressed/solved.
- The campaign team pointed out that the main barrier was internal coordination—once solved, all the past difficulties became knowledge that they used for the internal organisation of the cooperative.
- Fear of lack of communicative ability—thanks to Goteo's communicative drive and community, the project team reached a wide community of people both in Spain and around the world.

Depending on the total match-funding funds to be allocated and the total number of projects to be supported, Goteo recommends the matchers to allocate at least €4.000 per project during the first round. Conjuntament's objective is to support local initiatives of the City of Barcelona that contribute to the development of the neighbourhoods. The City Council Plan 2016–2019 establishes a strategic action to support the plural economy and options of sustainable consumption, including social and cooperative economy. In addition, Conjuntament is promoted by Barcelona Activa, which is the city council's organisation in charge of boosting economic policies and local development to ensure a better quality of life for Barcelona citizens, by promoting employment, encouraging entrepreneurship, and offering support to companies from the perspective of the plural economy. One of the lines of action of Barcelona Activa is to "Support for the plural economy, developing transforming economies, group entrepreneurship and social innovation."

An open application for projects was open during 6 weeks (from October 23 to December 4, 2017). During that period, Goteo promoted the opening in social media and other communications channels (media, blog, newsletter, etc.). They also organize in-person events and activities.

Success factors

- Legitimacy: Public institutions legitimize their budgets while allowing citizens to decide and prioritize how public money is used. Creating a space for participation where citizenship promotes and supports initiatives coming from below.
- Participation: Citizens decide to launch projects and to choose which projects to support.
- Sustainability: Projects come from neighbourhoods and organized citizenship, in opposition to top-down policies. Communities are behind these projects and they want to make them alive.
- Transparency: Citizens audit the whole process, as they can check and visualize instantly how the money is used.
- Success: The success rate rises until more than 90%, when a public institution multiplicates the donations made to the crowdfunding campaigns.
- Learning lab: While they are in campaign, projects learn as they work collectively, making the project stronger and growing their network.

Do's & Don'ts

- Ignorance (if not trivialisation) of the funding mechanisms and the tools themselves, evident in massive practices far from open standards, strongly profit-focused, and with no priority given to social impact or transparency.
- Interest but relative disconnection of many solidarity and philanthropic organisations that respond to the guiding principles of crowdfunding, especially its potential for nurturing a culture of open participation that goes beyond the occasional donation.
- There is plenty of room for improvement in the way people support social causes online, as well as in the way the impact of this is measured and how these initiatives can obtain resources in a more interactive, social process.
- *More "delocalisation:"* Only 50% of crowdfunding campaigns promoted by European organisations and projects use platforms based in a EU member country or Europe. Therefore, 50% of them chose the two global (and North-American) platforms—Kickstarter or Indiegogo. This points to the necessity of establishing local crowdfunding platforms.
- *Necessary improvements*: Apart from advancing in the study of the match-funding model behaviour, several reflections should be made about how current public policies—which encourage cultural, creative, and social initiatives—could develop a better fit with this hybrid funding model.
- *More flexibility*: Public organisations lack the agility and flexibility needed to operate with a new model of distribution of public funds, both on the operative and legal side.

Lessons learned

The match-funding programmes have a clear "multiplier" effect. To start with, the credibility of the crowdfunded project increases when it has the institution's hallmark. It receives the trust of the community and, thus, has a better chance of attracting crowd donations. The more crowd funds received, the higher the matching contribution by the institution and, as a result, the bigger the total project budget becomes. Data show that a crowdfunding campaign with institutional support receives on average

180% more from crowd donations than a campaign without institutional support. Also, its success probability (reaching the minimum set campaign/project budget) increases by up to 90% (in comparison to 71% success rate for campaigns without match-funding).

On the User Experience level, we've collected feedback over the years from the user communities that have participated in our calls and have performed a comparative evaluation of several European match-funding platforms.

Early findings revealed the need to develop additional improvements if we wanted to encourage co-responsible public and private contributions to projects that have the support of civil society. We saw, for example, that relations between individuals, communities, and institutions could be redesigned by visualising crowdfunding projects' needs (from projects carried out as part of match-funding programmes). This can be done based on data taken from their budget descriptions, their non-monetary needs, and their aims within their respective areas of influence.

This benchmarking study also identified which elements and changes had to be incorporated into the new design to enhance the user experience of citizens and institutions alike, and hence, the action and interaction of users in each of their roles become more intuitive, visual, and easier to understand.

With all this in mind, we presented in 2017 a better-organised new site, packed with new visual elements including interactive maps, information, a timeline showing dates and significant milestones within the call, new icons related to different areas of social impact, structure by phases showing the state of the call, and many other implementations that have been documented in Goteo Foundation's Github and the API entry point for developers interested in match-funding.

In addition, Goteo's match-funding site includes a map of influence of the call, showing the geographical location of the proposals received, the proportion of participation by gender, the expected scope of influence, and the location and method of enrolment for the training sessions.

The map of influence becomes, at this stage, a map of contributions to the call, showing in real time and interactively, the flow of contributions and their origins (geographically and by type of user, citizen, or institution).

We trust it will make it easier to understand this hybrid funding model which brings about a shift in the way resources are allocated to projects supported by the most active and engaged communities. And we hope to inspire more institutions and entities to use this powerful tool as part of their social responsibility lines of action.

All ideas and improvements outlined here aim to inspire institutions to release new crowdfunding calls and to encourage potential campaign initiators not only to apply but also to detect the priorities and interests of their communities. Institutions can reach sectors and audiences who do not usually apply for grants or public funding; they can identify new topics and specific types of projects that attract the interest of their communities. A study to define the needs related to a specific topic or community, or section of a community, can also be conducted by Goteo to inform future calls and tenders of a match-funding institution.

2.3.2. Startnext and the city of Hamburg

Startnext is Germany's first crowdfunding platform, founded in 2010. It offers entrepreneurs, innovators, and creative people the opportunity to present their ideas, raise the necessary funds, and build a community. Today, Startnext is the largest crowdfunding community for creative and sustainable projects and start-ups in the German-speaking countries. Nearly 6,000 projects have been successfully funded via the platform and over €50 million have been raised. Startnext is a social business and part of the Benefit Corporation network.

'Nordstarter' ('nord,' the German word for 'north') is an initiative of the City of Hamburg to foster its creative industry. Founded in 2011, it is a long-standing partner of Startnext: www.startnext.com/pages/nordstarter. Nearly €2 million have been collected by Hamburg's creatives via the regional crowdfunding network for over 500 projects. To foster crowdfunding, the city offers three types of support: (i) regular match-funding call for projects, where winners are awarded a prize, (ii) a dedicated contact person for one-to-one coaching on crowdfunding, and (iii) regular workshops, events, and meetups.

The strong community that was created around the initiative is remarkable. Regular events for everyone interested in crowdfunding as well as for former crowdfunders contributed to that. This strong regional crowdfunding network helps to spread the word about crowdfunding and creates a space for exchange. Based on this example, Startnext started a train-the-trainer programme to transfer knowledge and establish a local network of experienced crowdfunders in other German cities as well.

Name of platform: Startnext	Year of beginning of operations: 2010	Crowdfunding model adopted: Rewards	Country/ies of operation: Germany, Austria, Switzerland, and Lichtenstein	Total funds collected on the platform since launch of operations: €52,242,259
Services offered: Webinars for project starters; 'Train-the-trainer'— programme to establish a network of crowdfunding coaches in Germany; partners organise crowdfunding coachings and meetups in various German cities.	Total backers on platform: 900.000 registered users; +750.000 not registered users **Outreach:** Facebook: 70.000 Twitter: 10.000 Instagram: 3.500 Newsletter: 50.000 Website access: 2.5 Mio users/ month	Areas of activities: Not limited, everyone (matching Startnext's guidelines and categories) is welcome to start a campaign. Majority of projects: creative or sustainable.	Number of applicants to the match-funding call for projects: 43 **Number of selected projects:** 37 **Success rate:** 63% successful vs. 55% on average on Startnext	

(Continued)

(*Continued*)

Reasons for setting up the partnership:	Public resources committed:	Impact of the initiative:
Hamburg Kreativ Gesellschaft is an initiative of the City of Hamburg for the city's creative industry; it aims to be the primary point of contact and support for workshops, coachings, and networking in the creative industry; events and coachings on crowdfunding are available to offer an alternative financing option for artists, musicians, etc. and to enable more projects in the creative and cultural field.	Regular call for projects where winners are awarded a prize.	Support offered: Participants of the match-funding initiatives are offered to participate in different crowdfunding workshops, e.g. overview on crowdfunding, storytelling, and creating pitch videos. Outcome: The match-funding facilitated convincing people to start a crowdfunding campaign, to draw attention to crowdfunding, and to collect the necessary funds for the projects; media partnerships helped to spread the concept of crowdfunding.

Costs borne by public authorities in addition to the pledged resources—Funding for a part-time employee, who conducts one-to-one coaching sessions and workshops and organises events (free of charge for the audience).

Name of project: Schaluppe Hamburg (www.startnext.com/schaluppe)
Location: Hamburg
Sector: Culture
Proposing organisation: Verein für mobile Machenschaften (NGO) (www.mobilemachenschaften.de/schaluppe)
Duration of campaign: 30 days
Collected funds: €20,076
Number of backers: 442

Aim

Create an open-air cultural space on Hamburg's canals.

As a port city, Hamburg is surrounded by water and has two lakes and plenty of canals. The "Schaluppe" was built to create a non-commercial cultural space on the water. It is a 15-meter-long, 5-meter-wide raft, offering space for around 100 people. The project started in 2016, with more than 50 volunteers constructing the raft. Every year now, from May to October, a socio-cultural programme is offered, including theatre, circus, workshops, cinema, readings, or concerts.

A non-profit association is responsible for the project, with the purpose to promote and support art, culture, cosmopolitan attitudes, tolerance, and international understanding. They cooperate, for instance, with a refugee association with the aim to actively integrate them into the project structure. The "Schaluppe" pursues the following goals:

- a non-profit platform for newly arrived Hamburgers;
- promoting a transnational exchange of arts and culture;
- support for cultural participation;
- to discover and provide access to Hamburg's waterways.

The crowdfunding was conducted right at the start of the project to gain enough funds for the construction of the raft. The materials, as well as the rent for the building site, electrical power, safety equipment, insurance, and food and drinks for the volunteers were covered.

Outputs and impact

The crowdfunding campaign allowed the development of a community around the project which donated money and more importantly, time to realise the founding members' vision of creating a mobile space for art, culture, and civic engagement. The group organises regular events from May to October every year.

Success factors

- One contact person for all matters related to crowdfunding in Hamburg; insights and understanding of both sectors, the city and the creative industry, and committed to push crowdfunding in the region to build a strong network.

- Sufficient budget for a dedicated position.
- Incentives for projects to start a crowdfunding-campaign (in this case, prize money, coaching, communication support, and continuous point of contact).

Lessons learned

- Participation rates in the initiative (ongoing for 7 years now) correlate to the time invested by the partner; high effort, e.g. due to a contest, meant more projects starting a crowdfunding campaign.
- A good level of supporting activities is required to get started, e.g. how to do a video, what communication strategy to choose.
- The city has no specific framework for crowdfunding in place. Its support to Nordstarter fits into a larger support programme for cultural projects within the city of Hamburg.

2.3.3. Voor je Buurt and South Holland province

Voor je Buurt is a Dutch foundation and one of the first civic crowdfunding organisations globally. Its mission is to assist initiators of local social projects with collecting funds, volunteers, expertise, and materials to make their project successful. Projects range from the odd children's playground to social innovation projects like new concepts for local healthcare and projects to counter loneliness. Voor je Buurt manages over 10 national, thematic, and local civic engagement platforms and one global platform (Onepercentclub) as part of its programmes. All platforms allow initiators to start (donation and reward based) crowdfunding and/or crowdsourcing campaigns.

Voor je Buurt was founded in 2012 and launched its first platform in January 2013. Thus far, our platforms hosted nearly 2,400 local initiatives in the Netherlands, Flanders, and the developing countries, which raised about €7 million in donations and actively engaged over 75,000 people and organisations. Voor je Buurt has a paid staff of seven qualified professionals and has partnerships with over 40 municipalities, regional governments, foundations, educational and research institutions, and companies. Voor je Buurt also runs several "matchfunding" programmes, through which foundations and local governments co-finance local projects that use one of the online platforms.

Name of platform: Voor je Buurt	Year of beginning of operations: 2012	Crowdfunding model adopted: Rewards and donations	Country/ies of operation: Belgium, and the Netherlands	Total funds collected on the platform since launch of operations: €3.25 million
Services offered: Platform management and activation; community management; coaching and training; matching support and programmes; online helpdesk and support; software as a service.	**Total backers on platform:** 40.600 **Outreach:** Facebook: 52.000 Twitter: 22.000 Newsletter: 3.700 Website access: 50.000 unique visitors.	**Areas of activities:** Local social and cultural projects; social entrepreneurship; environment; healthcare; and urban development	**Number of applicants to the match-funding call for projects:** 26 **Number of selected projects:** 24 **Number of successful projects:** 18 (6 still ongoing); 100% success rate in previous editions of match-funding. General match-funding success rate: 90% vs. 83% average on Voor je Buurt.	

(*Continued*)

(Continued)

Reasons for setting up the partnership:	Public resources committed:	Impact of the initiative:
The provincial government had been running a subsidy programme for small-scale green projects for a couple of years. Participation is one of the main criterions used to assess project proposals. The officials expected that crowdfunding (and crowdsourcing) would help strengthen the participation element in the projects. Also, by running a crowdfunding campaign, the project initiator would prove in advance that people (and organisations) were willing to participate in the project.	€50,000 annually	Number of people trained: 50 Number of people reached with information about the match-funding possibility: at least 16,500. Number of public officials trained: +/−25 Number of municipalities involved: 12 Number of backers in campaigns that received match-funding: 1,006. Total amount of funding: €50,000 from the Province of South Holland and €70,655 from the crowd. Number of projects realised: 18

Costs borne by public authorities in addition to the pledged resources—Training and workshops, PR and communication, and administrative costs.

Name of the project: Natuurspeeltuin Voorschoten (Natural Playground Voorschoten) by Foundation Natuurspeeltuin Voorschoten
Location: Voorschoten (the Netherlands)
Sector: Nature, Health
Duration of campaign: 54 days
Collected funds: €6,862.02
Number of backers: 47

Aim

To create a nice, sustainable, and natural playground for children from Voorschoten, accessible for everyone.

Description of the project

After extensive preparation, the crowdfunding campaign voor Natuurspeeltuin Voorschoten was launched on May 29, 2017. The initiators raised €6,862.02 (114% of the target amount) in 54 days.

The playground should be a safe, sustainable, and natural playground where children, parents, and grandparents can meet. It was modelled as a meeting place, where children will have fun in direct contact with nature and will be introduced to local flora and fauna. It was also designed to serve as a playground where children will have the possibility to explore and push their boundaries.

Besides that, the playground needed to be accessible for everyone, especially for *disabled* children and elderly. The playground should be wheelchair accessible by creating paved and wide paths and bridges. Most of the parts that can be operated, such as the water pumps, will be at wheelchair height.

Outputs and impact

- Extensive collaboration between different parties. Project owners worked together with the municipality, primary schools, and the local Lions Club.
- A large financial investment was also needed. Project owners started by addressing their own network during the campaign.
- During the campaign, project owners raised 114% of the target amount. Part of the target amount was funded through match-funding by the Province of South Holland, which donated, together with another local party, half of the target amount. This is an example that cooperation between local initiatives and various institutions can lead to great results.

Barriers and bottlenecks

- Most of the donors are people known to the fundraiser or who are enthusiastic about the initiative, like family, friends, neighbours, and colleagues. Convincing unknown people to donate is more difficult.

The project initiator has to raise at least 50% of the target amount. The contribution by the Province is first shown on the platform once the initiator has raised 20% of the target amount. The minimum number of backers is 20. The Province of South Holland doubles the amount raised by crowdfunding up to a maximum of €5,000 per project if the project initiator is recognised as a Public Benefit Organisation (ANBI) and up to €2,000 per project if the initiator does not have an ANBI status.

Conclusions

- A new and fruitful way to activate communities and support project initiators.
- A promising solution for local authorities who are looking for new ways to participate in bottom-up activities.
- Clear commitment from the side of the participating government.
- Clear goals for the match-funding programme formulated.
- An enthusiastic and motivated programme manager from the participating government.
- Not just match-funding but also a training programme for project initiators as well as government professionals.
- Clearly formulated and simple criterions.
- Bringing project initiators together increases the chance of cross-overs and stronger projects (and is great fun).
- Make match-funding visible on the online project page early in the campaign (when 20% of the target amount is raised). Great stimulus for the project initiator and works well to communicate to (potential) backers that the government is supporting the project.
- Match-funding increases the chances of success (success rate for projects that received match-funding is significantly higher than that for all projects regardless of match-funding).

3. Conclusions

The case studies presented in this report highlight how crowdfunding platforms can become a reliable partner and a valuable asset to public administrations that face a reduction in budget availability and aim at increasing communication, promoting engagement, and channelling funds into initiatives that are perceived as priorities by stakeholders in their territories.

At the same time, the above-listed experiences are proof of the adaptability of crowdfunding, not only in terms of how they can respond to public authorities and citizens' agendas but also of how projects developed in many different sectors can obtain funding and acceptance by involving a larger crowd of supporters.

Reported experiences of partnership between a public authority and a crowdfunding platform have been implemented in Belgium, Germany, Italy, Spain, and the Netherlands. Each of the case studies has been developed in a different administrative and political context and has successfully achieved not only economic leverage but also unprecedented levels of participations both in terms of submitted ideas and in terms of contributors and beneficiaries of the funded initiatives. Furthermore, each public administration was able to select the form in which the partnership would be formalised, from setting up their own local platform, to offering dedicated support and consultancy services to set up a crowdfunding campaign, to allocating shares of their public resources that would be awarded to successful crowdfunding campaigns in certain areas. The flexibility of crowdfunding and its digital nature therefore allow public administrations not only to choose from different levels of commitment when implementing a crowdfunding or match-funding scheme but also to receive concrete feedback and results of the performance of the partnership in a short time—i.e. the period in which crowdfunding campaigns are active and citizens can support them.

Case studies also highlight that citizens' ideas, contributions, and participation can be triggered in many different sectors through the implementation of match-funding schemes: environment, social inclusion, entrepreneurship, agriculture, cultural and creative industries, and health research have all been topics in which crowdfunding has helped citizens and public authorities join forces, but the same partnership models could be replicated to include other sectors that have not been covered in this report.

When looking into the European relevance of existing experiences, several aspects need to be underlined.

In the first place, three of the crowdfunding platforms discussed in the case studies (Streekmotor, Growfunding, and Crowdfunding Südtirol) have been set up, thanks to the funding received either from ERDF or ESF, meaning that the concept of including crowdfunding in the economic and social development of regions has already found its way into the broader European cohesion policy framework.

Furthermore, such platforms have been providing economic and social returns to their communities, positioning themselves as multipliers and amplifiers of the initial economic investment undertaken by the managing authority. In addition to direct funding received for the set-up of a platform, two other experiences presented in the case studies are directly related to ESIF investment priorities and regional operational programmes (ROPs) as follows:

- the KICK-ER service has been explicitly developed as to be in line with Axis 1 Research and Innovation of the ERDF ROP 2014–2020 of Emilia-Romagna;
- Nordstarter has been implemented as a sub-initiative of a larger programme in support for creative industries in Hamburg, where "support to creativity" is listed among the regional Smart Specialisation Strategy priorities.

The paradigm shift that crowdfunding has been able to trigger at the local and regional level promises to become a game changer also in the way public authorities and citizens perceive cohesion policy in their lives. Furthermore, the European Union would likely benefit from increased visibility and recognition of its efforts through the intense communication activities that crowdfunding campaigns entail.

Despite the many advantages highlighted in the case studies and in Section 3, dynamics of civic crowdfunding and match-funding partnerships with public authorities are still far from unleashing their full potential. Section 4 is therefore dedicated to the identification of the most common obstacles that hinder the development of such innovative funding schemes and provides some insights on how these challenges have been successfully addressed.

4. Policy Recommendation

Taking into consideration figures, findings, impact, obstacles, and solutions analysed in this report, we recommend that further action is carried forward in the following policy areas:

- Explore synergies with thematic priorities and resources aimed at funding initiatives that might be suitable for a crowdfunding campaign, as to enhance the overall outreach and increase financial availability. This process would not only entail all the above-mentioned benefits deriving from a crowdfunding campaign developed in partnership with a public authority but also enlarge the number of projects that could receive funding without increasing the amount allocated from public budget. In addition to this, such partnership would grant increased visibility to the EU and cohesion policy and would allow managing authorities to be more responsive and better equipped when investing in territorial and citizens' needs.
- Mainstream trainings on crowdfunding education, by implementing trainings, workshops, and dissemination events that can be delivered to different societal groups and economic actors. Increased knowledge about crowdfunding is especially needed as to unlock the full potential of the funding scheme, since the main strength of this partnership relies on the concept that it can be a tool that is effectively accessible to any citizen, organisation, or enterprise.
- Expand the scope of public partnerships with crowdfunding, designing secure, transparent ways in which to scale-up the match-funding mechanism so as to also include lending and equity crowdfunding. The introduction of financial models of crowdfunding in the partnership might prove more challenging for public authorities, as any investment carries a certain level of risk. However, innovative funding schemes that might blend public grants with financial crowdfunding models might allow not only social and societal returns for citizen, but also financial returns and a renewed economic dynamic in the territory.
- Identify areas in which a pilot programme can be carried out in partnership with a crowdfunding platform. Subnational governments and civic

crowdfunding platforms should work together in specific areas, so as to implement innovative matching schemes that can reach a higher investment impact. Priority areas such as those identified in regional Smart Specialisation Strategies might be a solid starting point for the design of the matching scheme, since the listed sectors have already been identified as key in the economic territorial development and provide a long-term framework in which pilots can be tested.

About the Author

Francesca Passeri is a Head of Advocacy at the European Crowdfunding Network. She has a solid background in regional development and European Affairs. She previously worked for the Emilia-Romagna Delegation to the EU and has a strong interest in dissemination of knowledge about crowdfunding through regional and local authorities. She is currently developing efforts within the crowdfunding industry to enable the integration of crowdfunding into European Structural and Investment Funds. She holds a Master's degree in International Relations from the University of Bologna, where she focused on researching the potential of crowdfunding for regional development from a European perspective.

Endnotes

1. This article is an edited version of the report first published by the European Crowdfunding Network: Francesca Passeri (ed.) (2018), Triggering Participation: A Collection of Civic Crowdfunding and Match-funding Experiences in the EU, Brussels. The content has been developed as part of the Crowdfunding for European Structural and Investment Funds working group of the European Crowdfunding Network, a multi-stakeholder working group established with the aim of exploring the potential of crowdfunding as a blending and match-funding mechanism within the context of the European Cohesion Policy.
2. Rodrigo Davies (2014), Civic Crowdfunding: Participatory Communities, Entrepreneurs and the Political Economy of Place.
3. Modified to ease readability. Cfr Davies (2014) defines the same concept as "subnational governments create a crowdfunding platform."

4. Some examples: "Un passo per San Luca," Bologna; Luchtsingel pedestrian bridge in Rotterdam, The Netherlands; the Peckham Coal Line, London, UK.
5. https://bit.ly/2ysY77K
6. https://www.miyuca.it/laab
7. https://streekmotor23.be/project/6659
8. http://www.kick-er.it
9. https://www.ideaginger.it/progetti/trace.html
10. http://goteo.cc/sommobbcn
11. https://www.goteo.org/project/compartim-vehicles-electrics
12. http://www.themobilityfactory.eu

Chapter 11

Civic Crowdfunding in Milan: Between Grass-Roots Actors and Policy Opportunities

Carolina Pacchi and Ivana Pais

Abstract

In this chapter, we present an analysis of civic crowdfunding promoted by the Milan Municipality: it is the most systematic experience conducted in Italy thus far and the one that permits to draw some reflections, potentially useful also for other local authorities. The chapter is organised into paragraphs: the first one is an overview of civic crowdfunding in Italy; the second one discusses the crowdfunding initiative promoted by the Milan Municipality and the main elements emerging from the campaigns conducted so far; in the third one, we propose some interpretative hypotheses moving from fieldwork conducted through interviews to some actors involved; finally, the conclusions aim at reconnecting some evidence emerging from the Milan case to the wider international debate on civic crowdfunding as a form of social, economic, and political innovation.

1. Introduction

In Italy, crowdfunding started in 2005, promoted by *Produzioni dal Basso*, a leading actor in this sector till date. In more than 10 years, following similar growth worldwide, supply has also grown significantly in Italy.

A yearly mapping of crowdfunding platforms[1]—which is the source of data presented in this paragraph—can help to reconstruct the main diffusion trends in our country. In November 2016, there were 68 active platforms; it is the same figure as that in 2015, but with a relevant birth and death rate: 17 new ones have been opened, the same number of platforms have failed.

Among active platforms, 23 have adopted a *reward* model, 18 are *equity* ones, regulated by Consob (the financial markets regulation authority), 13 collect *donations*—usually with no reward—for socially oriented projects, four are based on *peer-to-peer lending,* and 10 have adopted mixed *reward* and *donation* models.

Four platforms out of six are generalist; the remaining ones are specialised, not only in the arts/music/culture sector and in the social sector but also in energy efficiency and environmental sustainability, in the agro-food and sports sectors. There is just one platform exclusively devoted to civic crowdfunding (*Planbee*); at the moment, it has hosted a limited number of projects (17 campaigns, one with the involvement of a municipality) The research by De Crescenzo and Pichler[2] shows that in Italy civic crowdfunding campaigns are hosted mainly on reward platforms, either pure reward or a reward–donation hybrid: 80% of the platforms involved in civic crowdfunding have a business model of this type; the percentage of platforms using the pure model is also high (60%).

Italy shows a peculiarity on the international stage: the presence of five area-based platforms; this is a significant element for civic projects, which tend to be locally embedded. Notwithstanding the high number of active platforms, the ones showing high volumes of users and transactions are very few: one out of four goes beyond 10 users and has more than 1 monthly transactions.

Since the very start, more than 100 campaigns have applied and 23 have been published. One out of three reaches its financial goal: reward campaigns are the better performing ones (52%), while *donation* campaigns are the worst ones (8%).

In 2017, the overall value of funded projects has been around €133 million, with an increase of 45% compared to 2016.[3] *Peer-to-Peer lending* platforms contribute 59% of such amount, equity platforms provide 15%, reward and donation ones collect 27% of the total.

The platforms, which are potentially more interesting for civic crowdfunding (reward and donation ones), are thus the largest group, but they collect proportionally less funding.

Looking at campaigns, 23% of them are social, 22% cultural, 11% are linked to scientific and medical research, and just 22% are product campaigns. Civic campaigns in 2016 were 18% of the total, according to platforms themselves. This figure includes all the campaigns concerning common goods, while the focus of our chapter is more limited: the ones that Lee, Zhao & Hassna define as "public crowdfunding:" "one in which the final product or service is going to be utilized by the public and where the initiator or the owner of the crowdfunding campaign or capital-seeking request is a public or governmental agency."[4]

In Italy, they are as a rule quite recent projects, but they are fast growing. The first campaign of this kind, *Un Passo per San Luca* (One Step for San Luca) took place in 2014. In this campaign, the Bologna Municipality acted as a "promoter," which is the category used by Davies to identify cases in which "the organization acts as the main fundraiser, taking responsibility for receiving and spending the money, using an existing platform [...] The benefit of the promoter approach is that it gives the organization complete control over the messaging associated with the campaign. The drawbacks are that the organization also bears all of the risks of failure, and organizations with large existing funding sources will need to justify carefully why they're asking for additional money."[5]

In the same year, the Milan Municipality launched the first initiative identifiable as one promoted by a "curator" in Italy "when the organisation draws attention to crowdfunding projects that support its values or benefit its community, most likely by hosting a page on an existing crowdfunding website. [...] As a curator, the organisation can leverage the best grass-roots ideas in the communities they serve. It is also presumed to have exercised some degree of due diligence on the campaigns it promotes, and so is likely to be held partly responsible if campaigns fail to reach their funding targets or to fulfil their projects."[6]

This initiative, which will be discussed more in-depth in the next paragraph, has been followed by other ones, which can be distinguished on the basis of three criteria as follows:

- *ex ante* or *ex post* match-funding;
- targeted to just one campaign or to complex projects, involving different actors and campaigns;
- one-off campaigns or projects lasting over time.

The most relevant initiative is the one promoted in 2018 by ANCI, the association of Italian Municipalities. It promoted a call for match-funding projects able to reach at least 50% of their goal through crowdfunding, with a maximum threshold of €150 for repairing damaged infrastructures and €40 for durable goods. Out of the 21 projects uploaded on the platform, 18 reached their goal for an overall value of more than €1.3 million. From a first analysis of the underlying data, two issues emerge, which we will also find in the Milan case: the fact that all the funded campaigns reach exactly their goal, without exceeding it, and the small number of funders (even the most successful campaigns never reach 100 donators, the average donation is €1,454).

Region Piemonte launched a similar project, aimed at supporting initiatives in the field of arts and culture, including the renewal and preservation of buildings. Three projects have been published and of the two already closed, one was not successful, while the other collected €25.130 for the renewal of the foyer of the Faraggiana theatre.

The public–private partnership among Region Umbria, the Ministry of Agriculture, and Forestry and the Perugina firm promoted the campaign "Castelluccio Renaissance," for the renewal of a village damaged badly by an earthquake; it collected almost €150, thanks to 556 donations. It is interesting to observe that even very small municipalities promoted complex and relevant projects. Soliera, a province with 15.000 inhabitants near Modena (Emilia Romagna Region), since 2016 promotes a public participation initiative that foresees the co-funding of four projects every year (from 60% to 80% of the total, depending on the total sum), one for each neighbourhood, with a total investment of €80.000. Projects are pre-selected by neighbourhood councils, then voted by citizens, and finally published on a crowdfunding platform, selected through a public call; only those who reach the funding goal are then co-funded by the municipality. In this way, projects have been funded for the equipment of a park with sports gear, the digitalisation of a school, the implementation of a river park, street art, etc.

The Nonantola Municipality, also near Modena, with around 15.000 inhabitants, launched a one-off campaign, in the framework of a participation process for the renovation of a community room, for a total amount of €12.700, 10.000 paid by the municipality. In the same way, the much larger Taranto Municipality (200.000 inhabitants, Puglia Region) launched a campaign for the implementation of a Mini FabLab, which collected €15.000.

On the contrary, the *Falling Star* campaign, aimed at implementing a lit path within the Mincio Park (Lombardia Region), did not reach its objective; in case of success, it would have obtained matching funding from the Co-Mantova sharing economy project. The objective was not reached in the case of the campaign promoted by the Lodi Municipality as well (Lombardia Region), for the renovation of a monument to the Italian Resistance during World War II. In this case, the municipality funded renovation works anyway. It is interesting to highlight that in the campaign page there is an interview with the Mayor, who when asked about the motivation for citizens to fund a project which would have been implemented anyway, answered: "In order to favour the development of an increasingly active citizenship towards the common good, in order to give to all the possibility to be part of the renovation of a monument which joins artistic and social value."

In general, in Italy most cases can be defined as involving a "facilitator's role," in which "the organisation supports its members, subsidiaries or partners who want to use crowdfunding to fund their projects by providing training and expertise. The benefit of this approach is that it has the potential to scale the use of crowdfunding and enable a larger pool of individuals to use it, and reduces the responsibility carried by the organisation."[7] This is the case for ASTER (Consortium for Innovation and Technology Transfer of Emilia-Romagna) which set up in 2015 "a dedicated crowdfunding guidance service that ensures independent guidance and training to the regional start-up ecosystem, as well as general education and awareness raising activities about crowdfunding to citizens. In this specific case, projects ideas that meet a set of requirements can benefit from the KICK-ER support service without incurring in any additional cost."[8]

On the contrary, there are no cases yet in which local governments adopted a "platform" model, one that "creates its own crowdfunding platform and hosts campaigns entirely on its own site;"[9] probably this is due to the fact that it is too costly compared to the advantages of using an already existing platform.

2. Civic Crowdfunding in Milan

In line with experimentation carried out in many urban contexts, in the last few years, the Municipality of Milan started projects on the frontiers of innovation: active citizenship and cooperation, local urban creativity, networking, multicultural city, etc.

Among other policy lines, with the "Milano Smart City" project, launched in April 2013, the municipality aimed at "building and sharing with all the relevant stakeholders the Smart Milan Strategy and to encourage the creation of a governance model for the smart city."[10]

Among the initiatives within this project, in December 2014 the deputy mayors for economic development and social and health policies approved the "Guidelines for the testing of crowdfunding actions" with the following objectives: to test innovative funding models for public interest projects; to favour the mix of public and private funding to support high social impact projects; to contribute to the diffusion of new fundraising models for the third sector; to support alternative modes of access to credit and, indirectly, to offer a level to economic development; and to promote active citizen participation in local government choices.

The municipality earmarked €400.000 for an 18-month test. After choosing *Eppela* as the support platform through a public tender, at the beginning of 2016 the municipality launched a call for projects in the following five areas: accessible city, with special attention to the most fragile segments of the population; technological innovation to support urban connectivity; innovative information systems for mobility, culture, and quality of life; innovation in care services and life–work balance; and valorisation of territorial resources in a sharing perspective. Overall, 56 projects have been submitted, out of which 21 possible civic crowd-funding projects were selected. In the end, only 18 effectively launched their campaign.

As in other cases discussed in the previous paragraphs, the campaigns are based on a match-funding mechanism: once each project collects pledges for half of their target, then the municipality would contribute the other half (up to a total of €50.000 for each project). From April to December 2016, the projects have been presented in our subsequent rounds in which some projects in turn remained open and visible on Eppela for 50 days; 16 out of 18 reached their funding goal.

Looking at the actual projects, it clearly shows that in many of them there is a shift from spaces to physical and digital places, quite dense from a relational point of view. The projects range from the promotion of local welfare networks for disadvantaged social groups, to local food systems and community gardens, to the reuse of abandoned buildings, to innova-tive forms of neighbourhood welfare provision, with funding goals rang-ing from €5.000 to €100.000. They show some interesting elements, from different points of view. Firstly, promoters belong almost exclusively to the third sector, with previous experience in the relationship with the municipality. Even so, only two projects proposed a target near the maxi-mum funding threshold set by the municipality: as an average, the others are quite smaller. This shows a pattern very different from the usual ones in public tenders, and this in turn may show how crowdfunding projects may be driven more by project needs than by funding opportunities. On the contrary, since this was a pilot experience, it is possible that promoters decided to test their ideas on rather small scales. The second element concerns match-funding. Overall, €328.000 have been collected through crowdfunding, and excluding the two non-funded projects, the others reached exactly 50% of their goal, necessary to obtain matching funding: this on the one hand highlights the likely difficulties in reaching the target, and on the other maximises the behaviour focused on receiving co-funding, rather than on projects themselves.

As far as supporters are concerned, more than 1.600 people have been involved. The average value per donation (more than €200, in some campaigns more than 500) is explained by the presence of 'big donors.' We will come back to the point, but it is interesting to highlight here the high level of donations collected through the "open field" in the platform (79% as an average); donations in this field do not give rewards, and the largest donors typically choose those.

3. Reviewing the Experience

After a description of the process and mechanisms which characterised the civic crowdfunding support path implemented by the Milan Municipality, it is possible to propose some cross-cutting reflections, impinging on a detailed analysis of each project and some in-depth interviews to project promoters.[11]

3.1. Promoters, Between New Ventures and Innovation Strategies in Traditional Organisations

An interesting first aspect is connected to the reasons why different actors decided to promote a crowdfunding campaign. In general, from the interviews, it emerged that—in the Milan case—crowdfunding has not been an occasion to identify new "purpose-driven urban entrepreneurs."[12] On the contrary, this tool has been an agent of transformation of traditional modes of interaction between the third sector and local governments in the promotion of civic project proposals. At the same time, a strategy such as crowdfunding, based on the collection of small or very small sums on the part of a high number of donors, is more difficult to implement for more structured third-sector organisations because they are accustomed to different funding modes.

Another interesting element concerns the scale of intervention: in Milan, almost all the funded campaigns are project-based initiatives, activated at the micro-scale (neighbourhood), rather than ventures operating at a meso-(city) level or platforms operating at urban global level. One important characteristic of civic crowdfunding is its "place-based" nature[13]: it is often illustrated through large-scale projects such as bridges or buildings; however, many projects use civic crowdfunding on a smaller scale, to the point that new definitions are gaining ground, such as those of "community-focused crowdfunding," or "hyper-local crowdfunding."[14]

3.2. Funders: Crowd or Network?

One of the central elements of crowdfunding—together with funding—is the crowd. The specific asset of crowdfunding is that it can act as a lever to modify the forms of such networks, by making them denser, increasing the ties among existing nodes or by extending them, through the creation of new ties including new nodes.

The first polarity—the strengthening of existing ties—tends to concern not particularly successful cases: those in which crowdfunding has just been a different option to channel funding that would have been collected anyway. On this very point, we can highlight the case of a campaign that, after engaging its existing network with unsatisfactory results, a few days before the closing of the campaign promoted a sort of "compulsory self-taxation" among project partners. As we have seen, the success rate of civic crowdfunding campaigns is higher than the average crowdfunding campaigns in Italy. It is clear, though, that match-funding offered by the municipality has been a strong incentive, which may have created some distortion.

In general, we found that campaigns aiming at extending existing networks are more complex to design and manage, but they seem to produce more satisfactory results. One of the campaigns we analysed has been promoted within a much larger strategy of organisational innovation, in which the association involved aimed to prove itself in areas different from the usual ones, to show a 'different face' and to open up to younger generations. This campaign connected two different communities: the potential users of the project, who are the residents of the area, and the members of the promoting association. The campaign even allowed the promoters— even if in a lesser measure—to get in contact with people external to either of these communities. The project promoters directly contacted all the 'unknown' donors in order to understand the reasons behind their donations and thus new exchanges and collaborations emerged.

The interviewees all agree in highlighting another aspect: the organisational and design commitment implied by crowdfunding is justified if the objective is not exclusively to raise funds, but if it involves other resources also. In more than one campaign, together with those funding the project, there have been others who contributed, e.g. through the organisation of events. Their voluntary contribution, in many cases, has had the same, if not a higher value, than that of donations, also given the specific features of each project. Another interesting contribution is the design and strategic one: in one of the cases, in parallel with the crowdfunding campaign, the promoters have been running an open consultation on participatory decision-making platform Oxway about ideas for possible functions for the target area. In this case, the project has been able to use online and offline tools and to successfully involve both already existing networks and people who did not know the project nor the promoters.

The importance of non-monetary contributions is confirmed by the fact that in one of the projects, when there has been the listing of contributors' names in the very space created thanks to the campaign, also the names of those who contributed resources other than funding (time, labour) have been listed. The valorisation of non-monetary contributions is a central issue for civic crowdfunding, in particular in order to include disadvantaged segments of the urban population; at the moment, though, it is not specifically supported by existing platforms.

In general, it is possible to single out crowdfunding from other peer-to-peer platforms because it pays scarce attention to the acknowledgement and valorisation of individual contribution, e.g. in terms of reputation. Almost all platforms enabling peer-to-peer transactions make an important investment in certifying users' identity and in building up profiles which enable people to trust strangers. This does not happen in crowdfunding: the promoter's reputation is largely external to the platform—also because the same promoters seldom promote recurring campaigns—and the funder's one is not considered particularly relevant. It is thus interesting to follow the ongoing—and up to now exclusively theoretical—debate on the potentialities and risks of "civic reputation" systems, for instance, through digital badge systems.[15]

3.3. The Role of the Public Administration

If the already mentioned Davies' typology helps us to distinguish the role of local authorities as promoters or supporters of a campaign, an interesting element emerging from the Milan experience is that, during the campaign itself, the public administration always plays a wider role. This may imply some negative effects: such as in the case of a project for the funding of a community vegetable garden, selected by the municipality for a crowdfunding campaign, but which couldn't obtain from a different office in the same administration the authorisation to connect to the water network. On the contrary, the hybridisation between different functions and sectors of the same administrative actor can create new value.

Promoters in some cases manifest the expectation to be more actively involved on the part of the crowdfunding platform and of the municipality itself in project communication. A request shows how the administration is considered as a "project partner" rather than a funding provider. The potential room for intervention of the administration is quite large, also because it could play an innovative 'bridging' role towards new publics and communities.

Much in the same way, promoters urge the administration to play a more active role in the valorisation of horizontal collaboration among different campaigns. In this pilot experience, forms of networking and collaboration emerged spontaneously, from reciprocal funding to the start-up of integrated projects. The interviewees acknowledged the value of such collaboration and wished an intentional intervention could be implemented for the build-up of such relationships, also looking to knowledge transfer between those who already implemented their campaigns and those who are designing them.

3.4. From Online to Offline and the Other Way Round

Crowdfunding emerges as a digital platform, but it does not end online. This is due, in a first instance, to a lack of expertise on the part of the third sector in the management of digital communication. The review of project pages and social media channels done by Stiver *et al.*[16] suggests social media use is less active, or more hidden, for civic crowdfunding projects. There are three causes identified by Stiver *et al.* as follows:

(1) The financial request is already tied up in local partnerships, creating a lesser need to use social media for wider reaching networks.
(2) Interactions between stakeholders take place offline through events such as town hall meetings, thus, minimising the need for online supplement.
(3) Civic crowdfunders are less accustomed to online engagement.

To these three reasons, we can add, in the Italian case, the fact that third-sector organisations are lagging behind as far as the promotion of social media-based communication strategies is concerned. It is a structural gap of the Italian third sector—more used to act than to communicate its action. On the contrary, crowdfunding is based on communication, both of the project and of the results, impinging also on innovative tools. Together with the skill updating, interviewees raised the question of the recognition of the legitimacy of the activities implemented in connection with crowdfunding, because they are so different from the traditional ones that in some cases the personnel of the associations had to work as volunteers, outside their working times, in order to bring the campaign to a conclusion.

Finally, all the interviewees, being locally rooted actors, accustomed to direct social relations, had difficulties in interacting with the relatively rigid and standardised mechanisms which characterise crowdfunding platforms. The usual construction modes of a fundraising campaign seem to be in general scarcely flexible and, thus, unable to adapt to the specific features of promoters. The exclusive use of the platform as a payment mode has been a problem for many promoters, because their local networks and communities are formed by elderly people, who do not own a credit card, neither are accustomed to online payments. In this sense, a relevant digital divide emerges, which is particularly relevant for those projects explicitly aiming to social inclusion and to work in disadvantaged neighbourhoods.

The consequence is that "civic communions"—defined by Procter[17] "the rhetorical processes and cultural performances that function to build community"—are almost exclusively built through face-to-face relationships. If the transition from online to offline community is a distinctive feature of civic crowdfunding, in the Milan projects offline activities played an important role in sustaining communities online,[18] rather than the other way round. This is a paradoxical result if we consider that at the launch of the campaigns offline collection of funding was forbidden. This restriction has been lifted in the course of the process, when it clearly emerged, for the reasons we discussed before, that exclusive online collection would have hindered the success of the funding goals.

3.5. Rewards: Limits or Opportunities?

Another example of the rigidity of the infrastructure concerns rewards. In the preceding paragraph we already noted how a large part of the collection has been implemented by circumventing the platform's socio-technical system, through the use of the open field. Interviewees confirm that, in general, donors were not interested in rewards and considered the superfluous. Many who made a donation with a reward, never claimed it. We can certainly discuss the quality of the proposed rewards, but it is an interesting element to be taken into account, also given the effort in organising them. Moreover, this lack of interest characterised both campaigns involving already known donors and campaigns aimed at enlarging support networks. In both cases, we can certainly say that donors had other types of motivation, in some cases a very strong value alignment, in others probably connected to the projects' potentialities: the funding model has been in some cases closer to the donation model, and in others, it was more similar to a small investment. Anyway, as underlined also by Stiver *et al.* "as civic crowdfunding outputs are for public, and often repeated, use (e.g. green space, education programs, technology access), there is less potential for a sole point of exchange (money for reward) and more encouragement towards longer term involvement in the project."

4. Conclusions

Moving from the critical elements emerging from our field research, it is possible to summarise the main policy implications of the Milanese case.

The first issue revolves around if, and to what extent, civic crowd-funding can be considered as a participatory practice. In general, it seems that civic crowdfunding, even when it is strongly supported by local governments, does not entail a significant redesign of local governance arrangements or of the interface between public actors and civic ones, and Milan has been no exception. This is partly due to the small dimension of the first experiments, still at a niche level, but it is probably also intimately connected to the nature of this tool.

The second one looks at the actual degree of openness and at the inclusionary/exclusionary potential of such practices. A critique frequently raised to civic crowdfunding campaigns is that they tend to produce benefits only for a tiny minority of the urban population. Local governments have a chance to rebalance the situation, addressing problems arising in more fragile areas within the urban contexts. While already addressing some of them, the Milan Municipality has expressed a strong intention to more specifically target such fragile neighbourhoods through future crowdfunding campaigns. Another angle useful to examine the inclusion/exclusion issue investigates if actors taking part in publicly supported civic crowdfunding projects are already known to local authorities. The Milan experience offers mixed results from this point of view: while in a minority of cases the promoters of projects were establishing for the first time a structured relationship with local government, in many other cases, the promoters were well established and structured third-sector organisations. In general, municipal offices have noticed that when launching projects directly aimed at final beneficiaries, they are more likely to get in contact with new actors, than when working on projects involving intermediation actors.

The third aspect proposes to reflect on the risk that the diffusion of civic crowdfunding entails a real risk of disinvestment and retreat of the public actor from the supply of critical services. What emerges from the Milan case, as from many others, is that, on the contrary, the level of engagement of the municipality has been higher, rather than lower, in order to effectively select, support, and match-fund civic crowdfunding campaigns. Local governments, as it appears also in the London case,[19] are typically extremely engaged and need to deploy specific competencies and skills. When local governments play the role of sponsors, but even when they act as curators, the project pre-selection mechanisms and the subsequent match-funding do in fact ensure a strong presence of the municipality both in the strategic design phase and in the feasibility phase, and this can in turn help the strengthening of public–private partnership not only with large third-sector organisations but also with smaller and more diffused societal actors.

About the Authors

Carolina Pacchi is an Associate Professor at Politecnico di Milano, where she teaches Planning Theory and Practice and Local Conflict Resolution. She is involved in research on the transformation of urban governance in European cities, forms of alternative politics and grassroots activism at local level and policy strategies based on stakeholder involvement. She has been involved in a number of EU funded research projects on governance in urban, environmental and local development policies and she is currently engaged in research on urban social segregation. She has been visiting researcher at TU Berlin (2016). She publishes at national and international level on her research topics, some recent publications include "Coworking Spaces in Milan: Location Patterns and Urban Effects," *Journal of Urban Technology* (with S. Di Vita and I. Mariotti, 2017); "Crowdfunding civico tra reti, comunità e ruolo del governo locale" In: R. Lodigiani (ed.) Milano 2017. *Una metropoli per innovare, crescere, sognare, Angeli, Milano* (with Ivana Pais, 2017).

Ivana Pais is an Associate Professor in Economic Sociology at the Faculty of Economics, Università Cattolica del Sacro Cuore in Milan, Italy. Her research interest focuses on the role of social capital and social networks in the digital economy. She has been mapping the Italian crowdfunding platforms since 2015 (http://www.crowdfundingreport.it/). Her publications about crowdfunding include: Bonini, T. and Pais, I. (2017). Hacking public service media funding: A scenario for rethinking the license fee as a form of civic crowdfunding. *International Journal on Media Management*, 19(2), 123–143; Balboni, B., Kocollari, U. and Pais, I. (2016). Crowdfunding for social enterprises: An exploratory analysis of the Italian context. In: Meric, J. Brabet, J. and Maque I. (eds.), International

Perspectives on Crowdfunding: Positive, Normative and Critical Theory (pp. 65–79). Emerald Group Publishing Limited; Guerzoni, M., Peirone, D., Pais, I. and Miglietta, A. (2016). The Emerging Crowdfunding Market in Italy: Are "the Crowd" Friends of Mine? In: Brüntje D. and Gajda O. (eds.), Crowdfunding in Europe (pp. 87–96). Springer, Cham; Pais, I. and Castrataro, D. (2014). Crowdfunding and free labor: Gift, exploitation or investment? *Sociologia del lavoro*, 133, 183–195.

Endnotes

1. TRAILab and Collaboriamo, *La mappatura delle piattaforme italiane di sharing economy e crowdfunding 2016*, http://sharitaly.com/la-sharing-economy-in-italia_la-mappatura-delle-piattaforme-collaborative-e-di-crowdfunding/
2. De Crescenzo, V. and Pichler, F. "December. Platforms business model in the Italian civic crowdfunding market: An empirical analysis." In: *Toulon-Verona Conference "Excellence in Services"* (2017).
3. Starteed, *Il crowdfunding in Italia: tutti i numeri e le piattaforme*, https://blog.starteed.com/il-crowdfunding-in-italia-tutti-i-numeri-e-le-piattaforme-aggiornato-a-gennaio-2017-e12535b00542#.i41sms846
4. Lee, C.H., Zhao, J.L. and Hassna, G. "Government-Incentivized Crowdfunding for One-Belt, One-Road Enterprises: Design and Research Issues." *Financial Innovation*, 2(1) (2016), 4.
5. Davies R. *Civic Crowdfunding: Participatory Communities, Entrepreneurs and the Political Economy of Place*, 2014, https://papers.ssrn.com/sol3/papers.cfm?abstract_id=2434615, p. 140.
6. *Ibid.*
7. *Ibid.*
8. ECN (European Crowdfunding Network) (2018), *Triggering Participation: A Collection of Civic Crowdfunding and Match-funding Experiences in the EU*, http://eurocrowd.org/2018/07/04/cf4esif-report-triggering-participation-collection-civic-crowdfunding-match-funding-experiences-eu-published/
9. Davies R. *Civic Crowdfunding: Participatory Communities, Entrepreneurs and the Political Economy of Place*, 2014, https://papers.ssrn.com/sol3/papers.cfm?abstract_id=2434615
10. Comune di Milano, *Linee guida Milano Smart City*, Milano, 2014, http://www.milanosmartcity.org/

11. We selected the interviewees on the basis of an internal heterogeneity principle and—clearly so—of their availability, after a first contact mediated by the Milan Municipality.

12. Cohen, B. and Munoz, P.A. "Toward a Theory of Purpose-Driven Urban Entrepreneurship." *Organization and Environment*, 28(3) (2015), 264–285, ISSN 1086-0266.

13. Charbit, C. and Desmoulins, G. Civic Crowdfunding: A collective option for local public goods? OECD Regional Development Working Papers, 2017/02, OECD Publishing, Paris, 2017, http://dx.doi.org/10.1787/b3f7a1c5-en

14. Stiver, A., Barroca, L., Minocha, S., Richards, M. and Roberts, D. "Civic Crowdfunding Research: Challenges, Opportunities, and Future Agenda." *New Media & Society*, 17(2) (2015), 249–271.

15. Bani, M. and De Paoli, S. "Ideas for a New Civic Reputation System for the Rising of Digital Civics: Digital Badges and their Role in Democratic Process." In: *ECEG2013-13th European Conference on eGovernment: ECEG 2013,* 45, 2013. Picci, L. Reputation-Based Governance and Making States "Legible" to Their Citizens. *The Reputation Society: How Online Opinions are Reshaping the Offline World*, 2012, p. 141.

16. Stiver, A., Barroca, L., Minocha, S., Richards, M. and Roberts, D. Civic Crowdfunding Research: Challenges, Opportunities, and Future Agenda." *New Media & Society*, 17(2) (2015), 249–271.

17. Procter, D. E. *Civic Communion: The Rhetoric of Community Building,* Rowman and Littlefield, Latham, MD, 2005.

18. Lin, H.F. "The Role of Online and Offline Features in Sustaining Virtual Communities: An Empirical Study." *Internet Research*, 17(2) (2007), 119–138.

19. Gullino S., Seetzen H., Cerulli C. and Pacchi C. "Citizen-Led Micro-Regeneration: Case Studies of Civic Crowdfunding in London and Milano." In: Fisker, J.K., Chiappini, L., Pugalis, L. and Bruzzese, A. (eds.), *The Production of Alternative Urban Spaces. An International Dialogue*, Routledge, 2019.

Part 3.3

State and Government Level

Chapter 12

Crowdfund Angus: The Impact of Public-Level Crowdfunding

Shelley Hague

Abstract

The demand for our service at Angus Council to provide support and assistance to access potential funding streams for individuals, community groups, and small businesses was becoming difficult to manage. Angus as an area tends not to regularly qualify for large amounts of funding from external sources due to the stringent criteria set by the European Commission, Scottish Government, or other external funders. Our team became frustrated at not being able to support local people and business, so we decided to create a bespoke crowdfunding platform that allowed businesses, individuals, and community groups to campaign for funding to achieve their goals. This was launched in August 2015 and is called Crowdfund Angus.

Our local authority was the first in Scotland to use crowdfunding, and it encourages more creative ideas from communities and businesses. Developing Crowdfund Angus has improved relationships between the council, local agencies, business, and individuals and has also introduced an important new direct contact between established businesses and economic development. This is vitally important to ensure our services are supporting the needs of the business community.

To date, we have ran over 30 campaigns, presented to over 4,000 people, and conducted a sold-out crowdfunding workshop in Angus House, in June 2017 in partnership with Crowdfunder UK. We have levered in over £760,000 in funding for projects and have directly benefited over 240 people in 2017/2018 alone.

After attending an event on crowdfunding organised by NESTA, the National Endowment for Science, Technology and Arts in the UK, there was a concept that a local authority could facilitate a portal that could host local projects, promote them, and work with communities to build their capacity. This was developed to empower communities and get the funding they needed while future-proofing the skills locally. Budgets and funding options for local councils have been cut significantly following the economic crisis of 2008. Yet, many of the crowdfunding portals operated in a very commercial manner and didn't offer space for a more holistic approach that would fit the philosophy of a public service. However, following a tweet from Crowdfunder, everything changed for Angus.

Following the initial contact with Crowdfunder, the pace of development intensified, and within a month, we had set up a steering group with representatives from the public, private, and voluntary sectors to gauge views and develop the portal. It was important that people had bought into this concept as it was very new in Angus. The vision was to create a bespoke crowdfunding platform that allowed businesses, individuals, and community groups to campaign for funding to achieve their goals.

Development sessions were held where key elements were discussed including the design of the portal, how it felt as a user, and what needed to be done in communities to ensure that they knew what it was and how to develop a campaign. This was the area that we focused on the most to ensure that the portal would be a success. The group created a bespoke crowdfunding platform that allowed businesses, individuals, and community groups to campaign for funding to achieve their goals. Crowdfunder and Angus Council provided workshops to the community to ensure the product being offered would meet the particular needs of their community. Following initial development, information sessions were held.

The feedback from these events was overwhelmingly positive with individuals, businesses, and third-sector groups eager to become involved. Crowdfunder also provided comprehensive training to the council's funding, policy, and projects team to ensure that support would be available at the grass-root level. The portal was launched in August 2015 and is called Crowdfund Angus.

Our portal fits with the Angus Community Planning vision—Angus is a great place to live, work, and visit. Under the Scottish Governments Economic Strategy, our portal supports the four I's, which are as follows:

- Investment—encouraging more funding into Angus and Scotland as a whole.
- Innovation—our local authority was the first in Scotland to use crowdfunding and it encourages more creative ideas from communities and businesses.
- Internationalisation—funding is sourced globally into Angus and statistics show that a significant amount is coming from overseas. Our portal also supports businesses to trade internationally through developing global supporters and increasing their presence overseas.
- Inclusive growth—through providing an outlet for projects, they have a greater chance of achieving their plans and growing.

In the lead up to the launch, the staff team within Angus Council worked with four projects that had come forward to be the first pilots. These included The Memory Box, which was a dementia project looking to fund dementia gardens; Following Dreams, which was a wheelchair racer with a dream of becoming a Paralympian; Fireworks for Arbroath, which was a campaign to get funding to deliver a local fireworks event; and the Bon Scott Statue, which to date is our most successful campaign, raising over £50,000 in 4 weeks.

The Bon Scott campaign is a great example of how to run a campaign. In 2015, the initiators of Bon Scott Statue, led by community group DD8 Music, decided to crowdfund for a statue of AC/DC legend Bon Scott for their hometown of Kirriemuir, Angus. People were quick to get behind the idea. Kirriemuir already has a few things honouring Bon Scott in the town, including Bon Scott Place, and an annual music festival, Bonfest, but looked to add the statue to their repertoire. After raising over £50,000 from 117 backers, the statue was given the go-ahead by local councils in January 2016. The rewards that they had on offer were a critical factor and ranged in value from £10 to £2,500. The project gained mass exposure featuring in newspapers like *The Guardian* and *Kirriemuir Herald*, which both celebrated the success of the crowdfunding campaign. More recently, the story has been featured on the *BBC* and in *The Scotsman*. The statue was unveiled on Saturday 30th April, as part of Bonfest 2016. This marked the 10th anniversary of the rock music festival in honour of AC/DC's Bon Scott.

Angus Council was the first local authority in Scotland to use crowd-funding and it encourages more creative ideas from communities and businesses than any other funding scheme we have delivered to date. Developing Crowdfund Angus has improved relationships between the council, local agencies, businesses, and individuals and has also intro-duced an important new direct contact between established businesses and economic development. This is vitally important to ensure our services are supporting the needs of the business community.

Crowdfunding for business is an area of development in Angus as the benefits are considerable. Angus is a rural authority area, which consists mainly of micro-businesses (89.4%) and small (9.2%) businesses. In these times of diminishing resources, one of the challenges we face is to provide support and opportunities for growth to the business community, while also encouraging further enterprise from new and emerging talent. Developing Crowdfund Angus has improved relationships between the council, local agencies, business, and individuals and has also introduced an important new direct contact between established businesses and eco-nomic development. This is vitally important to ensure our services are supporting the needs of the business community.

Within Angus Council, the economic development unit has the exper-tise of business advisers, employability skills advisers, and marketing sector-specific officers to advise and support the local economy. Employers can often feel bombarded by agencies all looking for their time and attention. An unexpected consequence of our activities with the portal is to have employers approach us; either to find out more about our other services or to discuss sponsoring projects that are on the portal.

When considering success factors, it would definitely start with the steering group who were very enthusiastic and positive about the development of Crowdfund Angus. They acted as ambassadors for the project and helped gain approval through the council's committee to develop the portal. Another key element is supporting our communities, since the launch in 2015 the awareness of Crowdfunding in Angus has grown significantly and those campaign holders now support new projects acting as a peer mentoring support.

Following the initial year, Angus Council agreed to move the Community Grant Scheme onto the portal to act as a match-fund. This was following a need for more community-led projects and demand-led events that the community was fully committed to. It was also an aspiration that moving from a traditional funding scheme which involved a lot of staff time to a more digital inclusive programme would be more efficient. It was agreed by the committee that this would be the way forward for community grants in Angus in 2016 and has been delivered in this way to date.

Embedding the grant scheme on to the crowdfunding portal was really straightforward with similar criteria and guidance being used; however, now campaigns going live on the portal could apply to be match-funded up to £1,200 by the Angus Council Grant Scheme. Initially, there was a lot of development work required, but once in operation, officer time could be dedicated to supporting projects and campaigns rather than administration of the scheme and writing reports. Another key aim for us was to deliver better value for money, and within the first year, for every £1 the Council put in towards a project, the community/business put in £6 (£1:£6 ratio). This was a great success for the council as not only were we empowering communities to fundraise but our grant scheme could also support more projects. Initial resistance came from groups that had become very reliant on the grant scheme and didn't want to engage with digital tools like crowdfunding or utilise social media; however, it would have been detrimental to the groups if we hadn't moved in this direction as now other funders are following suit.

Successful campaigns on Crowdfund Angus have included a wide range of rewards that appeal to everyone with a plan in place to ensure that throughout the campaign, messages are going out through social media and by any means. The key to a successful campaign is realising that the real work is done before the campaign launches. You've got to know your audience and think about rewards that are going to appeal to them. In the Bon Scott Statue example, they were very fortunate, in that they had a worldwide audience (AC/DC fans) that would support the campaign. Every campaign, irrespective of being local, national, or international, has a crowd out there somewhere!

A challenge for the project was managing the expectations of the groups embarking on a crowdfunding campaign and ensuring that they had all the right tools in order to achieve success. There have been some campaigns that haven't engaged with the Council service and haven't raised any funding. This is an ongoing challenge for us—keeping people engaged and interested in Crowdfunding; however, with the new development through European funding to develop Crowdfund Scotland, there is now a greater awareness of the work that is ongoing.

Over the past 3 years, there have been some interesting unintended consequences from Crowdfund Angus. The biggest success has been up-skilling the communities. This has been essential as more and more funders are now using crowdfunding to distribute their funding. Through our information, development, and awareness-raising sessions, our communities are now able to use more digital tools and social media to raise the funding they need for projects through crowdfunding. Another key message for us when working with potential campaigners is that this method of fundraising is more than just money. The additional benefits groups have had from having a campaign have helped them develop and grow. This has included increased levels of volunteering, new skills for the board and members, offers of free supplies, and offers to promote the local cause.

Throughout the process of developing a portal, Angus Council have always been keen to share the good practice and support the crowdfunding philosophy. This has led to presentations and meetings along the length and breadth of the UK. The next incarnation of the project is also in place to start developing a Crowdfund Scotland portal, which will focus on raising awareness, building capacity, and exploring options for the future. This is being developed through a cooperation project funded by the EU rural development programme.

In terms of success, to date the portal has hosted over 60 campaigns, the team has presented to over 4,000 people and has levered in over £760,000 in funding for projects and have directly benefited over 240 people in 2017/2018 alone. The influx of funding to the Angus economy has encouraged growth, employability, tourism, business, entrepreneurship, community capacity, and a new way of thinking towards funding. The Angus community continues to embrace the crowdfunding model with the figures showing the reach, multiplier effect, and additional leverage it can bring, and the Angus Council will continue to build on this.

About the Author

Shelley Hague is the Strategic Policy and Planning Manager within Angus Councils Strategic Policy, Transformation and Public Sector Reform Service. Due to the nature of Strategic Policy and Planning every day is different but Shelley spends a lot of her time supporting groups and businesses to find the funding they require to take forward their idea/ project. The Crowdfunding portal has been a massive piece of work for her and the team as an innovative funding solution for communities, individuals and businesses and she hopes it will change the way we look at funding.

Chapter 13

Initiative Comes at a Cost: Russian Experience of Crowdfunding for Policymakers

Evgeny P. Torkanovskiy

Abstract

Crowdfunding has gained ground in Russia as the means to facilitate the resolution of the most urgent problems at the local and regional levels as well as the means of communication between the authorities and civil activists. This chapter analyses different forms of civic crowdfunding and entrepreneurial crowdfunding for civic purposes in different regions of Russia. The research confirms the readiness of the local governments to contribute a share of necessary resources from public funds and organise online portals and voting systems and perform an expert study of the suggested projects as crowdfunding allows to finance projects that otherwise have deficits. The original feature of Russian civic crowdfunding turns out to be the obligatory share of expenses to be carried by the project's initiators along with budgetary grants. This suggests that civic crowdfunding can be recommended as the means to make the civil activists understand and act with responsibility as well as define the most urgent local problems and collect lacking funds for their resolution from the broad group of funders.

The idea of collective investing in projects through crowdfunding has found many fans in Russia and elsewhere in the world. Russian policymakers, especially at the regional and municipal level, have embraced crowdsourcing and crowdfunding as the means of interaction and cooperation with the civil society as well as financial support for local projects and social development.

1. Overview of Crowdfunding in Russia

Due to economic downturn, Russian crowdfunding, unlike its international peers, underwent optimisation in which only two competing national players survived—Planeta and Boomstarter. Despite subsiding optimism, crowdfunding has become a significant part of the creative class economic activity with an average investor's ticket ranging from RUR 1,000 to 1,500 (US\$15–25 equivalent) and the number of active investors reaching 0.03% of the total population.[1] On the pessimistic side, crowdfunding needs well-performing and trustworthy moderating platforms for success. However, practically all Russian crowdfunding platforms as businesses remain largely loss-making. It's true not only for pure-plays crowdfunding platforms as Planeta[2] and Boomstarter[3] that try to raise additional money through other activities and channels (selling tickets, organising events, issuing notes, etc.) but also for crowdinvesting and crowdlending start-ups like StarTrack[4] and Gorod Deneg (Town of Money).[5] Penenza crowdlending platform[6] (offers business loans for up to 60–90 days for companies participating in tenders) is a venture of an investment group and was opened to individual investors at the end of 2017 (only legal entities and clients of investment groups could invest before that), whereas Alfa Potok[7] crowdlends to small companies—customers of Russian AlfaBank—and is a fee-based service of the latter. The recent crowdinvesting newcomer Gosstart[8] is the subsidiary of Rusatom State Corporation. The major problem for all the platforms is the scope of business. The fee generated is insufficient to maintain platforms and thus platforms have to develop additional services or simply "burn" the money of founders or sponsors in the hope of attaining critical mass of investors and projects on the platform. Investees also express disappointment with the services of platforms as they expect to raise money from investors present on the platforms and in reality platforms offer just a legal way of raising money from investors already knowledgeable about investees or brought to the platforms by investees from other channels or through advertising activity outside the platform. In contrast, ICO market in Russia that represents a derivative of crowdfunding showed many new arrivals and successful ICOs, especially prior to the bitcoin's rapid descent[9]. ICO contains elements of various forms of attracting capital: IPO—as a result of investments in a project, an investor receives an asset (token) that is traded on public exchanges; crowdfunding—emission of tokens is similar to advanced sale and is associated with an active PR campaign; venture capital—the project is financed at an early stage.

By November 2017, the total amount raised through ICOs globally exceeded US\$ 3.7 billion and outpaced venture investment. The largest ICO amounts were raised in the USA, Russia, Singapore, and Hong Kong. Whereas globally, ICO market is tiny compared to other fundraising activities, in Russia ICO amounts to being comparable to the US or Hong Kong and occupies a prominent place in the local economy due to the small size of the Russian market.

The Russian state has also supported crowdfunding through budgetary investment in StarTrack crowdinvesting platform. The financing allowed the platform to cover its managerial costs during the first few years of existence. However, the platform has not yet reached break-even.

In 2016, the Russian state financial authority—the Central Bank of Russia—initiated the creation of regulatory sandboxes for new financial technologies. The regulatory sandbox or platform of the Russian Central Bank is a mechanism for steering new financial services and technologies that require changes in regulation. The platform is used to model the processes of introducing innovative financial services, products, and technologies and to test the hypotheses about the positive effects of their implementation. According to the results of steering, a financial service or technology may be approved with the subsequent formation of a plan to develop the necessary regulatory framework, or the initiative may be deemed inappropriate. Steering on the regulatory platform can be initiated by any organisation that has developed or plans to use an innovative financial service or technology. The initiator should just submit an application to the Russian Central Bank.

At the end of 2017, this regime has been applied to crowdfunding and crowdlending operations while the project of federal law on crowdfunding is under consideration in the national parliament and is expected to be signed into law in 2018. The regulation looks to provide guarantees for investors and reduce fraud possibilities. The draft, though bearing the name "On crowdfunding," seeks to regulate mainly crowdlending and crowdinvesting activities. The significant part of the draft deals with the requirements for the platforms and investors. The moderating platform should be incorporated in Russia, have capital of at least 5 million roubles, and be included in the Russian Central Bank's register of crowdfunding platforms. The activities of any platform not meeting these requirements are illegal. The Central Bank also has the right to set a ceiling for individual investments and total amount of investments by individual investors. However, many experts note that though the regulation is supported by major players, it may prove to be too onerous for a nascent industry and become an obstacle for the development of crowdfunding in Russia. Though it does not address traditional crowdfunding, the draft is insufficiently clear in this respect and may be applied to pure crowdfunding platforms thus making their functioning more onerous and scaring off investors. The draft also does not directly regulate municipal and regional initiatives of civic crowdfunding, and it may represent a problem for civic crowdfunding in the future. It is also useless with regard to ICOs as the majority of platforms are located abroad.

Crowdfunding has also been used for political fundraising similar to the US practice. Thus, opposition politician Alexey Navalny makes regular use of crowdfunding to finance his political campaign and initiatives, and many local politicians replicate these methods for their own campaigns. However, political crowdfunding in Russia is under close scrutiny not only from political opponents but also from the election officials and government authorities. The transparency of crowdfunding plays against many investors as they prefer to keep their profile low or choose anonymity.

We have found it useful for the purposes of our research to divide crowdfunding experience in Russia into three broad categories depending on the funds' final use, moderating organisation, and sources of financing.

The first category (we have chosen to call it civic crowdfunding) represents public–private partnership and/or participatory budgeting and implies the participation of both administration and civil society including socially responsible businesses for the benefit of local community. The funds are sourced partially from independent donors related to the local community and partially from municipal/regional budget. Unlike the following category, the moderating organisation or platform is not independent but is usually part of public administration (municipal or regional portal).

The second category is entrepreneurial crowdfunding—therein fall crowdfunding projects initiated by different economic agents who look for money to realize their own ideas. Among these initiators are individuals and organisations pursuing not just for-profit ends but also social goals important for the local communities. The funds are sourced from many independent donors who like the project's idea and choose to support the initiator. A good example of such crowdfunding is the Cocco Bello project.[10] Several crowdfunding campaigns allowed to develop a viable creamed honey-producing business in a small village of Maly Turysh in Urals (about 2,000 kilometres from Moscow) creating 18 jobs in a village of 50 inhabitants. The project allowed to reach both commercial and social sustainability purposes creating the only workplace in the faraway village and offering jobs to the otherwise ignored elderly population, "babushkas" thus achieving inclusive development goals in this particular community. Four successive crowdfunding campaigns allowed to raise more than RUR 4.5 million (US$80.000 equivalent) to acquire equipment and provide for the construction of the necessary infrastructure.

The third category is political as crowdfunding develops into an instrument for raising money to support political initiatives or certain candidates at the elections. The funds are sourced from independent donors sharing similar political views. The funds are raised through independent platforms or collected directly to a certain internet account.

Our focus will be the first category and the second category inasmuch as it relates to funding with the help of municipal authorities though we intend to give a broader picture to place public–private or civic crowdfunding in the context of national regulation, customs, and market.

Our research also includes examples of successful crowdfunding campaigns benefitting local communities without budgetary participation.

2. Civic Crowdfunding

Civic crowdfunding develops quickly along with participatory budgeting. For some time, participatory budgeting has been advanced by budget practitioners and academics as an important tool for inclusive and accountable governance. It has also been implemented in Russia with certain success.[11] Through participatory budgeting, citizens have the opportunity to gain first-hand knowledge of government operations, influence government policies, and hold governments accountable. In Russia, participatory budgeting (first initiated in Brazil in 1980s) started to develop in 2007 along with the World Bank project for Russian municipalities and local initiatives development. The main focus of the participatory budgeting, which is traditionally aimed at inclusivity and offers access to public finance for otherwise largely ignored strata of the population (poor, elderly, unemployed, and youth), has shifted in Russia. To reflect the change of focus, the Russian version of participatory budgeting got the name of "initiative budgeting." The particular characteristic of the Russian version of participatory budgeting is practically the obligatory share of projects' co-financed by the citizens. Another important feature of Russian participatory budgeting is the emphasis on the implementation of exclusively grass-roots people's initiatives aimed at addressing issues of local importance with no access to discussion of more general issues.

During the period 2007–2017, over 35,000 projects of initiative budgeting were implemented, and the number of beneficiaries of the programme exceeded 2 million people. More than 5,000 people were the most active members of the initiative groups, involved in all stages of the programme, and were popularizers and disseminators of the experience.

The projects for initiative budgeting and civic crowdfunding are selected from the lot submitted on relevant local portals through the iterative communication process involving active citizens, socially responsible businesses, experts, and local authorities.

The main idea that encourages crowdfunding to solve social problems at the municipal or regional level is the premise that creative people live on the relevant territory, who can generate ideas, solve problems, conduct research, contribute time and money, and coordinate their actions. Mostly, these people do not wait for rewards and derive moral satisfaction from their activities.

The main strategic objective of the municipal authorities to develop their cities is to raise the standard of living of the population, provide services to meet the needs of citizens, and ensure sustainable development of the local economy.[12] The following general model of interaction is widely used in Russian regions and municipalities.[13]

The population of municipalities—as a consumer of municipal services—is a powerful source of ideas in many areas. When an individual sees a problem, he/she instead of approaching to the local authorities, describes the problem and possible ways to solve it on a specially created site (for example, special part of the regional portal). The portal allows other citizens to evaluate the significance of the problem, its acuteness, as well as to invite experts to evaluate the suggested solution. Such interaction allows to select the best projects for crowdfunding as well as to ameliorate them in the process of discussion with other citizens and experts. As a result of the discussions, the best projects are selected that enjoy both the financial and moral support of the population. The selected projects receive financing from municipal and/or regional authorities with a part of the budget being earmarked for such civic crowdfunding projects. After the project is realized, an important issue is monitoring and control (assessing the speed, completeness, and effectiveness of solving a problem). A major role is played by the authorities who ensure the moderating role as well as provide organisational, technical, and informational support. It is also vital to get citizens to give the assessment of the way in which the problem is solved.

The distribution of projects implemented in Russia reflects the most pressing problems of population. The most supported and co-financed projects were road construction/repairs and water supply. Such projects together account for 51% of the total number of all implemented initiatives. They are followed with great gap by the renovation of cultural infrastructure (14%) and the improvement of the municipal territory and public services (10%). The projects aimed at arranging recreational facilities and children's playgrounds (7%) and sports infrastructure facilities (6%) represent the final best use of funds.[14]

With the help of crowdfunding, it is possible to collect funds to resolve problems and issues by uniting the resources of the locals. Often, problems such as arranging recreational areas, city parks, playgrounds, river and lake embankments, greening the streets, and arranging parking lots are not top priorities for municipal authorities that are short of budgetary funds, but they need to be solved in order to make life in the city better and more convenient. Users and experts should monitor the progress of work to eliminate the problem or to reduce its negative impact. After carrying out work to solve a socially significant problem, the citizen who initiated the project or detected the problem should give an assessment of the degree of elimination of the problem. It is worth noting that social problems in the field of population protection cannot be neutralized in the near-term perspective. Policymakers need to be aware that crowdfunding is a method that does not provide a guarantee of solving all problems. To implement such a model, it is necessary to clearly define the goal and to find out whether crowdfunding is the way to achieve it. This is a nascent method, which does not exclude errors. However, it has great potential and social significance, since it leads local authorities and citizens to mutual recognition and understanding. The academics and practitioners have already noted the long-term effect of participatory budgeting and crowdfunding on the level of social interaction in the community, civic activism, and level of interaction between the population and administration wherein administration becomes more open and responsible before the local community.

It is the responsibility of regional administrations to determine the individual cost of eligible projects and the size of the budget participation (subsidy). The size of the subsidy may vary from RUR 500.000 (US$8.000 equivalent) and can reach RUR 3 million (US$50.000 equivalent). The number of projects that can be implemented in one municipality also differs significantly from region to region. The total amount of allocations provided annually from regional budgets as a programme of participatory or initiative budgeting varies between RUR 60 and 600 million (US$1–10 million equivalent).

Local portals and platforms allow authorities and activists to signal problems, find like-minded people, and organize polls on different local issues.

One of the most publicized examples is the Active Citizen platform launched in 2014 in Moscow which allows city dwellers to express their opinions with regard to different issues related to the city functioning (ecology, transport, communal services, etc.). Polls are organized by Moscow government and provide quick means of measuring citizens' reaction to the administration's or other citizens' initiatives. However, Moscow government chose not to introduce crowdfunding on this platform, developing online voting instead.

In Tula region (200 kilometres from Moscow), thanks to the programme called People's Budget the region has seen more than 4,000 projects realized since 2011 worth about RUR 3.4 billion.[15] The People's Budget programme operates on the principles of co-financing, where inhabitants of the region can apply to participate in it, attract sponsors, or fund the projects directly. The share of civic co-financing depends on the total cost of the project and varies from 8% to 25%. Each project application is assessed by an expert commission of the regional government with funds being also allocated from the budgets of municipal districts. Over the past few years, the share of the regional budget expenditures amounted to RUR 2.4 billion, municipalities' share amounted to RUR 500 million, and residents' and philanthropists' share amounted to another RUR 500 million. As of 2018, similar programmes based on the pattern of Tula People's Budget were initiated in more than 20 regions of the Russian Federation. Annual contribution from Tula regional budget amounts to RUR 500 million (US$8 million equivalent). The principle of co-financing realized through crowdfunding, as an obligatory element of the regional programme, ensures an almost twofold increase in the project budget. Obligatory element of crowdfunding means that each project should be at least 10% crowdfunded. For example, in 2014, regional budget allocated RUR 377 million, whereas co-financing amounted to RUR 238 million, of which municipalities invested RUR 121 million, population invested RUR 72 million, and business sponsors invested RUR 45 million. This co-financing allowed the programme budget to be increased to RUR 616 million.

In Tatarstan, the Internet service "Map of Initiatives" was launched in 2017, designed to stimulate civil activity of the residents of the republic. In fact, this is an independent crowdfunding platform, where one can find like-minded people and provide support for the implementation of their initiatives.[16] The Internet platform allows to submit projects for review, discussion, and support and form teams for their implementation. The initiative can come from an individual, NGO, or business. After registration and approval of the project by the expert council, the initiative appears on the site. The colour of the projects on the map reflects the assessment that experts put to each initiative. "Green projects" are the most important, relevant, and easy to realize. Projects can be supported not only with money—for different forms of assistance, the user receives a reward in the form of points (for example, repost in a social network— 1 point). The most active participants will subsequently be able to exchange points for remuneration from the partners of the project. The authorities of the republic or municipalities can also support initiatives— individual projects have already received micro-grants amounting to RUR 30.000–50.000. Organizers believe that this should be enough to implement a good local project. During the 4 months of the platform's operation, 29 projects were supported.

The Government of Yakutsk, one of the northernmost cities in the world, has gone even further: uniting crowdsourcing and crowdfunding on its portal, One click Yakutsk.ru. Eight portal services provide citizens with various tools for participating in the life of the city, realising their ideas and providing mutual assistance. The portal is not only a simple and convenient tool for direct interaction between citizens, officials, public organisations, and municipal services but also a single technological platform that has proved itself as an effective mechanism for involving all residents in the management of the city. The site has become not just a technical tool but also an association of citizens who want to make their hometown better. With the help of the platform, powerful projects are implemented that directly affect the quality of life of citizens. A total of 32,449 users registered on the portal, with portal attendance being 7,000 unique visits a day.[17]

Besides polls, citizens can suggest initiatives for which up to 80% may be financed by the municipal government if at least 20% is collected by active citizens. This is one of the main conditions—participants must themselves invest at least 20% of the project cost.

The total People's Budget financing from city budget earmarked for civic crowdfunding purposes amounts to RUR 30 million (US$500.000 equivalent) annually and relates mainly to projects associated with local charities, cultural or educational events, improvement of public services, or repairs or building of small infrastructure objects. Since 2013, People's Budget programme allowed to realize 129 socially significant projects initiated by the citizens, with co-financing from the city budget. Thus, crowdfunding plays a central role in the public–private partnership where private initiative is supported by public (municipal) money.

The updated version of One Click Yakutsk portal includes a specialized subsystem—"City of Good." This functionality is also based on the principles of crowdfunding and is intended to help particular people in need and raise funds for the concrete projects of assistance from the locals without participation of the municipality. Thus, crowdfunding proved to be a popular mechanism for attracting funding from the locals in Yakutsk. The project allows to participate in civic crowdfunding not only for individuals but also for Yakut companies that would like to support any social project or help people in need. For Yakutsk administration, the main advantage of crowdfunding is that it is free of commissions and offers solutions to local problems.

The People's Budget programme implemented in many regions of Russia is the development of a social partnership between local government and society in the budgetary sphere as a mechanism for increasing civic engagement, as well as creating conditions for the development of civil society institutions. It is called upon to solve the following tasks: stimulation of direct participation of the population in the local self-government, activation of its participation in solving issues of local importance; increase of openness of activity of local government; attraction of public attention to the budget and budget process; and expansion of legal knowledge of the public in the sphere of finance. Jordan Raynor, one of the co-founders and director of Citizinvestor, explains that the main thing is not to pay for public services; the main thing is to want to know where exactly your money is going and to take part in running the city.[18] Initiative budgeting, Russian version of participatory budgeting, as a set of diverse, based on the civic initiative, practices to address issues of local importance with the direct participation of citizens in determining and selecting objects for spending budget funds, as well as subsequent monitoring of the implementation of selected projects, reflects this idea and allows with the help of crowdfunding to anchor the citizens' interests to the issues of primary importance to the local community.

3. Entrepreneurial Crowdfunding for Civil Projects

Many Russian municipalities and municipal organisations consider crowdfunding as an additional source of financing, a kind of money not associated with budgetary regulations and limitations.[19] Municipal budgets often lack money to provide for the needs of the local citizens and make appeal to private funds. They make use of crowdfunding as an alternative solution to their budgetary problems.

A vivid example is offered by Krivskoe municipality in Kaluga region. Municipality created its own social crowdfunding site to post the projects. Municipal administration put on this website several well-worked through ideas with full description and documentation. According to the administration's practice,[20] most sponsors never decline to donate if there is a project that is ready. The success of such municipal crowdfunding is related to two basic things. Firstly, it is the transparency of the project from concept to implementation, since all information about the project is posted on the open Internet site. Therefore, every potential sponsor has the opportunity to get acquainted with the idea in advance and follow the progress of the project. Secondly, if the project is focused on a large number of people, then businesses are even more willing to invest money in it.

In the case of Krivskoe, crowdfunding helped to solve a very problematic issue for the village—water supply. The village has three wells. One is old, there is not enough water in it for the whole Krivskoe, while in the other two, the quality of water is worse. Therefore, the proposed solution was to modernize the main well, to deepen it, and install modern pumps with a self-cleaning option. The project cost amounted to over RUR 2 million (US$35.000 equivalent). The realisation of this project was mainly financed by the local company OAO Krivskoe. At the same time, not a single rouble was spent from the municipal budget or the regional budget.

The intention of Krivskoe municipal authorities is to bring the volume of crowdfunded projects to 30–50% of the local budget and thus finance the communal projects for which there is no money in the municipal budget.

A similar example is offered by the municipal kindergarten in the southern city of Sochi. The teachers of the kindergarten have developed an innovative programme that involves the creation of a unique hybrid of a counselling centre for parents and a child support centre.[21] All services of such structure are to be provided free of charge to children who attend and do not attend kindergarten and their parents. The team submitted its project on Planeta crowdfunding platform to raise RUR 300.000 (US$5.000 equivalent) to finance the necessary infrastructure works. Municipal authorities did not have money to finance this programme from the local budget and the kindergarten made an appeal to private funds.

In both of these and many similar examples, the municipality takes on the role of an organizer or facilitator (preparing the project, offering its technical and administrative support) but not spending any money from the local scarce budgets. Moderating agent might be both independent and proprietary (municipal website or portal).

4. Critique of Crowdfunding for Community Purposes

In Russia, the real participation of citizens in the budgetary process through crowdfunding has significant successes and may serve in many respects as an example of the best practices in the field. However, it is not universally accepted in all Russian regions and municipalities. It is also limited by the lack of legal, financial, and other infrastructure at the federal and regional levels, based on proven mechanisms resulting from Russia's own experience, institutions for organising participation, which ensure maximum transparency and convenience for citizens. The successful practice of crowdfunding for communal projects necessitates an active informational campaign in their support. Such campaign needs both money and active experienced managers to gain the necessary momentum.

Many regional authorities in Russia do not yet understand what their interest is in implementing projects through crowdfunding. Nevertheless, civic crowdfunding has already gained ground and is one of dominant topics for the civil society and authorities' interaction.

Local authorities are often unable to assess the potential of crowdfunding, both civic and entrepreneurial, and they agree to participate in these procedures only as a result of "pressure from below" initiative groups. But even after this, the lack of skills to hold meetings of initiative groups, to design project documentation becomes an obstacle for crowdfunding development. Officials often lack or ignore project management techniques that are necessary to make crowdfunding projects successful. Russian practice also showed that the most important problem might be the lack of common language for communication between officials and participants in initiative groups. The training of officials to work with initiative groups and the involvement of external consultants to accompany projects are still episodic.

To overcome the lack of experience of the majority of citizens in the actual participation in public events (meetings, hearings, gatherings, etc.), there should be education of civil activists and population, especially in the field of crowdfunding, and active exchange of the best practices as well as encouragement of all forms of civil activity on the part of the local authorities. As for the authorities to negotiate the distrust of the population towards official innovation, the policymakers should have the sufficient knowledge and skills as well as financial resources to support the civil initiatives in order to increase the participation of the population in the practice of crowdfunding for local projects.

Crowdfunding cannot be considered as panacea for all municipal problems. Participatory processes run the risk of capture by interest groups.[22] Captured processes may continue to promote elitism in government decision-making especially when financial participation is necessary which can often not be provided by the most needy strata of population. In Russia, financial participation on the part of an initiator is the initiator's obligation without which the project is not eligible for financing from the budget. In the view of the local governments, such a clause precludes from considering projects that are not so vital for the initiators as they are not ready to have their money at stake along with state financing. It also increases financial discipline in the disbursement of the money obtained from all sources.

5. Conclusion

Russian practices of crowdfunding for municipal projects make use of both international practices and accumulated Russian experience. The major feature of Russian practice is the emphasis on the financial aspect of crowdfunding through obligatory financial participation of interested parties in the project. This is both due to the deficit nature of municipal budgets that plainly lack resources to finance the requirements of the population and the idea to have skin in the game for individual project initiators, thus increasing their responsibility for promoted projects.

In our opinion, gradual expansion of crowdfunding practices in Russian regions may lead to the appearance of special crowdfunding line in local budgets with funds allocated according to the local law and not at the will of the local mayor or governor. This will encourage the development of local social enterprises and elevate the profile of civil activists as well as the visibility of the problems at the government level. Another important feature may become the activisation of working collectives as the subject of crowdfunding activity. Due to the weakness of the Russian trade unions, the working collectives of enterprises, both small and large, may become the source of civic crowdfunding initiatives and partial financing. Also, modern IT will make crowdfunding projects more representative through development of polling techniques and applications. Such polling tools will not depend on the access or participation of the person in local affairs but will make the polling regular as well as informative and quasi-obligatory, thus making the engagement universal and expanding the reach of civic crowdfunding beyond activists.

About the Author

Evgeny P. Torkanovskiy is a Research Fellow with Russian Academy of Sciences Institute of Economics. His principal topics of research include crisis, innovation, and other special situations (M&A) management, collective investment vehicles (investment companies, funds) as well as crowdfunding and crowdlending. Besides research he pursued successful career in the financial sector putting his ideas into practice. He managed the first ever public tender offer in Russia as well as successful merger of the largest Russian confectionery producers. His experience as a fund manager was essential for public service where he started and led the Russian Industrial Development Fund offering loans and subsidies for innovative companies. Evgeny invests his time and efforts in promoting the climate change agenda for SMEs in the countries of the former USSR.

Endnotes

1. https://www.rbc.ru/own_business/15/06/2017/594006559a794715c9d5819c
2. https://planeta.ru/
3. http://boomstarter.ru/
4. https://starttrack.ru/
5. https://townmoney.ru/
6. https://penenza.ru/
7. https://business.potok.digital/
8. https://gosstart.ru/
9. The Token Spring of Central and Eastern Europe. International research by EWDN and ICO Bench http://cee-tokens.ewdn.com
10. History of Cocco Bello, family bee-garden and socially responsible project. http://project178122.tilda.ws
11. http://budget4me.ru
12. Decree of the Government of the Kirov region dd February 24, 2015 No 26/104 "On the implementation of "People's Budget" projects. http://www.Kirovreg.ru/publ/AkOUP.nsf/62bd0840256f32bcc3257149006da49f/E78DA3492FD0499043257E28002F626D
13. Pahomova O.A. "Crowdsourcing as a Means to Solve Social Problems of Small Towns. *Diskussia*, 66(3) (2016).
14. Vagin V.V., Gavrilova, N.V. and Shapovalova, N.A. "Initiative Budgeting in Russia: Best Practices and Directions for Development. In: Scientific-Research Financial Institute." *Financial Journal*, 4 (2015), 94–103.
15. http://tass.ru/ekonomika/4989461
16. https://ikarta.tatar/
17. https://www.exo-ykt.ru/articles/08/576/19927/
18. https://www.youtube.com/watch?v=9Vw37XmL_EM

19. Petrushenko Y.N. and A.V. Dudkin: Crowdfunding as an innovative tool for financing projects for social and economic development. Marketing and management of innovations, Moscow, 2014.
20. http://pressaobninsk.ru/nedbalfull/11237
21. http://www.ug.ru/news/19084, https://planeta.ru/campaigns/sochi28
22. Shah, Anwar. *Participatory Budgeting. Public Sector Governance and Accountability.* World Bank, Washington, DC, 2007.

Chapter 14

From Crowdfunding Initiative to Fintech Hub: Lithuanian Case

Arvydas Paškevičius, Leva Astrauskaitė
and Sigitas Mitkus

Abstract

Financial technologies (Fintech), expanding sharing economy, changing consumers' behaviour, and decentralisation in business models create new business possibilities. Those developments have put consumers' interest in the centre of financial services. The Brexit process has been another factor for increasing the attention of policymakers of European countries on Fintech development. Policymakers have to take into account new realities and adopt their policy following the trends. The ability to take a political leadership, involvement of political establishment, creating formal and informal administrative structures, and a spirit of cooperation among stakeholders are the main elements for the success of any political initiative. This chapter evaluates what kind of factors could define the country's choice to make Fintech as one of the policy priorities and how Fintech could trigger reforms in other policy areas such as immigration, education, risk management, taxation, improvements in business climate for start-ups, ICT infrastructure, and financial literacy. Lithuanian experience has been taken as a case study for the analysis due to the pace of the developments of the Fintech eco-system. Since 2015, a number of Fintech companies in the country has doubled, Fintech programmes have been introduced in universities, business associations, and innovation, including blockchain technology, centres have been established, crowdfunding platforms have been included into public–private financial engineering schemes, and, last but not least, up to 4% of the world's initial coin offerings (ICOs) have been initiated by Lithuanian entrepreneurs in 2017. Those facts prove that right political and administrative decisions could create a good basis for the development of a completely new eco-system of the economy.

1. Introduction

Financial services are at the edge of a huge transformation that has been driven by technological development and innovation. In countries with high predominance of traditional banking sector, the changes in financial services sector can be triggered by the effective policy of the governments. The traditional banking in Lithuania accounts for 79.2%[1] of the financial system with three dominant banks accounting for almost three-fourths of the entire sector in terms of assets. The deposits in the banks amounted to €19.8 billion in the first quarter of 2018.[2] The Lithuanian Government and Bank of Lithuania have included issues related to development of the financial system, access to finance, financial inclusion, competition, and consumer protection into their daily agenda.

Since 2015, the Government has been taking actions to become one of the best jurisdictions for financial services and financial technologies in the European Union. Despite the changes in the composition of the Government and Parliament of the Republic of Lithuania after the general elections in 2016, continuity of the policy has been ensured by introducing two concrete measures (No. 219.4 and No. 192.10) on financial services and financial technology industry development into the Government programme.[3]

Three interinstitutional working groups on Alternative financing (2014), Developments of Fintech industry (2016), and Financial Education (2017) have been created, three projects with the European Commission and EBRD have been planned (on Covered bonds and Securitisation (2015), SME Equity Listing Support Instrument (2017), and Institutional investors investments (2016)), and two national projects on Initial Coin

Offering (ICO) guidelines (2018) and revision of regulatory framework for investment funds (2018) have been initiated.[4] Those actions have had a major positive impact on the development of whole financial eco-systems, especially fintech, crowdfunding, and peer-to-peer (P2P) lending eco-systems, in Lithuania.

2. Crowdfunding and Peer-to-Peer Lending as the First Starting Point

Alternative financing, especially for SMEs, was taken as one of the main Lithuanian Government priorities in 2014. In 2015, a package of legal acts enhancing the efficiency of Lithuanian capital market was adopted. One of the elements of the package was adoption of Laws on Crowd-funding and Consumer credit that contains provisions on P2P lending[5] (Figure 1).

The decision to have two laws separating crowdfunding and P2P lend-ing activities was determined by a need to have the expeditious decisions correcting the situation in the consumer credit market and ensuring the consumer interests. The Parliament decided not only to reduce ceilings for

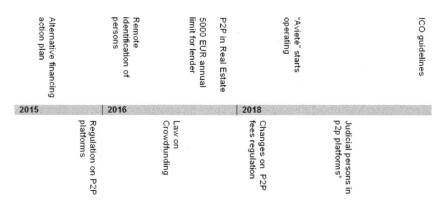

Figure 1: Policy actions on crowdfunding and P2P lending activities in Lithuania during 2015–2018.

Note: *To be adopted.

Source: Compiled by authors.

total cost of the credit to the consumer but also increase competition in the consumer credit market by creating a legal framework for P2P platforms to operate in Lithuania. The amendments to the Law on Consumer credit were adopted by the Parliament on November 5, 2015.

Lithuania was the eighth European country that introduced bespoke regime for crowdfunding activities on November 3, 2016. The Law had been designed taking into account the best practice of other EU member states such as Italy, UK, and Spain. On October 26, 2016, the Government approved the Resolution[6] that regulates remote identification of persons allowing not only using a qualified electronic signature but also by applying two supplementary methods, that is, by using the means of electronic identification issued in the EU under notified electronic identification schemes and by electronic means allowing video transmissions. Those provisions were incorporated into the Law on The Prevention of Money Laundering and Terrorism Financing[7] adopted by the Parliament on June 29, 2017.

The adoption of the Law on Real Estate Related Credit[8] establishing regulatory framework for P2P lending platforms operating in real estate market followed on November 10, 2016.

In addition, a €5,000 annual limitation for a lender for all loans in one platform was adopted by passing amendments to the Law on Consumer Credit on November 10, 2016.[9]

In order to ease P2P lending activities on June 15, 2018, the Parliament adopted amendments to the Law on Consumer credit[10] allowing P2P platform operators to take 50% of the fees at the moment of the signing of the agreement with a consumer. In addition to this measure in June 2018, the Government prepared draft amendments to the Law on Consumer Credit that would allow judicial persons to provide loans to individuals via P2P platforms. Those decisions have been made taking into account a sufficiently low ratio of non-performing loans issued by platform operators[11] and willingness to increase competition in the lending market.

On April 11, 2018, the guarantee institution Investment and Business Guarantees (INVEGA) introduced a new €4,615 million financial instrument for SMEs called "Avietė"[12] (Raspberry) that grants loans for SMEs through crowdfunding platforms. The maximum loan amount is €10.000 for up to 36 months and funding can be provided for up to 40% of the total amount to be used to finance both investments and circulating capital, except for refinancing of financial obligations, financial activities, and real estate.

Regulatory framework for crowdfunding can be used for equity ICO projects. The ICO Guidelines were approved by the Ministry of Finance of the Republic of Lithuania (Ministry of Finance) on June 8, 2018.[13] The initiators of ICO projects where security tokens are issued (granting the right to ownership, management of the company, or granting other rights to shareholders such as the right to receive part of the company's profit in the form of dividends or another form or providing for payment of interest or redemption of tokens) should follow the requirements listed in the Law on Crowdfunding and the Law on Securities.

DESICO[14] has been the first company that has announced its readiness to use the Lithuanian regulatory framework for Crowdfunding for establishing a global EU-based platform offering a fully compliant way for businesses to issue ICOs, and for token buyers to acquire and trade these tokens. Currently, the project is under implementation.

3. Fintech Strategy as the Main Driver for Fintech Development in the Country

There is not a single factor that defines a success and pace of fintech development in a country. It is important to look at whole composition of factors such as favourable public policy and legal environment, access to finance depending on company's phase of development, demand for fintech products and services, as well as supply of talents.

The evaluation of a possibility to include financial technologies into the broader agenda of Lithuanian public authorities started in the Ministry of Finance of the Republic of Lithuania in the end of 2015. A decision to raise Fintech development as a priority to political level was made on the basis of analyses in four areas: ICT infrastructure, existing talent pool, geopolitical trends (BREXIT process), and readiness of public institutions to be engaged in making regulatory changes.

Followed by the analyses, an inter-institutional working group[15] led by the Ministry of Finance and consisting of representatives of relevant public institutions was set up on July 25, 2016. Financial and business associations and academia representatives have been invited as associate members. Ministry of Finance, Bank of Lithuania, and a public agency responsible for attracting foreign investments "Invest Lithuania" allocated the necessary human and financial resources for this initiative.

An inter-institutional group in cooperation with business representatives was instructed to prepare "Action plan on Fintech industry development in Lithuania" (Action Plan). After public consultations, during which 70 proposals had been received, the Action plan was presented at the Cabinet of Ministers meeting on April 12, 2017 and approved by the Order of the Minister of Finance on May 8, 2017.[16] The Action Plan consisted of 34 concrete measures covering four main areas such as creating favourable public policy and legal environment, increasing access to finance depending on the company's phase of development, promoting demand for fintech products and services, and focusing on the supply of talents.

The first report on implementation of the plan was published in January 2018. It was stated that 90% of 34 measures had been implemented.

The main focus of the inter-institutional working group was on improving the regulatory framework for Fintech companies and preparing a broad public information campaign about jurisdiction (Figure 2).

The adoption and revision of innovation-oriented legislation on crowd-funding, P2P lending; "specialised" bank licence; remote identification of persons; payment services; regulatory initiatives, such as "Newcomer Program,"[17] regulatory "Sandbox," and technological blockchain sandbox "LBChain;"[18] and introduction of e-licensing tool[19] were the main initiatives implemented during 2016–2018.

Access to finance depending on Fintech company's phase of development was one of the major themes discussed within the Fintech community. Two-thirds of all fintech companies in Lithuania had less than 10 employees.[20] There is no support programme specifically created for Fintech companies, but the companies can apply for a support from the financial instruments such as guarantees, soft loans, venture capital funds, and global grants designed by INVEGA.[21]

Evaluation phase | Fintech Conference | Newcomer program, Regulatory sandbox, LB Chain project* | Specialised bank licence | Fintech Conference | Law on payments (PSD2 transposition) | International financial markets conference | 16+1 Fintech coordination Centre | E-licencing tool* | Fintech programs start at KTU, VU, VGTU

| 2015 | 2016 | 2017 | 2018 |

Inter institutional working group | Regulations on Crowdfunding, remote identification of persons* | Fintech Action plan | International financial markets conference | Associate member of Hyperledger** | ICO guidelines | 500 letters initiative | Virtual currency exchange operators

Figure 2: *Policy actions taken towards implementation of Fintech strategy.*

Visits*** abroad

UK, Belgium	UK, Belgium, France	France, UK, USA, Sweden, Singapore	France, UK, USA, China, Austria, Jordan, Belgium

Private initiatives

	Conference: *Sharing economy in Lithuania 2019: Current state and new opportunities.* Lithuanian P2P Lending and Crowdfunding Association established.	*Insurtech* Conference *SWITCH* Conference *6th ECN Crowdfunding Convention.* Association: "Fintech Lithuania" established.	"Fintech Hub LT" association, Crypto Economy Organization, Blockchain Vilnius Centre established.

Notes: *Initiatives of Bank of Lithuania, **Ministry of Finance, *** Visits of the Minister of Finance or the Governor of the Bank of Lithuania.

Source: Compiled by authors.

As for public information campaign, efforts have been concentrated on promoting Lithuania as a country that has future-ready payment (non-banks can have direct access to Single European Payment Area), world-class Information Communication Technology (Lithuania is among the 20 best performers in the world in terms of broadband speed[22] and public Wi-Fi[23]) infrastructure, as well as a business-friendly environment. The fourth lowest profit tax in the EU and three times deduction of research and development expenses have been introduced.

A talent pool is one of the most important elements in the fintech ecosystem. Around 52% of the Lithuanian population speaks at least two languages, there are 26,000 IT professionals, 11,000 developers, and 7,600 IT students in the country.[24] On September 1, 2018, Kaunas University of Technology (KTU) and Vilnius Gediminas Technical University (VGTU) started a Bachelor's degree programme, meanwhile Vilnius University (VU) started a Master's in Fintech study programmes.

Fintech movement has contributed to discussions on easing immigration policy too; two initiatives—Bluecard[25] (special visas for highly skilled professionals from non-EU/EAA countries who want to work and live in Lithuania) and Startup Visa[26] (special visas for start-ups founders from non-EU/EAA countries who want to run innovative businesses in Lithuania)—have been launched to attract highly skilled professionals and entrepreneurs.

Due to orchestrated efforts of public institutions and strong engagement of the Fintech community, the number of Fintech companies doubled in 2017.[27] Political leaders, heads of the main public agencies and four associations (Fintech Lithuania association, established in July 2017, Fintech hub LT (2017), P2P lending and crowdfunding association (2016), Crypto Economy Organisation (2018)), and legal firms have been taking an active part in promoting Lithuania as a Fintech Hub.

One of the most important measures of promoting the country has been a conference diplomacy. A scope of activity and level of engagement have been coordinated by relevant public institutions focusing on the main Fintech events such as Money2020, Paris Fintech forum, Innovate Global Summit, and others. Since 2015, Lithuanian officials have been regularly visiting the world's financial centres such as London, Paris, Singapore, Beijing, Tel Aviv, Pretoria, Frankfurt, Stockholm, and New York.

On July 7, 2018 in Sofia (Bulgaria), prime ministers of Central and Eastern European Countries (CEEC) and People's Republic of China supported the Lithuanian initiative to establish the 16+1 Fintech Coordination Centre in Vilnius and organise the 16+1 High Level Fintech Forum in Lithuania in 2019. The Fintech Coordination Centre will act as a facilitator of discussions on innovation in financial services, consumer and data protection, regulatory practice, risk management, and cybersecurity.

Since 2016, the Ministry of Finance in cooperation with the Bank of Lithuania has set up a tradition to organise an annual Fintech conference in November. The first Fintech conference gathered about 200 participants. The number of participants at the conference in 2017 reached 1,000. In addition to Fintech conference in November, the Ministry of Finance dedicates one session on Fintech development at the international financial markets conference organised in May.

In June 2018, the ministers of Finance and Economy of Lithuania sent 500 letters to the world's biggest Fintech companies offering Lithuania as jurisdiction for their activities.

3.1. Understanding Risks

Fintech innovations create not only opportunities but also challenges related to consumer and data protection, cybersecurity, anti-money laundering, fight against the financing of terrorism, as well as managing reputational risk.

According to the Law on the Basics of National Security[28] of the Republic of Lithuania, finance and credit sectors of the economy are of strategic importance for national security. All applications for licences to provide financial services in Lithuania should be considered by the Governmental National Security Commission.

In order to establish a coordinated policy towards cybersecurity in the country, the Law on Cyber Security was adopted on December 19, 2017. The National Cyber Security Centre was appointed as a single entity that ensures overall security of the public and private IT networks as well as coordinates actions of public institutions in the fields of cybersecurity.

On June 15, 2018, amendments to the Law on Bank of Lithuania that strengthen powers of the Bank of Lithuania to carry out effective supervision of financial markets and consumer and investor interests came into force. All financial market players are subject to harmonised supervision rules, fine calculation, and imposition system for infringements based on uniform principles.

Despite the above-mentioned measures, very low level of non-resident financial activities, high quality of supervision of the financial sector, and risk management should remain the top priorities for the Government and Bank of Lithuania.

4. Crypto Economy and Blockchain Technology: Lithuanian Phenomenon

Since 2016, ICOs have been seen as a new and rapidly expanding fund-raising instrument. It was mainly due to the ease of raising money and openness to masses that funds are much greater than those in crowdfunding, and last but not least, lack of regulation is a main factor.

Depending on different sources, since the beginning of 2017, around 800–1,800 ICO projects have been initiated and US$12–12.5 billion[29] has been raised.

Since the beginning of 2017, around 30–40 ICO projects have been initiated by Lithuanian nationals in Lithuania and abroad.[30] They accounted for up to US$0.5 billion and amounted to 3–4% of the global market in 2017.

Fintech development and messages from policymakers and regulators on openness for innovation in financial services and new technologies such as blockchain have contributed to the development of crypto economy movement in Lithuania. The first messages from the Ministry of Finance and Bank of Lithuania were sent to the public at the *Bitcoin conference* in Vilnius on April 8, 2016 followed by an announcement by the Bank of Lithuania on creating the technological blockchain sandbox "LBchain" and joining the Hyperledger community by the Ministry of Finance in January 2018.[31]

A measure to evaluate a possibility to introduce regulation for virtual currency exchange operators was included into the Action Plan on Fintech industry development in Lithuania. During the third quarter of 2017, inter-institutional consultations and evaluation of practice in other jurisdictions such as in Japan and US had been carried out. After examination, it was decided to wait until the adoption of the 5th EU directive on Anti-Money Laundering in 2018.

Those actions have been contributed by private-sector initiatives such as creating informal blockchain clubs and opening the Blockchain Vilnius Centre[32] in January 2018.

Lack of clarity in regulating security and utility ICO business, risk management issues, and the fast-growing sector determined the need to discuss regulatory framework for ICO. The outcome of the discussion was a publication of ICO Guidelines by the Ministry of Finance covering Regulatory, Taxation, Accounting, and Anti-money Laundering aspects on June 8, 2018.[33]

Lithuania has been among the first few countries in the world that has published the most comprehensive guidelines on how to make ICO business in Lithuania.

As for anti-money laundering aspects related to activities of virtual currency exchange operators in July 2018, the Ministry of Finance initiated an interinstitutional consultation on amendments to the Law on the Prevention of Money Laundering and Terrorism Financing. It is expected that the amendments will be adopted by the end of 2018.

5. Methodology

The research methodology is introduced in order to evaluate the impact of political and administrative decisions on the development of a completely new eco-system of the economy.

Previous approximations to the topic were made using these methods as follows:

- surveys and observations (Ley and Weaven, 2011; Hui *et al.* 2012; Kim and Moor, 2017; Gerber *et al.* 2012);
- experiments and models (Lin and Viswanathan, 2013);
- correlations and regressions (Lu *et al.* 2014; Lee and Lee, 2012; Giudici *et al.* 2013);
- the chosen method is Pearson correlation (r) with statistical significance check at p-value; the main argumentation of the choice is as follows:

  o presumption of linear relationship of the variables is set;
  o size of data >20 observations;
  o variables are supposed to be normally distributed.

The Pearson Correlation Coefficient (PCC) is defined (Benesty *et al.* 2009) as follows:

$$\rho(a,b) = (E(ab))/(\sigma\_a\sigma\_b), \tag{1}$$

where $E(ab)$ is the cross-correlation between a and b, and $\sigma\_a^2 = E(a2)$ and $\sigma\_b^2 = E(b2)$ are the variances of the signals a and b, respectively.

One of the most important properties of the PCC is that $0 \leq \rho(a,b) \leq 1$. The PCC gives an indication of the strength of the linear relationship between the two random variables a and b. If $\rho(a,b) = 0$, then a and b are said to be uncorrelated. The closer the value of $\rho(a,b)$ to 1, the stronger the correlation between the two variables. If the two variables are independent, then $2(a,b) = 0$. But the converse is not true because the PCC detects only linear dependencies between the two variables a and b. For a nonlinear dependency, the PCC may be equal to zero. However, in the special case when a and b are jointly normal, "independent" is equivalent to "uncorrelated" (Benesty *et al.* 2009).

The P value, or calculated probability, is the probability of finding the observed, or more extreme, results when the null hypothesis (H0) of a study question is true—the definition of "extreme" depends on how the hypothesis is being tested. The p-value, or calculated probability, is the probability of finding the observed, or more extreme, results when the null hypothesis (H0) of a study question is true—the definition of 'extreme' depends on how the hypothesis is being tested. If p-value is less than the chosen significance level, then you reject the null hypothesis, i.e. accept that the sample gives reasonable evidence to support the alternative hypothesis (Benesty *et al.* 2009). The authors of this paper refer to statistical significance as $p < 0.05$.

The main aim of the research is to determine the impact of political and administrative decisions on the development of crowdfunding in Lithuania.

Variables were chosen according to the availability of the data.
Numeric of linear regressions equations was investigated.

The following were the dependent variables: (1) number of consumer credits paid out by P2P lending platforms over the period, pcs.; (2) amount of consumer credits paid out by P2P lending platforms over the period, million Euros; (3) average annual rate of total consumer credit prices in P2P lending platforms, %; (4) average annual consumer credit interest rate, %; (5) total cost of consumer credit paid by consumer borrowers during the reporting period, million Euros; (6) average loan duration, months; (7) number of consumer credit agreements terminated, when the consumer has not fulfilled his obligations, pcs.; and (8) number of complaints received, pcs.

The independent variable was a qualitative measure representing the political decisions and strategies: establishment of three interinstitutional working groups on Alternative financing (2014), Developments of Fintech industry (2016) and Financial Education (2017), implementation of three projects with the European Commission and EBRD related to the development of capital markets (on Covered bonds and Securitisation (2015), SME Equity Listing Support Instrument (2017) and Institutional investors investments (2016), initiation of two national projects on Initial Coin Offering (ICO) guidelines (2018) and revision of regulatory framework for investment funds (2018), and establishment of four Fintech associations (2016, 2017, and 2018). Giving values of 1 to interinstitutional working groups, 2 to projects, 3 to regulatory actions, and 4 to associations, the qualitative measure was created.

Data set and period. The data provider was the Central Bank of Lithuania. The earliest monthly data were taken (depending on variable) from April, 2016.

The Pearson correlation coefficients were calculated between variables using the Microsoft Excel toolkit. All coefficients were validated by p-value check.

Conclusions were made.

6. Empirical Results

According to methodology, the political and administrative decisions and crowdfunding indicators' correlations were formed as follows (see Table 1).

The results show the following:

- practically no correlation ($r = 0.009$) exists between the amount of consumer credits paid out by P2P lending platforms over the period and qualitative measure of political and administrative decisions; according to the values to given the categories, it could be concluded that all

Table 1:　Relationship between crowdfunding and political and administrative decisions.

	Number of consumer credits paid out by peer-to-peer lending platforms over the period, pcs.	Amount of consumer credits paid out by peer-to-peer lending platforms over the period, min. Eur;	Average annual rate of total consumer credit prices in peer-to-peer lending platforms, %	Average annual consumer credit interest rate, %	Total cost of consumer credit paid by consumer borrowers during the reporting period, min. Eur;	Average loan duration, months;	Number of consumer credit agreements terminated, when the consumer has not fulfilled his obligations, pcs.	Number complaints received, pcs.	Qualitative measure of political and administrative decisions
Number of consumer credits paid out by peer-to-peer lending platforms over the period, pcs.	1								
Amount of consumer credits paid out by peer-to-peer lending platforms over the period, min. Eur;	0.993931546	1							
Average annual rate of total consumer credit prices in peer-to-peer lending platforms, %	-0.961975164	-0.968642152	1						
Average annual consumer credit interest rate, %	-0.942921362	-0.954448596	0.997181164	1					

Total cost of consumer credit paid by consumer borrowers during the reporting period, min. Eur;	0.956481314	0.983226824	-0.963925798	-0.957710486	1			
Average loan duration, months;	0.901325658	0.936097558	-0.964789491	-0.975098275	0.969027053	1		
Number of consumer credit agreements terminated, when the consumer has not fulfilled his obligations, pcs.	0.903556267	0.918590235	-0.939977754	-0.942656713	0.918110607	0.907750874	1	
Number complaints received, pcs.	0.390132223	0.344425654	-0.333050922	-0.338998715	0.200741923	0.283438806	0.139287977	1
Qualitative measure of political and administrative decisions	0.063878567	0.008650975	-0.18346294	-0.178024643	0.086792738	0.091130853	-0.08100768	0.576228021	1

Sources: Compiled by authors.

political actions were less important on the growth of consumer credits paid out by P2P lending platforms;

- a very weak positive correlation ($r = 0.06$) exists between the number of consumer credits paid out by P2P lending platforms over the period and qualitative measure of political and administrative decisions; to sum up, all political actions were less powerful to the movements of number of consumer credits paid out by P2P lending platforms;
- a very weak positive correlation ($r = 0.09$) exists between the total cost of consumer credit paid by consumers during the reporting period and qualitative measure of political and administrative decisions; according to the values given to the categories, it could be concluded that political decisions were of very small effect to consumer creditworthiness;
- a very weak positive correlation ($r = 0.09$) exists between the average loan duration in months and qualitative measure of political and administrative decisions; to sum up, all political actions and loan duration do not show the overlapping connections;
- moderate positive correlation ($r = 0.58$) exists between the number of complaints received and qualitative measure of political and administrative decisions; according to the values given to categories, it could be concluded that political decisions empowered market participants to defend their rights and more confidence has been brought to the market;
- a very weak negative correlation ($r = -0.08$) exists between the number of consumer credit agreements terminated, when the consumer has not fulfilled his obligations and qualitative measure of political and administrative decisions; to sum up, strengthening the regulatory environment is the prevention of market shrinkage or the emergence of credit risk;
- a weak negative correlation ($r = -0.17$) exists between the average annual consumer credit interest rate and qualitative measure of political and administrative decisions; hence, political and administrative decisions could empower market attractiveness with competitive price ranges;
- a weak negative correlation ($r = -0.18$) exists between the average annual rate of total consumer credit prices in P2P lending platforms and qualitative measure of political and administrative decisions; therefore, the presumption on competitive price ranges under political decisions made is supported.

- It should be noted that these abbreviations are distant approximations to samples means according to statistically insignificant measures (all variables, except the *number of complaints received, pcs,* possess *p*-value greater than the 0.05 significance level).

7. Conclusions: Main Priorities for the Future

Recent political and administrative actions strengthen both consumer and business confidence on new financial instruments.

Consumer credit price levels are negatively related to those political decisions and administrative actions, meaning the creation of an attractive and competitive eco-system depends on more comprehensive actions (e.g. larger means were given to associations than interinstitutional working groups).

Unfortunately, there is a very slight positive effect between the size of the crowdfunding market (amount of consumer credits paid out by P2P lending platforms; number of consumer credits paid out by P2P lending platforms) and the political decisions and actions. These connections could be explained by a longer responsive period, which is needed by the market to react to political changes. There is still a lack of confidence in new financial instruments and awareness about them in the society.

The Lithuanian case proves that proper impact analyses, right political and administrative decisions, administrative structures, as well as signals to the market could create a good basis for the development of a completely new Fintech *ecosystem* in a country.

Evaluation of policy actions shows that Fintech ecosystem could increase competition in the highly concentrated financial market, provide better choice for consumers, create new jobs, and generate new business models.

In order to keep pace of the development, Lithuania needs to focus on (1) tightening Fintech's inter-linkages with sectors like logistics, e-commerce, sustainable finance, etc., that use Artificial Intelligence, Internet of Things, Big Data, Cybersecurity solutions; (2) increasing the collaboration among universities and business; (3) creating financial engineering instruments that include new forms of business models such as crowdfunding, ICO, and P2P lending; (4) creating networks among Fintech hubs and policymakers; and (5) strengthening cooperation among law enforcement institutions in order to cope with challenges such as risk management, consumer and data protection, cybersecurity, as well as money laundering and terrorism financing.

About the Authors

Arvydas Paškevičius, Doctor of Social Sciences, Professor, Head of the Finance Department, the Vilnius University, Lithuania. Research interests include issues of Lithuanian capital market development, Issues of the EU Capital Markets Union, and FinTech. He has also delivered talks on the following subjects: Management of Finance (in Lithuanian and English); Financial Management of Enterprises (in Lithuanian and English); Project Finance Management (in Lithuanian and English); Foreign Exchange Markets; and Management of Currency Risk. Qualification certificates: Certificate of Property Appraiser Qualification (business appraiser) No. A000161, issued on April 21, 2000.

Leva Astrauskaitė is an Assistant Professor as well as a Researcher at the Vilnius University, Lithuania. Teaching subjects include Financial markets, Financial analysis, Finance theory, International Banking, Money and Credit. The research interests include business analysis, corporate finance, capital market. Leva is specialising in various empirical models and techniques (e.g. correlations, regressions, time series, etc.). Research interests works haves been published revealed in 11 personal and co-author publications and in one book.

Sigitas Mitkus is a PhD candidate at the Vilnius University, in Lithuania. His research interests cover policy impact on the development of financial markets, creating new financial instruments for institutional and retail investors, Fintech, Crowdfunding and ICO regulation.

Endnotes

1.   https://www.oecd.org/finance/Lithuania-financial-markets-2017.pdf
2.   https://www.lb.lt/uploads/publications/docs/20050_74f56ecd7913f7a75db1
     6e563df50416.pdf
3.   https://lrv.lt/uploads/main/documents/files/XVII%20Vyriausybes%20
     programa_EN20170112.pdf

4. Information provided by the Ministry of Finance of the Republic of Lithuania.
5. The package consisted of Law on Reward Employees with Stocks (adopted in June 2017), Law on Legal Protection Regime for Bonds Holders together with the amendments to the Law on Companies that allows for Joint Stock Companies to offer bonds publicly (June 2016), amendments to the Law on Partnerships that bring improvements to the publicly disclosed information (June 2017), the Law on Accounting that simplifies application of business accounting standards (June 2017), the Law on Crowdfunding (November 2016).
6. https://e-seimas.lrs.lt/portal/legalAct/lt/TAD/a06f5e61a18e11e68987 e8320e9a5185
7. https://e-seimas.lrs.lt/portal/legalAct/lt/TAD/26a9f270fb6511e796a2c6c63a dd27e9?jfwid=2r1m4zwg
8. https://e-seimas.lrs.lt/portal/legalAct/lt/TAD/24bc9070a80111e68987e8320 e9a5185?jfwid=q8i88lpik
9. https://e-seimas.lrs.lt/portal/legalAct/lt/TAD/9b303402a80111e68987e8320 e9a5185?positionInSearchResults=0&searchModelUUID=f31bd730-04ef-4656-9271-68fbf1b5c426
10. https://www.e-tar.lt/portal/lt/legalAct/a2ff40f06f0f11e8b83be60b2e217f90
11. On December 31, 2016, non-performing loan ratio accounted for 12%. Source: Vartojimo kreditų rinkos apžvalga (2016). Bank of Lithuania.
12. http://invega.lt/en/solf-loans/crowd-funding-loans-aviete/
13. http://finmin.lrv.lt/uploads/finmin/documents/files/ICO%20Guidelines%20 Lithuania.pdf
14. https://www.desico.io/en#product
15. https://e-seimas.lrs.lt/portal/legalAct/lt/TAD/102538c054fe11e688d29c6e5e f0deee?jfwid=5sjolfzvo
16. https://e-seimas.lrs.lt/portal/legalAct/lt/TAD/6fe058823ccf11e79eb19446fc 7155e9?jfwid=q8i88lp51
17. https://www.lb.lt/en/newcomer-programme
18. https://www.lb.lt/en/news/bank-of-lithuania-calls-for-proposals-to-develop-a-blockchain-platform
19. https://www.lb.lt/en/authorisation-application
20. https://investlithuania.com/wp-content/uploads/2018/02/Lithuanian-Fintech-Report-2017.pdf
21. http://invega.lt/en/
22. http://www.speedtest.net/global-index
23. https://www.ooma.com/blog/best-worst-wifi-countries/
24. https://investlithuania.com/key-sectors/technology/fintech/
25. http://www.eubluecard.lt/

26. https://startupvisalithuania.com/
27. https://investlithuania.com/wp-content/uploads/2018/02/Lithuanian-Fintech-Report-2017.pdf
28. https://e-seimas.lrs.lt/portal/legalAct/lt/TAD/TAIS.265230?jfwid=nz8qn7zki
29. Investopedia, ICOdata, Coindesk, Coinschedule
30. CoFounder, POLITICO.
31. https://finmin.lrv.lt/en/news/lithuania-is-a-member-of-the-world-class-high-tech-blockchain-community
32. https://bcgateway.eu/
33. http://finmin.lrv.lt/uploads/finmin/documents/files/ICO%20Guidelines%20Lithuania.pdf

Chapter 15

Civic Crowdfunding in Germany—An Overview of the Developing Landscape in Germany

Karsten Wenzlaff

Abstract

This chapter provides an overview about the Civic Crowdfunding activities in Germany. Civic Crowdfunding is understood differently in the scientific literature, as I will discuss in a forthcoming conference paper. In this chapter, I start with the assumption that Civic Crowdfunding projects are aiming to establish a public good, which is essentially non-rivalrous and non-excludable. But as we can see in the examples below, such a definition does not always hold.

1. The History of Civic Crowdfunding in Germany

Civic Crowdfunding is both an old and a new phenomenon in Germany. In the late 19th century and the early 20th century, cultural institutions in the expanding cities were crowdfunded in the sense that cooperatives financed the building of theatres, parks, and swimming facilities. At that time, without the internet, citizens joined causes by establishing infrastructure open to the public. Infrastructure such as theatres, parks, and swimming pools are not in the strict sense non-rivalrous and non-excludable, but given the fact that cultural institutions were often owned and built for rich people, the so-called "Volkstheater" (folks' theatres), which emerged at that time allowed a certain group of poorer citizen to access cultural activities.

We could, however, trace Civic Crowdfunding to an even earlier origin, long before the financing of the often quoted the—Pedestal of the Statue of Liberty in New York. In 1745, the newspaper *Wandsbecker Mercurius* was established. Wandsbeck (today Wandsbek) is now a district of the city of Hamburg, but in the 18th century was placed just outside of the territory of the city of Hamburg, which at that time was part of the Kingdom of Denmark. The newspaper *Wandsbecker Mercurius* and the follow-up publication *Wandsbecker Bothe* was financed through a pre-purchase subscription model. The small circulation was however accessible to a wider public. Yet, it contributed to freedom of expression and freedom of press, both public goods in the strict sense, which were both scarce in the 18th century.

2. Reward-Based Civic Crowdfunding Platforms in Germany

Most platforms that allow Civic Crowdfunding projects have been established in the last 10 years. A wave of new reward-based platforms came alive in 2010, partially inspired by the success of Kickstarter and Indiegogo. Platforms like Startnext, Visionbakery, Wemakeit, and Betterplace allow the financing of Civic Projects and often have specific categories devoted to these projects. Friendraising platforms like GoFundMe and Leetchi are also active in Germany but have not featured Civic Crowdfunding Projects to a major degree (unlike, for instance, the platform GoFundMe in the US, which has been a platform for activists financing political causes on the platform).

Unlike in Italy, France, and UK, there are no specific platforms dedicated to reward-based Civic Crowdfunding. Place2Help was a Civic Crowdfunding Platform based in Munich, which tried to focus on Civic Crowdfunding. The business model depended on the cooperation with sponsors and it seems that this business model was not sustainable, because the platform in Munich is no longer active. The founders of

Place2Help opened a second platform in Frankfurt. This platform works as Meta-Platform for Regional Crowdfunding Campaigns in Frankfurt/Main. On the platform, a number of projects on campaigns such as Kickstarter, Indiegogo, Startnext can be found.

Without specific Civic Crowdfunding platforms, the reward-based platforms for the creative industries serve as hub for those projects. There are two major German reward-based Crowdfunding platforms: Startnext and Visionbakery.

Startnext is the biggest reward-based platform in Germany. The platform focuses on creative campaigns (Music, Movies), but they also have a category which is called "Community." In this category, Civic Crowdfunding projects such as the funding of a fountain in a Berlin Park can be found (https://www.startnext.com/fez-berlin-brunnen). Another important project which has gained a lot of traction recently on Startnext is the Pacific Garbage Screening Project (https://www.startnext.com/pgs). Both projects are perfect examples of non-rivalrous and non-excludable public goods.

Visionbakery, another reward-based platform, is much smaller not only in terms of overall volume but also quite active in the field of Civic Crowdfunding. They have a specific category for social projects. An example is this Civic Crowdfunding projects for a new kids playground space (http://www.visionbakery.com/teilhabe).

When it comes to Civic Actors operating a platform, we also see quite a few examples. Public Authorities are using the so-called White-label platforms. The platform is operated by a "normal" Crowdfunding provider, but the logo and the appearance are connected to a city or region. The first platform of this kind was called "Nordstarter," which was operated by the city of Hamburg. Further cities were Dresden, Munich, and Kassel. Sometimes federal states partnered with platform, often through the public state-owned banks, such as the L-Bank in Baden-Württemberg and the Investitionsbank Schleswig-Holstein.

This cooperation took place with the platform Startnext. The platform would "mirror" the projects on their own platform. For instance, all projects on the Nordstarter-Platform from Hamburg would also be featured on the Crowdfunding platform Startnext. The cities helped the projects with free consultancy and promotion. The disadvantage was that this model was quite costly—so the public authorities now switched to a different model called "Pages." On the platform Startnext, each of their partners has a sub-page with all projects from the area that they promote.

3. Equity-Based Civic Crowdfunding Platforms in Germany

Germany is one of the few countries in Europe which has a dedicated platform to Civic Equity-based Crowdfunding. The platform "Leih-Deiner-Stadt-Geld" (Lend money to your own city) is based in Frankfurt. It is currently not very active because cities have other means of financing their activities in Germany. But the platforms show that Civic Crowdfunding uses not only the donation- or reward-based Crowdfunding model but also equity- and lending-based Crowdfunding.

The platform is using a type of investment vehicle called Subordinated Loan. It is a loan which pays interest; however, the structure of the loan is that it functions like equity investments. When the project defaults, the investor does not receive any money back. The most notable project was an investment into Firearms Equipment at the city of Oestrich-Winkel near Frankfurt (https://www.leihdeinerstadtgeld.de/oestrich-winkel).

4. Donation-Based Civic Crowdfunding and Infrastructure in Germany

Public infrastructure is essentially a public good as well, with streets, bridges, or tunnels being prime examples. The important international examples (for instance, the Luchtsingel-Bridge in Rotterdam or the Tasmanian Beach in New Zealand, which was financed by the Crowd) are discussed elsewhere in this book. Infrastructure does not have to be physical; digital infrastructure also produces public goods.

In Germany, we have experienced some examples of financing infrastructure through Crowdfunding. The Freifunk movement used Crowdfunding for their purposes. The Wuppertal Freifunk movement collected €1,000 for a public Wireless Network in Wuppertal (https://www.startnext.com/de/wlan-fuer-das-luisenviertel/).

There are other examples as well, where playgrounds, schools, and churches were funded using Crowdfunding. Most of these campaigns have used a simple donation-based model. The majority of donation-based platforms have created a cooperation with a local bank. In Germany, unlike in other countries in Europe, most banks are run as cooperatives (Volksbanken) or local saving banks (Sparkassen). They have a strong affiliation with a region or a city. These banks are using White-label software by three major software providers: Startnext, Betterplace, and Table of Visions. We have counted almost 150 donation-based Crowdfunding platforms in Germany.

The banks provide the platforms free of charge to the projects. While most other platforms charge for operating and transaction costs, the donation-based platforms run by banks often cover all costs, thus providing a public good of essentially free Crowdfunding.

For the banks, using the donation-based Crowdfunding software is a great way to interact with their customers. They can channel requests for sponsorship to their platforms and co-finance those projects that receive enough attention. For the banks, the donation-based platforms are a form of corporate responsibility. Furthermore, the banks use the donation-based model to experiment with other forms of Alternative Finance as well. It is the author's opinion that these platforms, although small, are a very large driver of Civic Crowdfunding in Germany.

5. Civic Crowdfunding and Environmental Causes in Germany

Efforts to protect the natural environment, restore biodiversity, or fight climate change are another category of causes that create public goods, most of them non-rivalrous and non-excludable. Therefore, environmental projects can be considered Civic Crowdfunding as well. In Germany, we can find environmental causes both in equity- and in donation-based Crowdfunding.

German equity-based Crowdfunding Platforms such as Greenvesting, Bettervest, Ecoligo, and Econeers allow investment in Energy Projects. They not only generate profits for the investors but also generate profits for the environment because of reduced CO_2 emissions.

Donation-based Crowdfunding platforms focusing on environmental causes also exist. The first one in Europe was OneplanetCrowd in the Netherlands, which also operated in Germany.

A German platform with a similar objective is Ecocrowd. Ecocrowd supports sustainable projects in all kinds of areas and many of the projects can be considered Civic Crowdfunding. For instance, this campaign for a sustainable tourism agency for Africa combines normal Crowdfunding and Civic Crowdfunding (https://www.ecocrowd.de/projekte/mingle-africa-bildung-durch-reisen/).

Ecocrowd is run by the Deutsche Umweltstiftung (German Environment Foundation), a private foundation which receives grants from the German Ministry of Environment. The Ministry also supported the development of EcoCrowd; however, after the operation of the platform was paid for, the platform is now run based on membership, so citizens can become members of the platform in order to support the growth of the platform.

6. Civic Crowdfunding and Tourism in Germany

Tourism is a field where the definition of Civic Crowdfunding is not entirely applicable. In Germany, we do not find any dedicated platforms for tourism projects. However, the large reward-based and equity-based platforms have featured projects in the field of tourism, for instance, investments in a hotel resort or upgrading the energy efficiency of a hotel.

Hotels would not be considered a public good, given that they are often privately owned and the resources they provide (accommodation) are both rivalrous and excludable. However, promoting a region to tourists can generate spillover effects for other companies, such as restaurants. Therefore, one could argue that awareness of the touristic benefits of a region provide a public good to the companies in that specialisation.

The public Tourism Agency in Thüringen worked together with the German platform Startnext to promote tourism crowdfunding projects in Thüringen. Thüringen is a state in the South-eastern part of Germany. The agency developed a guidebook for projects as well and listed Crowdfunding projects in Thüringen through an API provided by Startnext (https://thueringen.tourismusnetzwerk.info/inhalte/produktentwicklung/crowdfunding/).

It that sense, it fits the Civic Crowdfunding definition put forward by some scholars that Civic Crowdfunding are also projects run by Public Authorities, even if they do not create a public good directly.

7. Civic Crowdfunding and Sports in Germany

Sport is another field where Civic Crowdfunding plays a role. Normally, sports activities create private goods which are both rivalrous and excludable—think of a major league football game and the high prices associated with entry. However, having a wide range of sport facilities, local clubs, and a healthy population can be considered a public good as well. Sports is a public undertaking in Germany. The German government supports local sports clubs through tax payments. Even commercially successful sport institutions like soccer clubs are required to invest in local sport networks and help to train the next generation of athletes. At the same time, there is a financing gap where neither private nor public money can support all activities which are needed. And this is maybe the case for Civic Crowdfunding in Sports.

In Germany, there are a few platforms dedicated to Sport Projects, but certainly, the platform Fairplaid is one of the biggest platforms in the field of Sport reward-based Crowdfunding. On the platform, it is possible to support local amateur sport teams, for instance, when they travel to tournaments or buy new uniforms. The platform is mostly used as a Friendraising platform—the team members advertise for the campaign within their circle of friends, parents, family, and colleagues.

8. Civic Crowdfunding and Public Support in Germany

A description of the Civic Crowdfunding landscape would not be complete without the activities of public institutions in this field, outside of hosting platforms themselves.

Because a number of platforms are active in Berlin, the city created a meta-platform called crowdfunding-berlin.com where all projects in Berlin on all platforms were featured. The city also ran a Crowdfunding competition which gave prices to the best crowdfunding campaigns in specific branches, for instance, design and technology. The meta-platform also offers consultants the opportunity to present themselves and has an FAQ for Crowdfunding supporters.

A similar model is taking place in Munich. An SME from Munich can apply to a subsidy for costs in preparing the campaign, for instance, in shooting a pitch video. The idea is to support innovative Crowdfunding campaigns by subsidising their initial costs.

Civic Crowdfunding has also been in the political debate. The regional Parliament in Baden-Württemberg has requested the Ministry of Finance in Baden-Württemberg to give feedback on Civic Crowdfunding and to develop proposals on how to support the Civic Crowdfunding in the state.

9. Conclusion

This chapter tried to outline the landscape of Civic Crowdfunding in Germany. It has established that commercial platforms have opened for Civic Crowdfunding projects, as well as Civic Institutions, such as City Governments, Regional Governments, Public Banks, and Public Development Agencies which are regularly interacting with the CF Ecosystem.

About the Author

Karsten Wenzlaff is the Founder of the Institute of Communications for Social Media (ikosom), a Berlin-based research facility for new forms of electronic technology. He has received an M.Phil. in International Relations with a thesis on international financial regulation from the University of Cambridge. Currently, he is a lecturer and PhD Student at the University of Hamburg at the Chair of Digital Markets.

Chapter 16

Crowdfunding Act: Accelerating the Growth of Crowdfunding Market in Finland

Ilkka Harju and Aki Kallio

Abstract

In Finland, popularity of crowdfunding expanded rapidly starting from 2013. The Ministry of Finance together with the Ministry of Employment and Economy conducted a survey in late 2013 in order to map the size of crowdfunding markets in Finland. It was seen as a fast, effective, and flexible funding channel, especially for start-up companies and new innovative enterprises. The Finnish Crowdfunding Act (734/2016, August 25, 2016) came into effect on September 1, 2016. The preparation work in the Ministry of Finance started in 2014 based on an order from the Minister of Employment and Economy in office at the time. The aim was to ease the strict interpretation made by the Finnish Financial Supervisory Authority (FIN-FSA) in mid-2014, basically categorising all investment-based crowdfunding as placing of financial instruments without a firm commitment basis as defined in Markets in Financial Instruments Directive (MIFID I, 2004/39/EC). The Ministry of Finance prepared the project in cooperation with market operators, interest groups, and the relevant authorities.

The law bill (HE 46/2016 vp) aimed to establish clear ground rules for crowdfunding at the legislative level in order to ensure that the business sector could operate and grow in Finland. The aim was also to clarify the responsibilities of various authorities in the supervision of crowdfunding, to improve investor protection, and to diversify the operability of financial markets. A further goal of the law bill was to increase the financing options of SMEs as well as small, innovative, and growth-oriented companies in particular. Investors were seen to receive higher risk and higher yielding investment opportunities. The Crowdfunding Act applies to mediation and acquisition of crowdfunding for financing business activities in Finland i.e. the Crowdfunding Act addresses only investment-based and loan-based crowdfunding with financial return. For operating in other EU member states, an intermediary should still apply, for example, an MIFID operating licence. The Crowdfunding Act specified the terminology and the responsibilities of the authorities in the business sector and made an effort to make clear a complex subject more understandable. It also strengthened the position of investors by specifying clear rules on investor protection and disclosure requirements. At the same time, crowdfunding became a recognised part of the regulated financial markets.

Due to changes in the Finnish political climate and, consequently, a more strict approach towards financial services regulation in general, the Crowdfunding Act was criticised by the FIN-FSA during 2017, in particular, as being too liberal and innovative. The central argument in the criticism towards original Crowdfunding Act was that not a single application to operate as a registered investment-based crowdfunding platform had been received by the FIN-FSA. An opposite argument was that the FIN-FSA had been against such light-touch regulation in the first place and had in practice discouraged eventual applicants from seeking such registration. Following this debate, the Ministry of Finance allowed the FIN-FSA to draft an alternative legislative approach to regulate crowdfunding in a more rigid and legalistic manner.

Despite the changes to the regulatory framework, the crowdfunding market in Finland has continued to grow. In Finland, the crowdfunding market has more than doubled in size from 2015 (€70.5 million) to 2016 (€153.2 million) according to a survey by the Ministry of Finance and the growth is expected to continue according to predictions of the Ministry. Based on the recent survey by the Bank of Finland, the growth has indeed continued. The crowdfunding market in 2017 was €246.7 million showing an impressive 61% growth rate compared to the last year.

The Ministry of Finance has previously stated that it expects the crowdfunding market to grow further. Factors contributing to this growth include the new Crowdfunding Act, which clarifies the rights and obligations of both market operators and investors. The Crowdfunding Act placed loan-based and investment-based crowdfunding within the sphere of regulated financial markets. However, based on the above-referred recent amendments (mainly related to strengthening of investor protection) made to both the Investment Services Act and the Crowdfunding Act in relation to transposition of MiFIR and MIFID II, it is unclear whether the growth of the alternative finance market will continue as strong as till date. Taking into account the contents of the recent Commission proposal, it is unfortunately becoming evident that new restrictions imposed in relation to transposition of MiFIR and MIFID II to both the Investment Services Act and the Crowdfunding Act were premature and unnecessary.

1. Background

In Finland, popularity of crowdfunding expanded rapidly starting from 2013. The Ministry of Finance together with the Ministry of Employment and Economy conducted a survey in late 2013 in order to map the size of crowdfunding markets in Finland. Based on the survey, which was published on March 13, 2014, high expectations were being placed on crowdfunding. It was seen as a fast, effective, and flexible funding channel especially for start-up companies and new innovative enterprises. Experts stated that crowdfunding would allow for more effective use of private individuals' funds and increase the overall availability of financing.[1] Under its structural policy programme, Prime Minister Jyrki Katainen's Government decided in late 2013—especially after publication of the survey results in a separate report—to explore the needs for development in the area of crowdsourced loans and investment.

The main potential problems with crowdfunding, according to the survey results, were seen in the risk of misuse, the lack of effective regulation, the absence of formal regulatory supervision, and inadequate investor protection, particularly in questions related to the disclosure and exchange of information. However, it was seen to be of paramount importance that the regulation would be kept administratively light in order to achieve sound objectives stated in the survey report, but for reasons of improved investor protection and preventing misuse, it might be necessary to introduce a unified set of disclosure requirements as well as mechanisms for operator registration.

2. Drafting of the Crowdfunding Act

The Finnish Crowdfunding Act (734/2016, August 25, 2016)[2] entered into force on September 1, 2016.[3] The preparation work in the Ministry of Finance started in 2014 based on an order from the Minister of Employment and Economy in office at the time. The aim was to ease the strict interpretation made by the Finnish Financial Supervisory Authority (FIN-FSA) in mid-2014[4] basically categorising all investment-based crowdfunding as placing of financial instruments without a firm commitment basis as defined in the Markets in Financial Instruments Directive (MIFID I, 2004/39/EC).[5] Additionally, the strict interpretations of the Money Collecting Act (255/2006) by the National Police Authority received negative feedback not only from the business community but also from several political parties and non-profit organisations. Money Collecting Act was widely seen as an obstacle to introduce especially donation-based crowdfunding in Finland.

The Ministry of Finance prepared the project in cooperation with market operators, interest groups, and the relevant authorities. The draft Crowdfunding Act was disclosed and circulated for comments in May–June 2015. Statements were received from 47 interested parties, most of whom responded positively to the proposal. The Ministry of Finance also arranged two public consultations with authorities and one public consultation with market operators. Those drafting the Crowdfunding Act met with key stakeholders dozens of times in connection with the project. Every effort was made to take the feedback from consultations and statements into account in the process of drafting the Crowdfunding Act.

Crowdfunding was seen as a new form of financing that complements traditional financing channels or even partly replaces them. It has been of particular interest in recent years, foremost to seed and growth companies as well as innovative enterprises and other SMEs both in Finland and internationally. For investors, it was assessed that crowdfunding would offer higher yielding and higher risk investment opportunities than traditional investments.

It is one of the business law enhancement priorities of the current Finnish Government lead by Prime Minister Juha Sipilä to eliminate bottlenecks in corporate financing. SME funding is considered important, both on shares and debt, with tailored thresholds and other legislative requirements which deviate from those applied in general securities markets legislation. The purpose of the Crowdfunding Act is to enhance especially start-ups, innovative enterprises, and other small-scale new and emerging businesses by providing possibilities of alternative financing to traditional financing sources via, e.g. lesser disclosure requirements than those set out in the EU Prospectus regime. Access to finance is one of the most important factors for innovative start-ups and scale-ups when it comes to their possibilities to enter the market and develop new innovative products and services.

Due to the recent international financial crisis, the SMEs' access to finance from banks had become more difficult, or at least more expensive, because banks' capital adequacy and solvency requirements had been robustly tightened. Companies were therefore in need of new alternative sources of funding. In Finland, the availability of financing from banks had not deteriorated as much as in many other EU member states. Nevertheless, crowdfunding had also become more popular in Finland, because research showed that banks had tightened their requirements relating to collateral and special conditions linked to them (covenants, e.g. with regard to cash flow) and had also increased the margins on their loans. Demand and supply of crowdfunding has also been increased by the fact that it is nowadays possible to reach large numbers of people cost-effectively online.

During the drafting process of the Crowdfunding Act, many risks were identified. Crowdfunding was typically seen to be used by seed and growth companies as well as SMEs to acquire financing. Investments in start-ups, in particular, are notoriously known to be of high risk. It is estimated that more than half of the start-ups go out of business within a few years. Of the surviving start-ups, only a few become big and successful, and most of them struggle daily to keep afloat and to continue in business.

On the contrary, in all investment activities there is a risk of losing some or all of the invested assets. The Crowdfunding Act aims to ensure that investors are informed of all relevant information to enable them to make informed investment decisions. Moreover, crowdfunding as a business sector is also in a rapidly evolving growth phase, and all procedures have not yet been established. In this respect, too, the aim of the Crowdfunding Act was to steer the sector in the right direction and to define the rights and obligations of intermediaries, investors, and companies seeking finance in a centralised and transparently structured way.

It was also noted that no regulated market recognised in MIFID or any other secondary marketplace exists for securities (e.g. shares of non-listed companies) given as consideration to investors for financing in crowdfunding. For this reason, they are usually not traded in the secondary market, which may make it extremely difficult to convert the investment into cash later.

3. The Goal of the Crowdfunding Act

The law bill (HE 46/2016 vp) aimed to establish clear ground rules for crowdfunding at the legislative level in order to ensure that the business sector could operate and grow in Finland. The aim was also to clarify the responsibilities of various authorities in the supervision of crowdfunding, to improve investor protection, and to diversify the operability of financial markets. A further goal of the law bill was to increase the financing options of SMEs as well as small innovative and growth-oriented companies in particular. Investors were seen to receive higher risk and higher yielding investment opportunities.[6]

The innovative elements of the law can be divided into two main objectives. First, the intention has been to seek possibilities in EU law to take advantage of member state options and other flexibilities in order to make the regime more lucrative to small businesses. Second, the intention has been to enhance small businesses' access to alternative financing sources. The Crowdfunding Act particularly eases the regulation of investment-based crowdfunding and correspondingly clarifies the basic rules for loan-based crowdfunding. It is the general approach of the Crowdfunding Act to be innovation friendly as it improves the opportunities for businesses to finance their growth. Thus, all single paragraphs of the Crowdfunding Act should therefore be seen in this context.

Particular focus of the Crowdfunding Act is to strengthen the possibilities of crowdfunding platforms operating in the internet to grow, prosper, and succeed. Innovation elements were examined by two surveys directed to market participants *ex ante* and one survey *ex post*. These surveys were conducted by the Ministry of Finance and reached the crowdfunding business community in its entirety. The results of the surveys were a valuable tool, which enabled to better adapt the legislative measures to the needs of the business requirements.

The Crowdfunding Act eased the regulation of investment-based crowdfunding (based on MIFID I article 3 optional exemption) and correspondingly clarified the ground rules for loan-based crowdfunding. In addition, the Crowdfunding Act clarified the responsibilities of various authorities in the supervision of crowdfunding, improved investor protection, and aimed to diversify financial markets by way of facilitating the entry of crowdfunding platforms therein.

The crowdfunding business sector's market operators are very small compared with established market operators, and they have only limited administrative resources at their disposal. In addition, the sector to date is responsible for only a small part of the money collected and mediated in the financial markets. Both these observations were seen to stress the need for proportionate regulatory approach.[7]

An operating licence requirement (authorisation from FIN-FSA) and the process associated with it, therefore, were seen to constitute an unreasonably high threshold for access to business and entry into the sector. For this reason, it was assessed that transposing MIFID I article 3 optional exemption into national law would be the most efficient way to promote the development of crowdfunding in the domestic market and its competitiveness in relation to established financial market operators. Inspiration to take advantage of MIFID I article 3 was based on a comparative survey including 10 European countries, of which France and Italy proved to be the most useful benchmarks. As it was foreseen that MIFID II[8] would extend the requirements set in optional article 3, those conditions were proactively transposed into the Crowdfunding Act. Thus, only technical fine-tuning was deemed to be necessary when transposing MIFID II into national law in 2018. For example, as no investment-based crowdfunding platforms sought for registration, this option was removed from the Crowdfunding Act and only the possibility to seek for a licence to operate as an investment firm is currently available.

4. Scope of the Crowdfunding Act

The Crowdfunding Act applies to mediation and acquisition of crowdfunding for financing business activities in Finland, i.e. the Crowdfunding Act addresses only investment-based and loan-based crowdfunding with financial return. For operating in other EU member states, an intermediary should still apply for an MIFID operating licence, for example. MIFID II provides for an opportunity for cross-border activity in accordance with the so-called single-authorisation principle (European passport). If a credit institution, management company, investment firm, or insurance undertaking has obtained an operating licence in a member state of the European Economic Area (EEA), it may engage in activity under its operating licence in all EEA member states. A company may either establish a branch or provide cross-border services through a separate notification procedure. The European passport provides an opportunity to expand crowdfunding mediation and other platform services to cross-border EU-level activity.

Non-interest bearing loans are not currently available in the Finnish crowdfunding market practice. However, if such loans would emerge, it could be argued that they would fall into the scope of the Money Collection Act. Donation-based crowdfunding is also covered by the Money Collection Act, which basically covers all collection of money without consideration. Investment-based crowdfunding and loan-based crowdfunding were carved out from the scope of the Money Collecting Act.

One Finnish-based crowdfunding intermediary, which is operating on the basis of an investment firm authorisation, has operations currently in the United Kingdom, Norway, Denmark, and Finland. There is also another one, which has just recently expanded its operation to Norway via setting up a branch in the country. According to a survey by the Ministry of Finance, which was published on December 28, 2016, crowdfunding intermediaries which were offering investment-based or loan-based crowdfunding in Finland gathered approximately from 2% to maximum of 10% of investment from abroad (mainly from other EU member states).

Under the Crowdfunding Act, crowdfunding for business activity can be collected by a business entity (limited liability company, cooperative, limited partnership, general partnership, European limited company, European cooperative, and association) or a foundation. The party collecting crowdfunding cannot be in bankruptcy or other insolvency procedures. On the contrary, a restructuring procedure does not prevent the collection of crowdfunding as long as investors are adequately informed about the issue and the associated risks. The party collecting crowdfunding cannot be a listed company, i.e. the securities of listed companies cannot be the investment objects of loan-based or investment-based crowdfunding investment. The party collecting crowdfunding is not required to be a legal entity; thus, a business name is also possible.

The Crowdfunding Act essentially applies to crowdfunding intermediaries. To ensure adequate investor protection, the Crowdfunding Act also lays down obligations on the party collecting the financing through crowdfunding and, correspondingly, certain rights for investors and customers. For crowdfunding intermediaries (platforms), the Crowdfunding Act replaced the administratively onerous, cumbersome, and the time-consuming operating licence process with a registration process that is less expensive, simpler, and faster. In addition, the Crowdfunding Act dismissed the necessity to join the Investor Compensation Scheme, reduced the minimum capital requirement from €125,000 to €50,000, introduced a concept of good crowdfunding practice, as well as self-regulation for the industry and created a crowdfunding exemption to Finnish prospectus rules (till €5 million instead of the €2.5 million threshold, which generally obliges to publish a prospectus for a securities offer).

Alternatively, the capital requirement can be replaced by a professional liability insurance policy, bank guarantee, or other corresponding collateral which the FIN-FSA deems to be sufficient. If that were the case, due account of the nature and scope of the notice provider's activities must be taken into account. The insurer or other collateral provider must be domiciled in an EEA state, unless an exception to this is granted by the FIN-FSA upon application. The insurance policy must also meet the following terms: (i) the insurance shall be valid for compensation of losses for which the crowdfunding intermediary is responsible under the Crowdfunding Act; (ii) the sum insured shall be a minimum of €1 million per loss and a total of €1.5 million for all losses in a year; (iii) if the insurance policy incorporates a deductible, the insurer shall pay the insurance compensation to the party suffering the loss without subtracting the deductible; (iv) the insurance shall compensate a loss that arises as a result of an act or omission that occurred during the period of insurance and for which a written claim for compensation is presented to the notice provider or the insurer during the validity of the insurance policy or within 3 years of the expiry of the policy.

This exemption supported the growth of crowdfunding, with the aim of facilitating the utilisation of crowdfunding in business financing more extensively than at the time without the obligation to prepare a separate and often rather expensive prospectus. Moreover, the much-debated exemption based on national discretion turned out to be a visionary one, while anticipating the European Commission's proposal of November 30, 2015 to increase the threshold to €8 million.

5. Crowdfunding Intermediaries to the Register

Most financial services provided on a professional basis require an authorisation from the FIN-FSA. However, in some regimes like payment services and alternative investment fund managers, where the service providers are often small in size, EU law also provides registration procedure as an alternative to authorisation. Registration procedure has proven popular in Finland, particularly among small payment service providers (40+) and alternative investment fund managers (70+). With regard to crowdfunding platforms, the practical experience is still thin, but so far, eight registration applications have been received and five registrations granted by the FIN-FSA. In addition, two large credit institutions and two small investment firms have launched a crowdfunding platform. Authorised financial market participants who may operate as crowdfunding intermediaries on the basis of their authorisation (if based on CRD, MIFID or AIFMD) would be allowed to operate cross-border after appropriate notification procedure. However, they need to follow certain rules of the Crowdfunding Act. The crowdfunding intermediaries seeking registration are not allowed to operate cross-border. Many credit institutions and insurance undertakings have also started to collaborate with crowdfunding intermediaries. So far, the news has been more or less expectant and cautiously positive.

It seems that rules related to cross-border investments or lending are more or less vague to national authorities and especially to retail investors. The legal framework does not hinder investors or lenders from cross-border transactions based on their own initiative. During one-to-one discussions with Finnish-based crowdfunding intermediaries, it has been stated that regulatory fragmentation based on different member states introducing their own bespoke crowdfunding regimes as well as non-consistent interpretation of the existing EU-rules by market supervisors have made cross-border capital-raising difficult. In addition, crowdfunding intermediaries in Finland have stressed that existing EU regulatory framework does not take into account the special characteristics of either loan-based or investment-based crowdfunding and hence it should be evaluated whether crowdfunding as a new way of finance could be supported via amendments to existing EU regulatory framework which could better promote cross-border capital-raising. A prime example is the interpretation of article 3(b) of Payment Services Directive (PSD II)[9] in connection with loan-based crowdfunding. It seems that market supervisors in different EU member states tend to interpret PSD in a way which does not promote regulatory convergence.

It is worthwhile noting that reference to article 3(b) of PSD II means that duly registered platforms may not currently manage customers' funds or assets. However, platforms can apply for a license based on the Act on Payment Institutions (297/2010), but they can also operate as registered payment institutions. Registration gives them a somewhat more restricted chance to offer payment services compared to the full license. If the crowdfunding recipient aims to administer or keep in its possession any financial instruments or cash assets of its customers (customer assets) or be in debt to a customer, it must apply for a registration or license based on the Act on Payment Institutions. This is based on interpretation of the FIN-FSA, although it is not clearly stated in the Crowdfunding Act.

Based on various stakeholder meetings, public hearings, studies, and finally wide-scale inception impact assessment, the European Commission, as part of its Fintech action plan, in March 2018, presented a proposal for a regulation on crowdfunding service providers. If adopted at the EU level, the new regulation aims to allow platforms to apply for an EU passport based on a single set of rules. This will make it easier for them to offer their services across the EU. The new rules aim to improve access to this innovative form of finance for small investors and businesses in need of funding, particularly start-ups. Investors on crowdfunding platforms will benefit from a better protection regime and a higher level of guarantees, based on clear rules on disclosure requirements for project owners and crowdfunding platforms, rules on governance and risk management, and a coherent approach to supervision. The Commission proposal only applies to those crowdfunding services entailing a financial return for investors, such as investment and lending-based crowdfunding. The Commission proposal as it currently stands will not have any effect to the Crowdfunding Act and its content. The Commission proposal and member state negotiations related thereto are currently pending.

One Swedish-based platform has tried to enter the Finnish markets but has not been able to gather solid ground. Based on the Crowdfunding Act, it should have applied registration unless it is operating in Finland based on MIFID or similar authorisation acquired in its home Member State. We have witnessed a number of court cases where another Swedish peer-to-peer lending platform advanced superfluous interest rate levels. One pilot court case entered into the Finnish Supreme Court which turned to the European Court of Justice for advice. However, the Swedish platform went bankrupt and the case died. The question raised by the Supreme Court was whether the platform operated under Consumer Credit Directive.[10] As far as we know, the question remains unsolved. Other foreign platforms have not tried to enter the Finnish market probably due to the relatively small size and existing competition in the market.

The Mortgage Credit Directive[11] was transposed in Finland as of January 1, 2017 including some national regulation concerning peer-to-peer consumer lending. However, taking into account the date of transposition of the directive, there is only limited experience on its application in practice.

Under the Crowdfunding Act, crowdfunding can be mediated only by businesses that are entered in a special register of crowdfunding intermediaries. The conditions for registration are specified in detail in the Crowdfunding Act. The register will be open to reliable operators that have adequate knowledge of the financial markets. The FIN-FSA will be responsible for the register and the supervision of the intermediaries.

Through registration, the aim was to create a sufficient threshold for entry into the market, which in turn is of great significance from the perspective of the credibility of investor protection and this form of financing (i.e. crowdfunding). In preparing the registration of crowdfunding intermediaries, the aim was to make it lighter than other financial market regulation (often heavy, with an expensive operating licence process), because the objective was to diversify the domestic financial markets by providing sustainable and predictable conditions for new, alternative forms of financing. The relative lightness of regulation is based on the business sector's small size, growth phase, risk profile, and other characteristics (like the proportionality principle recognised in EU financial services law which takes into account the size and complexity of the business and the risks involved in it).

6. Investor Protection is Improved

The Crowdfunding Act specified the terminology and the responsibilities of the authorities in the business sector and made an effort to clarify a complex subject more understandable. It also strengthened the position of investors by specifying clear rules on investor protection and disclosure requirements. At the same time, crowdfunding became a recognised part of the regulated financial markets.

In the Crowdfunding Act, no maximum investable amounts are set but a mandatory appropriateness test is required for investments above €2,000.

As concerns due diligence, for the purpose of a considered assessment of an issuer and the favourability of an offer, the issuer must disclose true and sufficient information about factors that are likely to materially influence a company's value or its repayment ability, before starting to acquire funds. However, the valuation always depends on parameters used. Crowdfunding intermediaries must ensure that issuers meet the obligation laid down in the Ministry of Finance Decree (1045/2016) on disclosure requirements set forth to issuers (not full but adequate due diligence).[12] MIFID conflict of interest rules also apply as well as rules on detecting and preventing money laundering and terrorist financing. Provisions on knowing the customer are laid down in the Act on Detecting and Preventing Money Laundering and Terrorist Financing (444/2017), and hence, duly registered crowdfunding operators are fully subject to AML/ CFT requirements.

Investor protection was additionally improved, for example, by extending under the Crowdfunding Act the obligation to disclose information also to situations in which securities other than those referred to in the Securities Markets Act (746/2012) are utilised in acquiring or mediating crowdfunding. At the time, securities other than those referred to in the Securities Market Act, essentially non-transferable securities, were not included within the scope of the Securities Market Act or the Investment Services Act (747/2012). They were, therefore, largely outside of regulation and thereby without investor protection. In addition, investor protection was specified in a separate Decree of the Ministry of Finance relating to the disclosure obligation of an entity acquiring financing through crowdfunding. The Decree on the contents and structure of the issuer's disclosure obligation include risk warning and detailed contents and form thereof. According to the Decree, an issuer (i.e. company seeking financing through crowdfunding) must disclose a document containing key crowdfunding information (Basic Information Document) available to investors throughout the period of validity of the offer. Basic Information Document must: (i) in terms of its contents and format, be clear, easy to read, and understandable; (ii) be no longer than six A4-size pages; and (iii) include information on: (a) the issuer, (b) the investment instrument and the offering, and (c) the guarantor and collateral (if applicable). In addition, the key risks associated with the crowdfunding recipient's business and the investment instrument must be listed on the Basic Information Document (risk warning). The Crowdfunding Act, for its part, has provisions on a crowdfunding intermediary's procedures and its obligations towards the investor. In addition, the Crowdfunding Act lays down provision on supervision and sanctions.

Sanctions in the Crowdfunding Act can be divided into penal provisions, administrative fines, penalty payments, and liability for damages. For example, anyone who wilfully or through gross negligence mediates crowdfunding without registration, must be sentenced to pay a fine or to imprisonment for no more than 1 year for a *crowdfunding offence*, unless the act is of a minor nature or a more severe punishment is provided for it elsewhere in the law. The neglect or violation of which a penalty payment is imposed are, e.g. the provisions on the duty of disclosure.

The rules will facilitate and clarify the obligation to provide information (including key risks), which contributes to higher efficiency and reduces the costs of the financing operations. It will be easier for the investors to compare the various purposes of investment when all issuers disclose the information in the same form when seeking finance through crowdfunding. In accordance with the Crowdfunding Act, no false or misleading information may be given in the marketing of crowdfunding. Information that proves to be false or misleading after it has been presented and which may be of material importance to the customer of a crowdfunding intermediary must be adequately corrected or supplemented without delay. Crowdfunding intermediaries must ensure that issuers meet their disclosure obligations laid down in the Ministry of Finance Decree.

The Decree of the Ministry of Finance of December 8, 2016 came into effect on December 15, 2016 and specified the obligations concerning the provision of information in the Crowdfunding Act. The Decree specifies the information to be provided on the issuer, the investment project, and the investment instrument when applying for crowdfunding. This enables the investor to understand better what the issuer, the investment project, as well as the investment instrument are really like. For example, the issuer must state clearly and in a readily understandable way the risks that may be involved regarding both the issuer and the investment.

The reform will facilitate and clarify the obligation to provide information, which contributes to higher efficiency and reduces the costs of the financing operations. This is based on the fact that the documents will be largely harmonised in terms of their content and structure. It will be easier for the investors to compare the various purposes of investment when all issuers disclose the information in the same form when seeking crowdfunding.

7. Crowdfunding Act is also Relevant for the Existing Market Operators

Authorised financial market operators may utilise the fact that the Crowdfunding Act has a lighter structure than other financial market regulations. They may, for example, set up a subsidiary to mediate loan-based or investment-based crowdfunding or link up a crowdfunding intermediary already operating in the market with their own authorised activities through an arrangement utilising a tied agent. A basis of the Crowdfunding Act is that credit institutions, payment institutions, management companies, investment firms, and alternative fund managers may also mediate crowdfunding by virtue of their own operating licence as long as they adhere to certain separately specified provisions of the Crowdfunding Act.

8. Transposition of MIFID II and the Crowdfunding Act

Due to changes in the Finnish political climate and, consequently, a more strict approach towards financial services regulation in general, the Crowdfunding Act was criticised by the FIN-FSA during 2017, in particular, as being too liberal and innovative. The central argument in the criticism towards original Crowdfunding Act was that not a single application to operate as a registered investment-based crowdfunding platform had been received by the FIN-FSA. An opposite argument was that the FIN-FSA had been against such light-touch regulation in the first place and had in practice discouraged eventual applicants from seeking such registration. Following this debate, the Ministry of Finance allowed the FIN-FSA to draft an alternative legislative approach to regulate crowdfunding in a more rigid and legalistic manner. MIFID II article 3, which includes a somewhat extended EU member state optional exemption, was therefore, on the basis of FIN-FSA drafting, transposed to Finnish law in a formal, robust, and legalistic manner. The optional exemption was thus interpreted as to allow only narrow national discretion. In this context, 6 of the 23 paragraphs of the Crowdfunding Act were technically amended. These amendments covered Scope of Application (Section 1), Definitions (Section 2), Obligation to register (Section 3), Right of crowdfunding intermediary to mediate loan-based or investment-based crowdfunding (Section 9), Crowdfunding intermediary's procedures and its obligations towards the investor (Section 10), and Administrative fines and penalty payments (Section 15).

The amendments reflect, first, an overall approach where MIFID is considered a framework EU regime which is above and supreme to other legal regimes in the field of financial services. Second, the amendments reveal the dimensions of the FIN-FSA's rare administrative position. As the FIN-FSA is located within the central bank, it enjoys an extraordinary independence which makes it possible to operate regardless of the Government's political goals and ambitions. The proposed amendments were met with hostility by the crowdfunding business community as amendments had been prepared without proper open consultation with the stakeholders. The debate was fierce and continued at the hearings of the Finnish Parliament's Commerce Committee. The spearhead of the debate was the FIN-FSA position where it objected to the light-touch regulatory approach that sought to enhance start-up businesses in financial markets. All in all, 36 stakeholders both from the public sector and the private sector participated in those hearings.

In the MIFID II transposition law bill (HE 151/2017), the most relevant change was the downgrading of the registration procedure. Thus, it is no longer possible to register an investment-based crowdfunding intermediary within the FIN-FSA. Instead, an investment firm authorisation is required. In this regard, the rules mean a complete turnaround and essentially lead back to square one, i.e. to a similar situation that was the starting point of drafting the Crowdfunding Act in mid-2014 with the exception that loan-based crowdfunding to businesses is now a regulated activity as set forth in the Crowdfunding Act. Indeed, the role of the FIN-FSA was crucial in drafting the new regime. Particularly, investment-based crowdfunding will be subject to heavy regulation, as the general financial market legislation was applicable to the provision of investment services through crowdfunding.

Contrary to the traditional Finnish approach in transposing EU financial services law, which aims to seek and follow the mainstream line of European thinking, the FIN-FSA position towards crowdfunding appears to be quite extreme in European Securities and Markets Authority (ESMA) comparison. Whereas in most EU member states crowdfunding is considered to constitute a business of reception and transmission of client orders, FIN-FSA approach views crowdfunding as placing of financial instruments. However, the FIN-FSA has not published observations on any wrongdoings with regard to crowdfunding. Nor were the impact assessment and the international comparison of the 2016 law bill updated in the MIFID II transposition law bill. Thus, the proposed amendments were to a great extent based on a more legalistic reading of the EU member state optional exemption in MIFID II article 3, and also reflected the FIN-FSA's negative approach to light-touch regulation of investment-based crowdfunding. This tension reveals the different views with regard to regulation of financial technology and, particularly, whether start-up businesses should be enhanced with light-touch rules or not. It is not surprising that the traditional financial services business community is against favouring such disruptions and frequently appeals for a level playing field.

9. The Size of the Market is Growing Rapidly

In Finland, the crowdfunding market has more than doubled in size from 2015 (€70.5 million) to 2016 (€153.2 million) according to a survey by the Ministry of Finance and the growth is expected to continue according to predictions of the Ministry.[13] Based on the recent survey by the Bank of Finland,[14] the growth has indeed continued. The crowdfunding market in 2017 was €246.7 million showing an impressive 61% growth rate compared to the last year.

The crowdfunding market in 2017 can be summarised as follows:

- investment-based crowdfunding: €63 million (year-on-year increase of 51%);

- loan-based crowdfunding for businesses: €75.8 million (year-on-year increase of 64%);
- peer-to-peer lending between private individuals: €106.8 million (year-on-year increase of 67%);
- reward-based crowdfunding: €1 million (year-on-year increase of 5%).

The Ministry of Finance has previously stated that it expects the crowdfunding market to grow further. Factors contributing to this growth include the new Crowdfunding Act, which clarifies the rights and obligations of both market operators and investors. The Crowdfunding Act placed loan-based and investment-based crowdfunding within the sphere of regulated financial markets. However, based on the above referred recent amendments (mainly related to strengthening of investor protection) made to both the Investment Services Act and the Crowdfunding Act in relation to transposition of MiFIR and MIFID II, it is unclear whether the growth of the alternative finance market will continue as strong as till date. Taking into account the contents of the recent Commission proposal, it is unfortunately becoming evident that new restrictions imposed in relation to transposition of MiFIR and MIFID II to both the Investment Services Act and the Crowdfunding Act were premature and unnecessary.

Finnish crowdfunding platforms have not so far utilised blockchain technology or virtual currencies. Regulation of virtual currencies is being contemplated in the context of amendments being made to the Act on Detecting and Preventing Money Laundering and Terrorist Financing and Money Collection Act. However, such amendments have not yet been entered into force in Finland (at the time of writing of this article, i.e. July 31, 2018).

About the Authors

Ilkka Harju is a Senior Legislative Counsellor at the Ministry of Finance of Finland. He has been in charge of many of the most complex projects related to the Financial Markets both before and after Finland joined the European Union. He acts as the Development and Quality Manager in the Financial Markets Department of the Ministry. He is also a highly valued and appreciated lecturer in the field of Financial Markets.

Aki Kallio is a Managing Compliance Officer (Corporates & Institutions) at Danske Bank A/S Finland Branch. He is currently in charge of managing and supervising a compliance team responsible for conducting control and monitoring activities with regard to wide range of business units operating under the umbrella of Corporates & Institutions as well as providing legal advice in various issues falling into the scope of Financial Markets. Previously, Mr. Kallio worked as a Ministerial Adviser in the Financial Markets Department of the Ministry of Finance of Finland, where his main responsibility area consisted of various assignments relating to divergent legislative and other actions of Financial Markets Department especially in the field of capital and bond markets. In particular, Mr. Kallio was in charge of assessing and developing both Finnish bond market (acting as a head secretary of the Finnish Bond Market expert group) as well as alternative finance market (responsible for drafting and entering into force the Finnish Crowdfunding Act and acting as a co-head of the Financial Market Department's Fintech -expert group). In addition, Mr. Kallio acted as a Finnish representative of the Commission's informal expert group European Crowdfunding Stakeholder Forum (ECSF) which assisted the Commission in developing EU-wide policies for crowdfunding and alternative finance. Mr. Kallio is a frequent lecturer in the field of bond market as well as alternative finance/fintech both in domestic and international events. Prior to joining the Ministry of Finance of Finland, Mr. Kallio worked for two leading Finnish law firms and as a corporate counsel.

Endnotes

1. Ministry of Finance and Ministry of Employment and Economy: Report on Crowdfunding Survey 13th March 2014 (https://api.hankeikkuna.fi/asiakirjat/a91a2dba-46f8-4497-8431-e3af46466fcd/10740ca7-b8b9-4533-ab7d-fe3c90726a4f/JULKAISU_20140327151501.PDF).
2. Unofficial English translation of the Crowdfunding Act available via https://www.finlex.fi/en/laki/kaannokset/2016/en20160734.pdf.
3. Ministry of Finance—Press release August 25, 2016: Crowdfunding Act enters into force in September (https://vm.fi/en/article/-/asset_publisher/joukkorahoituslaki-voimaan-syyskuussa).

4. For more about the interpretation, please see Borenius Attorneys Ltd., Legal Alert, FIN-FSA Makes Move to Regulate Equity Crowdfunding Platforms July 30, 2014 (https://www.borenius.com/2014/07/30/legal-alert-fin-fsa-makes-move-to-regulate-equity-crowdfunding-platforms/) and European Crowdfunding Network - Finnish Financial Authority Publishes Guidelines for Securities Crowdfunding July 22, 2014 (https://eurocrowd.org/2014/07/22/finnish-financial-authority-publishes-guidelines-securities-crowdfunding/).

5. Directive 2004/39/EC of the European Parliament and of the Council of April 21, 2004 on markets in financial instruments amending Council Directives 85/611/EEC and 93/6/EEC and Directive 2000/12/EC of the European Parliament and of the Council and repealing Council Directive 93/22/EEC.

6. Ministry of Finance—Press release April 7, 2016: Crowdfunding Act to provide new financing options for business growth (https://vm.fi/en/article/-/asset_publisher/joukkorahoituslaki-tuo-toimialalle-kevytta-saantelya).

7. For more information about the background concerning goals and assessment related to the Crowdfunding Act, please see Ministry of Finance—Press release April 12, 2016: Why is Crowdfunding Act needed? Questions and Answers (https://vm.fi/en/article/-/asset_publisher/miksi-tarvitaan-joukkorahoituslaki-).

8. Directive 2014/65/EU of the European Parliament and of the Council of May 15, 2014 on markets in financial instruments and amending Directive 2002/92/EC and Directive 2011/61/EU.

9. Directive (EU) 2015/2366 of the European Parliament and of the Council of November 25, 2015 on payment services in the internal market, amending Directives 2002/65/EC, 2009/110/EC and 2013/36/EU and Regulation (EU) No. 1093/2010, and repealing Directive 2007/64/EC.

10. Directive 2008/48/EC of the European Parliament and of the Council of April 23, 2008 on credit agreements for consumers and repealing Council Directive 87/102/EEC.

11. Directive 2014/17/EU of the European Parliament and of the Council of February 4, 2014 on credit agreements for consumers relating to residential immovable property and amending Directives 2008/48/EC and 2013/36/EU and Regulation (EU) No. 1093/2010.

12. Ministry of Finance—Press release December 8, 2016: Investor protection clarified in crowdfunding (https://vm.fi/en/article/-/asset_publisher/sijoittajansuojaa-selkeytetaan-joukkorahoituksessa).

13. Ministry of Finance—Press release December 28, 2017: Rapid increase in popularity of crowdfunding (https://vm.fi/en/article/-/asset_publisher/joukkorahoituksen-suosio-kasvaa-nopeasti).
14. Bank of Finland—Statistics March 29, 2018: Crowdfunding and peer-to-peer pending (https://www.suomenpankki.fi/en/Statistics/crowdfunding/).

Chapter 17

Crowdsourcing Ideas for Public Investment: The Experience of Youth Participatory Budgeting in Portugal

Susana Jacinta Queirós Bernardino
and José de Freitas Santos

Abstract

Participatory budgeting is a particular case of crowdsourcing that involves the participation of citizens in the political decision-making process regarding the distribution of a small part of the public budget among projects that were presented and voted by young citizens for public investment.

The objective of this chapter is to assess the content and results of the projects presented by young citizens in a participatory budgeting process that took place in 2017 in Portugal. Through the analysis of a database of 167 projects, this chapter aims to (i) understand the characteristics of the projects presented through this specific public crowdsourcing platform; (ii) study the profile of the youth participants; and (iii) analyse the voting results.

The results show that the projects were developed under the domain of different thematic areas, and most of them intend to be implemented at a regional or national level. The projects proposed in the crowdsourcing

CrowdAsset: Crowdfunding for Policymakers

platform benefit both littoral and inner territories and involved an average budget around €28.700, ranging from a minimum of €200 to the maximum allowed in the application (€75.000).

With respect to the proponents' profile, we observe that male youth citizens were slightly more active in the generation of ideas for projects than female youth citizens, whereas the applications were mostly made individually or by teams composed of a unique gender type.

Concerning the voting results, we observe that there is an intense concentration of votes in a few projects. Projects developed within the scope of inclusive sports and environmental sustainability gathered more votes, as well as those that have a regional or multiregional character or intend simultaneously to benefit both coastal and interior regions.

1. Introduction

Crowdsourcing platforms offer enormous potential for public managers to engage citizens in the policy decision-making process. One of the applications of crowdsourcing is the "idea crowdsourcing" that occurs when a public administration invites ideas for small projects from the crowd and then evaluates the proposed ideas through the votes of citizens. This is a particular case of participatory budgeting that is an alternative political process that involves the citizens in the distribution of a small part of the public funds that are attached to projects that were presented and voted by citizens.

Participatory budgeting is a new form of public financing that intends to engage citizens in the political decision-making process. This governance practice has its origins in 1989 in Brazil, more specifically in the Municipality of Porto Alegre, and after that has been expanded to other municipalities.[1] The primary aim of this movement was the desire to extend and deepen the democracy existing in the country.

Since the first experiences in Brazil, participatory budgeting has been implemented in thousands of cities all over the world. Its initial proliferation occurred mainly in developing countries, especially in Latin America, and since 2000, this participatory governance mechanism was also adopted in European countries and after 2008 has become increasingly popular in several municipalities worldwide.[2] After the initial development, participatory budgeting is currently embedded in a more complex participatory system that governments intend to build up.

Participatory budgeting is one of the most innovative practices of citizen participation that has become a reference worldwide and, due to its ability to get people together to a common cause, is also recognised as the most promising tool with respect to society's democratisation.

The virtue of participatory budgeting derives from its ability to blend in a balanced way both an institutionalised political participation and an autonomous citizens' participation. Participatory budgeting also has the potential to invigorate the political discourse, since it creates conditions for the construction of a new public arena for debate and negotiation, which take advantage of the participation and ideas collection of all the citizens.

Due to its potential to transform public management, participatory budgeting has been recommended by international organisations, such as the World Bank and UN-Habitat. Further, several authors have recognised the importance of involving young people in active citizenship and engaging them more actively in societal and political issues. These processes are also able to promote capacity building and the empowerment of young people. On the contrary, the involvement of young people could take advantage of their potential to national development, economic and social sustainability, and to the conception of solutions to attenuate some intergenerational problems.[3]

In Portugal, in line with these recommendations, the Portuguese government has implemented the Young Participatory Budgeting Portugal. This initiative was promoted at the national level, in partnership with the Portuguese Institute of Sport and Youth.

Here, we aim to analyse the content and results of the Portuguese young citizens in the participative budgeting process. To attain this purpose, the investigation draws on the study of the electronic platform used by the Portuguese government to "crowd" the ideas of young Portuguese citizens. As specific objectives, we intend to (i) understand the characteristics of the projects presented through this specific public crowdsourcing platform, (ii) study the profile of the youth participants, and (iii) analyse the voting results.

We will organise this chapter into four main sections. In the first part, a literature review of pertinent topics will be developed. Specifically, we will analyse the concepts of crowdsourcing and participatory budgeting. In this section, we examine the main advantages and limitations of participatory budgeting as well as the preconditions to its successful application. In Section 2, the case of youth participatory budgeting in Portugal is described. Then we explain our methodology. Section 3 presents the main results attained in the research. We end with the conclusions and final remarks.

2. Crowdsourcing and Participatory Budgeting

The concept of crowdsourcing was coined in 2006 by Jeff Howe in the article 'the rise of crowdsourcing' to include a diverse range of activities and projects involving public participation. In more specific terms, we can define crowdsourcing as an open call for anybody to participate in a task open online where the crowd is invited to participate by submitting information, knowledge, or talent.[4] The term crowdsourcing is also used in contexts in which the task is open only to a restricted group like the youth as a part of the participatory budgeting process. Crowdsourcing has become popular after the information technologies (smartphones, tablets) have enabled people to participate more easily in sophisticated processes of generation of ideas and collective co-creation. Crowdsourcing could be voluntary or paid, as people participate without receiving payment or if the tasks developed are paid. The participatory budgeting is a voluntary crowdsourcing process where participants collaborate to achieve a goal—the finance of a project proposed by them.

The increasing participation of citizens in the public life and the desire to become more active in the political decisions lead governments to open new ways of listening to citizens' opinions. The use of crowdsourcing platforms in a political context can be seen as a logical extension of the democratic process, engaging local citizens in idea generation activities to help public administrators plan investments that could be more widely accepted by future users. Also, the use of crowdsourcing can attenuate the difficulties in the offline traditional participation process (face-to-face public participation methods), bringing new insights and innovation to solve specific problems of the citizens at different territorial levels.

One application of the principles of the crowdsourcing is the participatory budgeting. According to a collective book published by the World Bank entitled Participatory Budgeting,[5] participatory budgeting is defined as "a decision-making process through which citizens deliberate and negotiate over the distribution of public resources (that) allow citizens to play a direct role in deciding how and where resources should be spent." A broad definition of participatory budgeting, according to the same publication, "usually describes it as a process through which citizens can contribute to decision making over at least part of a governmental budget." In a broad sense, participatory budgeting refers to "citizens' engagement with public budgets, including such mechanisms as analysis by civil society of spending policies as an input to public debate." Still according to the World Bank publication, during this participatory process that participatory budgeting encompasses, "men and women are invited to express their opinions about, and then vote on, the most important development projects for their neighbourhood, town or region. Once decisions are final, citizen oversight committees monitor project spending to prevent corruption and ensure accountability of their elected officials."

According to the book, citizens participation occurs "when citizens or their representatives (who are not elected officials) interact with and provide feedback to the government at the policy formulation or implementation stage of governance." Thus, participatory budgeting represents a mechanism through which individuals are encouraged to interact more closely and to get engaged in public policy and public decision-making. Also, "participatory budgeting aims to infuse the values of citizen involvement into the most basic and frequently the most formal procedure of governance—the distribution of resources through the budgeting process." In fact, as referred in the World Bank publication, "participatory budgeting is a process that is open to any citizen who wants to participate, combines direct and representative democracy, involves deliberation (not merely consultation), redistributes resources toward the poor, and is self-regulating, such that participants help define the rules governing the process, including the criteria by which resources are allocated."

The democratic advantage of participatory budgeting derives from the fact that it strongly reduces the perceived gap between the representatives and represented. Participatory budgeting also contributes for changing existing representative conditions, "by allowing citizens to move beyond simply electing politicians to make decisions about the allocation of local resources."[6]

One of the particularities of the participatory budgeting is its ability to enable any citizen (an "ordinary citizen") to be involved in the design of the proposal, the planning process, and the evaluation of the public policies.

Participatory budgeting takes advantage of the potential of bi-directional citizens' participation and involvement, and they include not only the idea generation process but also the appreciation of the ideas gathered via the crowdsourcing mechanisms.

Many potential advantages have been recognised to participatory budgeting. First, participatory budgeting has been recognised as an important tool for inclusive and accountable governance. As stated by the recent book published by the World Bank, "Participatory budgeting represents a direct-democracy approach to budgeting." The merits of the participatory budgeting also include reconciling the satisfaction of two different but interconnected needs: (i) improving the state performance while (ii) enhancing the quality of democracy.

Participatory budgeting processes could also foster citizens' knowledge about government operations, as well as to increase the citizens' requirements on the government's good governance practices. Thus, participatory budgeting strongly contributes to citizens' empowerment and the promotion of a more vibrant civil society. As stated by the book entitled *Participatory Budgeting*, participatory budgeting involves the decentralisation of the decision-making authority to citizens, "by moving the locus of decision making from the private offices of politicians and technocrats to public forums." Thus, based on this participatory process, governments stimulate citizens to be engaged in open and public debates, as well as increasing their knowledge of public affairs.

Authors even consider that participatory budgeting could play the role of "citizens schools" or to form "better citizens," since it leads individuals to have a clear comprehension of their rights and duties as citizens, as well as to exert an active citizenship and get involved in the selection of public policy priorities, where the scarce resources of the State will be employed. Participatory budgeting also contributes to raising individuals contentious and stimulus to involve them in civic life.

An empirical investigation in Central and Eastern Europe[7] finds that the introduction of participatory mechanisms has changed the citizens' participation and mind-set, which was initially characterised by a passive provision of information about their preferences. Also, participatory budgeting can produce an impressive change in the relationship between the State and the Society. This finding is observed by the research performed in Central and Eastern Europe, which reveals that participatory budgeting has contributed to an improvement of communication and comprehension between citizens and government.

In addition, participatory budgeting could foster the development of several citizens' capabilities, such as negotiation (between citizens and public entities), citizens' internal efficacy since they feel more confident about their ability to influence decisions, as well as decision-making capabilities related to the improvement of citizens' ability to make autonomous, reflective, and consequential decision-making.[8]

Participatory budgeting also contributes to more inclusive governance, since it gives a voice to all the citizens. In fact, in participatory budgeting, all individuals have the opportunity to influence the public decision-making, even those who typically are considered marginalised and excluded groups of the society. As stated by the publication made by the World Bank, "participatory budgeting programs provide poor and historically excluded citizens with access to important decision-making venues." Thus, participatory budgeting represents an important step towards reducing traditional and rigid social hierarchies.

As a consequence, the publication considers that participatory budgeting has the potential to make governments "more responsive to citizens' needs and preferences and more accountable to them for performance in resource allocation and service delivery." In accordance with the book *Participatory Budgeting*, "participatory budgeting also helps promote transparency, which has the potential to reduce government inefficiencies and corruption." Participatory budgeting could also contribute to a more effective and equitable resources distribution.

Further, the investigation pursued in Europe[9] found that the implementation of this participatory mechanism conducted led to an increase in the number of resources available for local development, by allowing governments to carry additional revenues for local development.

Current literature adverts for other motivations that governments could have to implement participatory budgeting, beyond those that are traditionally recognised. Within these motivations, the author refers to the enlargement of the government support, the weakening of the government opponents, the fulfilment of the ideological commitments, or even the enlargement of the government alliances.

Although there are vast potential benefits that have been assigned to participatory budgeting, some constraints have also been pointed out. First, despite the theoretical benefits, some authors are sceptical about the effective ability of participatory budgeting to be genuinely democratic and inclusive, as there is the risk of the voting process to be controlled by interest groups. As stated by the collective book published by the World Bank, when this occurs, participatory budgeting reinforces existing social injustices and facilitates the unjust exercise of power, as "giving the appearance of broader participation and inclusive governance while using public funds to advance the interest of powerful elites."

Other researchers,[10] based on an investigation conducted in Peru, find that some cultural, socioeconomic, and institutional factors could raise barriers to the participation of some groups, specifically women in the Peru case, and as consequence, participatory budgeting is not as inclusive as the theoretical arguments suggest.

Also, some disappointment could occur within citizens and the general society if they perceive that budgeting programmes function poorly, and they are not able to fulfil all its (theoretical) potentialities. As alerted by the World Bank publication, "if participation does not result in real change, it discourages future participation" and could threaten the sustainability of the participatory programmes. In fact, in order for citizens to continue to participate in these programmes, they must perceive that "there is a meaningful relationship between their input to political sphere and policy outcomes." The perception that their intervention really exerts an effective political influence is critical to keep their attention on the programmes.

A trend observed in recent years is the use of information and communication technologies to support participatory budgeting. The benefits that the digitalisation of the participatory processes could have are enlargement of the number and variety of individuals prone to participate, increasing the participation rate, and reduction of the costs involved. Digitalisation also brings new risks to participatory budgeting, such as the potential digital exclusion of some citizens and bias on more young, educated, and high-income people. Notwithstanding, the investigation of Sampaio[11] refers that this concern is being accommodated by most of the governments that typically have also made available physical public voting places and supporting citizens who have difficulties in these activities.

Another limitation of participatory budgeting is the participants' dependence on public entities, namely the mayor's office that figures as the leading actor in the process. The book published by the World Bank[12] also adverts that the time horizon of the participatory budgets typically does not have long-term planning, the proposals being restricted mainly to short-term programmes. Also, the scope of the proposals developed could not effectively address the citizens' needs, since most of the proposals have a local character and are not able to respond to some of the most pressing macro citizens' needs, such as unemployment or violence.

Thus, overall, literature review suggests some caution concerning the real benefits of participatory budgeting, since the materialisation of its potential benefits could not be accepted as guaranteed nor fully generalised. In fact, as referred by Wampler, the implementation of participatory budgeting has produced increasingly diverse results with high to very weak success rates. This suggests that participatory budgeting has a high potential, even though these benefits are not always able to be achieved.

Although participatory budgeting aims to involve a set of entities, such as citizens, non-governmental organisations, and civil society institutions, governments usually have a very central role, namely on the design and implementation of participatory programmes.

Participatory budgeting is usually implemented at the municipal level, although some initiatives are promoted at state or regional levels. Even so, as argued by the publication of the World Bank "Participation rises more quickly when the government commits significant support and resources to participatory budgeting." This pattern is justified by citizens' expectations, since they will be more prone to invest their time and effort in participatory programmes when they feel that the consequences of the programmes could be higher. Thus, public policy on the conception of the participatory programmes can constrain the opportunity and type of participation of all the citizens.

In fact, participatory budgeting is usually a bottom-up process, where public entities have a critical role on the design of the different stages of the programmes, as well as the rules that support it. Governments and public entities also have a significant role to play in the communication of the programmes that is critical for its success.

In order for participatory budgeting programmes to be implemented in a given country, the literature has highlighted certain preconditions. Shortly, these factors could be summarised as: (i) political will and commitment to the implementation of the programmes; (ii) civil society's willingness and ability to contribute to political debate; (iii) supportive political environment, including legislative support and political decentralisation; (iv) sufficient financial resources to support the projects selected by citizens (v) institutional design of the programmes that include the rules-based design; and (vi) bureaucratic competences to technically support the process.

The literature review conducted recently[13] highlights that the study of institutional design is critical for the success of participatory budgeting, especially on digital participatory ones.

Although the book entitled *Participatory Budgeting* recognises that the design for the participatory budgeting varies considerably between cities or countries, the author suggests some guidelines for its conception. Specifically, the author recommends the division by region within the programmes. This approach will facilitate the promotion of meetings and the resources allocation process. In this regard, the rules for attributing resources to different regions should also be previously defined and could include aspects such as poverty rates, population density, or the infrastructure that the region has. The programme design could include an agenda of priorities, containing some areas that are considered a priority.

The organisation of meetings with people is also indicated as relevant, to provide information, to foster proposal presentations, or to stimulus debates on proposals. Another decision that public entities need to make is to define who could vote on selected proposals and the type of voting that will be pursued (for instance, public or anonymous).

Although different types of participatory budgeting could be conceived, a typical participatory budgeting process includes the following stages[14]: (i) the city/state announces the participatory budgeting programme; (ii) citizens make their project proposals; (iii) the technical staff confirms and verifies the proposals and discloses the list of proposals considered eligible; and finally, (iv) the citizens vote for projects. These stages are usually implemented through meetings that are public and open to all citizens and which are mediated by the technical staff or by delegates elected by the participants. Even in the cases where electronic participatory budgeting is implemented, the (co)existence of face-to-face meetings is considered critical to the success of the programmes. The conjunction of these multichannel communication platforms—online and offline (face-to-face)—is observed in several successful cases.[15]

3. The Youth Participative Budgeting in Portugal

Portugal, following the trend observed in some of the most important countries in Europe, has also implemented participatory budgeting programmes, including the use of digital platforms. As revealed by a worldwide investigation conducted by Sampaio, Portugal is the country which has observed a high number of occurrences of digital participatory budgeting (44 occurrences found by the author) and which was implemented in many locations (25 communities according to the same research).

One of these participatory initiatives was the young participative budgeting. This is an innovative initiative, pioneering worldwide, that the Portuguese Government first implemented in 2017. The Youth Participatory Budgeting Portugal was approved by the Council of Ministers in September 2017 and aims to strengthen the quality of democracy and the involvement of young citizens in the Portuguese decision-making processes.

The programme, promoted by the Portuguese Institute of Sport and Youth, aims to encourage youth citizens to be more involved in public life and public investment decisions. Although some Portuguese's municipalities have already developed some initiatives for youth regarding participatory budgeting, they were isolated ones and had a different impact from the initiative that was promoted at the national level and under the supervision of the Portuguese Government.

In the youth participatory budgeting, young citizens aged between 14 and 30 years are encouraged to present and decide on investment projects. The aims are to involve younger citizens as an important group of Portuguese society, giving them a stimulus to be creative and to develop the entrepreneurial potential of the young Portuguese as well as involving them in the political decision-making process.

Either Portuguese or foreign citizens, legally residing in Portugal, are allowed to submit proposals, since their age is in accordance with the application's rules.

The application of the projects could be made in four main thematic areas: (i) inclusive sport, (ii) education for sciences, (iii) social innovation, and (iv) environmental sustainability.

The proposals are realised through the submission of a form in the online platform created on the Youth Participatory Budgeting[16] and could also be presented in person in the meetings promoted by the organisation throughout the national territory or even at the Portuguese Institute of Sports and Youth services.

To be eligible for the programme, the applicants should meet the following criteria:

- the idea has to be included in one of the four thematic areas of the programme (as previously mentioned, inclusive sport, education for sciences, social innovation, and environmental sustainability);
- not exceed the maximum amount of €75.000;
- not involve the construction of infrastructures;
- benefit more than one municipality;
- be well specified and located in the national territory;
- be technically feasible;
- not contradict the Government Programme or ongoing projects and programmes in the different areas of public policy.

Participants could propose more than one project, which does not have to be restricted to their residence area.

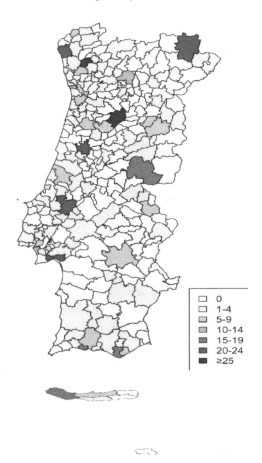

Figure 1: *Participants in the meetings according to the municipality of residence.*
Source: Portuguese Institute of Sport and Youth (2018).[17]

The edition started with the discussion and preparation of proposals, which included the promotion of participation meetings. The Participation Meeting aims to stimulate face-to-face presentation and discussion of the project proposals, as well as to inform the young applicants who intend to be engaged in the competition. The Participation Meetings were developed in 142 municipalities of the Portuguese territory, as shown in Figure 1, which corresponds to a coverage of almost half (46.1%) of Portuguese municipalities.

In 2017, a total of 4,245 proposals were submitted, although the technical commission approved only 167, once they have met the eligibility criteria.

After the pre-analysis of the application, a list of approved projects was made available in the platform and only these projects could be subject to the national public vote. The voting took place for about 1 month, between November 27 and December 22, 2017.

The decision-making process is democratic, since the projects to be supported by the Portuguese Government will be those that gathered a more significant number of votes. The votes can be made in two different ways: through a Short Message Service (SMS) or the online platform [opjovem.gov.pt], after a previous registration. The Government had committed itself to respect the participants' decision, by executing the winning investments.

In the 2017 edition, a total budget of €300.000 was available to support the competition, made on behalf of the budgetary allocation of the Portuguese Institute of Sports and Youth that corresponds to 10% of the total value of Participatory Budgeting Portugal.

Due to the success of the first edition of the competition, the organisations have decided to continue with the initiative and also to reinforce the available budget to €500.000.

4. Methodology

Our objective is to assess the content and the results of the projects presented by young citizens in a participatory budgeting process that took place in 2017 in Portugal. Through the analysis of a database of 167 projects that were included in the youth participatory budgeting, the aims of the chapter are threefold. First, we intend to understand the characteristics of the projects presented through this specific public platform, namely the thematic areas covered, the geographical scope, and the budget involved. Second, we aim to study the profile of the youth participants, specifically gender, and whether they are willing to participate as a team or individually. Finally, we intend to analyse the voting results, by understating the extent to which there is dispersion or concentration of the votes in some projects and, in such case, the characteristics of the projects that received more votes.

We constructed a database using the information available at the platform, which was consulted in April 2018.

The information was then categorised in order to allow the statistical treatment of the data. The codes used are summarised in Table 1.

Also, other continuous variables were included in the database such as the number of promoters involved as well as the number of votes received by each project.

We also have contacted the organising institution, Portuguese Institute of Sport and Youth, that kindly provided us with some more information about meeting the participants and voters.

Table 1: *Categorisation of the variables.*

Variable	Code
Thematic area	1. Inclusive sports
	2. Education for sciences
	3. Social innovation
	4. Environmental sustainability
Geographical scope	1. Local
	2. Regional
	3. Multiregional
	4. National
NUTS II territory classification	1. North
	2. Centre
	3. Lisbon
	4. Alentejo
	5. Algarve
	6. Islands
Territory characteristics	1. Littoral
	2. Interior
	3. Both
Promoters' gender	1. Female
	2. Male
	3. Both
Budget classification	1. €0–20.000
	2. €20.001–40.000
	3. € >40.000

Source: Authors' own elaboration.

5. Results

5.1. Characterisation of Projects

The data analysis shows that the projects were developed under the domain of the four areas defined by the leading public institution (Table 2), although a higher concentration on environmental sustainability (39.5%) and education for sciences themes (27.5%) was observed.

Table 2: *General characterisation of the projects.*

	N	Percentage (%)
Thematic area		
Inclusive sport	32	19.2
Education for the sciences	46	27.5
Social innovation	23	13.8
Environmental sustainability	66	39.5
Geographical scope		
Local	26	15.6
Regional	79	47.3
Multiregional	10	6.0
National	52	31.1
Location of the project (a)		
North	33	25.2
Centre	44	33.6
Lisbon	16	12.2
Alentejo	18	13.7
Algarve	13	9.9
Islands	7.0	5.3
Territory characterisation (b)		
Littoral	60	52.6
Interior	50	43.9
Both	4.0	3.5

Notes: (a) Some projects are located in two or more regions; this categorisation was made only for projects that do not work on a national basis ($N = 114$). (b) This categorisation was made for projects that do not work on a national basis ($N = 114$). *Source*: Authors' own elaboration.

Most of the projects have a regional nature (47.3%) or intend to be implemented at a national level (31.1%). Less common are the ideas generated by youth citizens to develop local (15.6%) or multiregional projects (6.0%). The submissions covered all the Portuguese regions, with particular emphasis on the Central (33.6%) and Northern (25.2%) regions. Multiregional projects covered essentially two regions of the national territory (70% of the multiregional projects) and only about a third (30%) benefits three regions simultaneously.

Interestingly, we observe that most regional and multiregional project proposals cover the environmental sustainability area (49.4% and 60.0% of the cases, respectively). The inclusive sports theme is particularly prominent for local projects, since this topic is explored in about a third of the local projects (a percentage higher than that observed in projects with a different geographical scope). We also observe that education for the sciences theme is mainly explored in national projects.

The projects proposed in the platform aimed to benefit both littoral (52.6%) and inner territories (43.9%), although only a small percentage of the projects jointly benefit the two types of territories (3.5%). Notwithstanding, we observe that local projects are mainly developed in littoral areas (88% of the cases), while regional and multiregional projects are divided into littoral and inner zones in a more similar way.

Concerning the budget amount that the project estimates necessary to be implemented, we observe that applications vary from €200 (minimum) to the maximum allowed in the application (€75.000, in 13 cases). Notwithstanding, the average budget was around €28.700, and just about a quarter of the youth citizens have proposed projects that involved €40.000 or more. Only three projects did not indicate the estimated budget that the project needs to be implemented.

On average, multiregional projects involve higher budgets (average of €34.575) than projects with a different geographical scope, including projects of national scope (average budget of €28.256). Further, projects that simultaneously benefit both coastal and interior areas involve a slightly higher budget (average of €33.937,5) than those that apply to only one of these types of territories (Table 3).

Considering the theme covered in the application, the results show that the projects in the social innovation area typically presented a

Table 3:   *Characterisation of the project budget according to the project features.*

	Average	Minimum	Maximum	25th percentile	50th percentile	75th percentile
Overall	28.712,5	200,0	75.000,0	13.125,0	20.000,0	40.000,0
Geographical scope						
Local	27.560,0	3.000,0	75.000,0	12.000,0	20.000,0	40.000,0
Regional	28.633,1	5.000,0	75.000,0	10.000,0	20.000,0	40.000,0
Multiregional	34.575,0	10.000,0	60.000,0	15.562,5	27.500,0	60.000,0
National	28.256,7	200,0	75.000,0	15.000,0	20.000,0	38.250,0
Characteristics of the territory						
Littoral	28.431,0	3.000,0	75.000,0	10.000,0	20.000,0	40.000,0
Interior	29.443,9	5.000,0	75.000,0	15.000,0	24.000,0	40.000,0
Both	33.937,5	10.000,0	60.000,0	11.437,5	32.875,0	57.500,0
Thematic areas						
Inclusive sport	21.230,7	1.500,0	71.500,0	7.000,0	15.000,0	30.000,0
Education for the sciences	23.178,3	200,0	75.000,0	12.000,0	20.000,0	26.250,0
Social innovation	49.565,2	20.000,0	75.000,0	35.000,00	60.000,0	75.000,0
Environmental sustainability	28.820,3	5.000,0	75.000,0	15.000,0	20.000,0	40.000,0

Source: Authors' own elaboration.

higher budget (average of €49.565.2). Projects that deal with inclusive sports are those that on average involve a lower budget (average of €21.230,7).

5.2. Characterisation of the Young Proponents' Profile

Concerning the proponents' profile, the investigation reveals that male youth citizens were slightly more active in the project's generation than female youth citizens (48.5% and 41.9%, respectively). We also observe that the applications were mostly made by individuals from a gender type,

Table 4: *Characterisation of the proponents of the projects.*

	N	Percentage
Gender		
Female	70	41.9
Male	81	48.5
Both	16	9.6
Number of proponents		
1	139	83.2
2	9.0	5.4
3	7.0	4.2
4	3.0	1.8
5	3.0	1.8
6	6.0	3.6

Source: Authors' own elaboration.

since only about one-10th of the projects (9.6%) were proposed by blended teams, composed of both women and men (see Table 4).

The analysis of the database shows that the applications were made between 1 and 6 proponents. Notwithstanding, most of the submissions were generally made individually, by a single promoter (83.2%), both in female (90.0%) and in male (93.8%) promoters. The second most frequent situation is the application by two (5.4%) or three promoters (4.2%). The constitution of teams with 4 or more individuals is less common. Multigender applications typically involve a more substantial number of applicants (4.3 on average).

Considering the proponents' gender and the characteristics of the projects, some slight differences are observed. The data analysis shows that compared to women, men tend to develop more projects in the environmental sustainability areas and less on social innovation. In inclusive sports and education for sciences themes, no significant differences were found. Female and multigender teams tend to be more prone to develop projects in littoral areas (66.7% and 61.5%, respectively), while the opposite is observed for male citizens (projects for inner regions accounts about 60.4% of the male proposals). Further, we observe that both male and female citizens tend to develop low-budgeting projects, as applications presented by women involve an estimated budget of less than €40.000 (54.3%), while men account for 56.3%. In contrast, multigender teams tend to make proposals with a slightly higher budget (projects involving less than €20.000 represent 42.9% of the applications).

5.3. Characterisation of the Voting Results

Concerning the voting results, 1.440 individuals were registered in the platform as voters. The voters were aged between 14 and 31 years and were close to the age profile allowed for the submission of ideas in the platform.

We observe that 6.739 votes were made during the voting process. The seven (7) winner projects received, respectively, 832, 796, 686, 441, 410, 403, and 379 votes. In contrast, 19 projects (11.4%) did not receive any votes. The percentile analysis (Table 5) also reveals that there is a strong concentration of votes in a few projects. The concentration of votes is also confirmed by frequency analysis, since most of the projects (53.9%) received 10 votes or less and a small percentage (6%) received more than 300 votes.

Further, the analysis also shows that projects developed within the scope of inclusive sports and environmental sustainability gathered more votes (on average, 67 and 61.6 votes, respectively). Also, projects that have a regional (average of 69.5 votes) or multiregional focus (average of 78.8 votes) received substantially more votes than those that are targeted to be implemented locally or nationally (average of 17.7 and 31.2 votes, respectively). Projects that will benefit coastal regions (average of 31.2 votes) were slightly less popular than those benefiting inner regions (average of 39.8 votes). Nevertheless, projects that simultaneously benefit littoral and interior Portuguese regions received more votes (average of 89 votes), although the variability of the votes attained is also higher in

Table 5: *Analysis of voting results received in accordance with the projects' characteristics.*

	Average	Standard deviation	Minimum	Maximum	25th percentile	50th percentile	75th percentile
Overall geographical scope	50.0	127,009	0	832	2.0	9.0	35.0
Local	17.7	22,361	0	74	1.75	8.5	30.5
Regional	69.5	152,997	0	832	2.0	11.0	48.0
Multiregional	78.8	213,601	0	686	1.5	11.5	23.25
National	31.2	82,169	0	441	1.25	6.0	20.75
Characteristics of the territory							
Littoral	31.2	82,169	0	441	1.25	6.0	20.75
Interior	39.8	84,005	0	379	1.0	5.0	30.75
Both	89.0	177,334	0	355	0.0	0.5	266.5
Thematic areas							
Inclusive sport	67.0	121,38	0	441	1.5	10.0	59.5
Education for the sciences	38.4	118,777	0	686	1.0	5.0	15.5
Social innovation	16.7	23,183	0	98	4.0	7.0	21.0
Environmental sustainability	61.6	153,136	0	832	2.0	13.5	46.25
Budget							
€0–20.000	28.4	63,296	0	410	1.5	7.0	33.5
€20.001–40.000	55.2	151,671	0	796	2.0	11.0	25.0
€ >40.000	93.2	193,000	0	832	3.0	9.0	66.8

Source: Authors' own elaboration.

involving more consuming budgets, as the standard deviation analysis suggests.

Finally, our findings show that the voting results are somewhat positively correlated with the project's budget, since projects involving higher budgets typically also receive more votes (Table 6).

Considering the promoters' characteristics, we observe that projects proposed by female citizens, although less frequent, on average received a higher number of votes (average of 72.9 votes) than those proposed by male (average of 37.8 votes) or multigender teams (average of 12.5 votes). Further, typically teams composed of 2 or 3 members attracted more attention from voters (average of 92.7 and 138.6 votes, respectively) than those who have a single proposer (average of 46.5 votes). Projects involving many proponents (equal to or more than 4) are those who received the worst voting result (average of 7.8).

According to the information made available by the Portuguese Sports and Youth Institute, the proponents of the (7) winning projects typically have high academic qualifications.

Table 6: Analysis of voting results received in accordance with the promoters' characteristics.

	Average	Standard deviation	Minimum	Maximum	25th percentile	50th percentile	75th percentile
Overall	50,06587	127,0009	0	832	2	9	35
Gender							
Female	72.9	158,650	0	796	2	10	40
Male	37.8	103,933	0	832	2	6	32
Both	12.5	18,640	0	74	1.25	4	15
Number of proponents							
1	46.5	116,347	0	796	2	7	33
2	92.7	136,883	3	410	11.5	40	148
3	138.6	306,596	0	832	4	31	63
>= 4	7.8	9,864	0	30	1	2.5	14.75

Source: Authors' own elaboration.

6. Conclusions and Final Remarks

Participatory budgeting has been one of the political options of the Portuguese government. The main purpose is to enhance the involvement of the Portuguese citizens in public decision-making process. A proportion of the resources available to participatory budgeting, about one-10th, was assigned for the implementation of the youth participatory budgeting. This was a pioneering initiative worldwide, which was implemented for the first time in Portugal in 2017.

Among the characteristics that make it unique is the fact that Portuguese Youth Participatory Budgeting makes use of a digital platform and is promoted at a national level, although there is a breakdown by regions, either in the definition of projects or the promotion of discussion meetings. Another peculiarity is the fact that it has been approved institutionally in the maximum organs of government (Council of Ministers), although its implementation has been promoted through a partnership with the Portuguese Institute of Sports and Youth.

The combination of new technologies, through the construction of a digital platform, with face-to-face meetings, was able to attract the attention of many young people, not only in the process of idea generation but also in its discussion and voting.

In total, the young participants generated 4,245 ideas for new projects. Although only a small number of these ideas (just 167) were considered eligible, by observing the rules defined in the competition, this demonstrates the high capacity of this tool to mobilise young people to participate and to stimulate them in the generation of new ideas. Each young person could present more than one idea and develop proposals for territories outside their area of residence. This option carried out in the design of the participatory budgeting programme allowed to enrich its potential as an instrument able to encourage young citizens to present solutions for the problems that the neighbourhood, region, or country face. Further, the young participatory budgeting also has an essential contribution for the development of ideas that consider the specificities of the territory, and for which young people could be especially sensitive (more than the government probably are).

The Portuguese young participatory budgeting still involves a considerable amount of money, which allowed the implementation of seven projects. The proposals presented during the process covered the four different thematic areas envisioned in the design of the programme. The proposals have different characteristics and distinct geographic scope, including programmes with a national, multiregional, regional, or local scope. The projects presented also have a widespread territorial coverage including both coastal and interior regions. As argued by a book published by the World Bank,[18] the implementation of participatory mechanisms, such as the young participatory budgeting, increases the resources available for the reinforcement of local and regional projects and contributes to the development of the territory.

Nevertheless, the programme was mainly anchored around a digital platform, although face-to-face meetings also had a significant adherence and were developed in about half of the country's municipalities. This fact confirms the capacity of the participatory budgeting to promote debate among younger audiences, which also allows the development of their capacities as argued by the relevant literature on the topic.

The Portuguese youth participatory budgeting was initially developed through a top-down process since the design and promotion of the programme was held by the Portuguese government, but which, after that, has fostered bottom-up innovation and the generation of ideas. Through the participatory budgeting, young people were heard (and also encouraged to express their ideas) about the problems that concern them and for which they could have a different view on the solution that might be more favourable.

Because it is open, public, and based on democratic processes, participatory budgeting assumes an inclusive character, not only among young people but also between generations, once it gives attention to people who are generally not as well represented in the national decision-making systems as the mature citizens.

It should be noted that, in its design, the youth participatory budgeting assumed the commitment to effectively fundraise the implementation of projects that had received a higher number of votes (subject only to the requirement of being considered as eligible for the programme). This issue could be important for motivating young people to participate, since they feel that they really have an active voice and a participation in the decision-making process.

Although young men are a little more actively involved in the idea generation process, young women also have been significantly engaged in the programme, either individually or in multigender teams. The research also revealed that most of the individuals had participated individually in the proposal development. Proposals submitted by large teams, composed of four or more proponents, were not frequent.

Participatory budgeting was also able to attract the young citizens' attention for the analysis of the project, since more than 1,440 individuals had expressed their preferences during the voting process, through 6.739 votes made. Despite the high number of young citizens involved in voting, we observe a certain concentration of results on a relatively small number of projects, suggesting some consensus between the participants on projects that seem to be more interesting. The analysis of the voting results suggests some slight preferences from the young citizens, namely its tenuous predilection on regional and multiregional projects, as well as projects benefiting simultaneously coastal and interior regions. The slight preference on regional and multiregional projects could be explained by some emotional propensity that could be involved in the decision-making process that lead young citizens to elect projects that could benefit territories for which they have a higher affinity. Also, this pattern could derive from the facility to explain in a clear and persuasive way the benefits of a project that has its geographical aim well defined, rather than broader projects with ill-defined focus.

The proponents' profile also seems to have some impact on the voting process, namely the gender and the dimension of the team. Teams with two or three proponents tend to receive more votes than the others. Two possible reasons could justify this pattern: the first, due to their enhanced ability to use larger personal networks in the voting process and the other could derive from its capacity to conceive a more complex and robust solution as they are generated by collective processes within the team. Under these circumstances, it is recommended for Portuguese entities to implement mechanisms able to stimulate discussion and teamwork among Portuguese youth citizens, through training processes, workshops, or even team-building activities.

The investigation pursued allowed to produce a description of a programme that could have a high potential for young citizens' empowerment and social and economic development. Through the analysis of this case, a better knowledge of the way the programme was developed, the characteristics of the projects, as well as its proponents' profile and the voting results were achieved.

Some practical implications could be derived from this case study. First, it could help Portuguese entities to understand better the relevance and impact of the programme. Also, a better comprehension of the features of the projects and citizens involved could help to leverage the design of the programme. Also, the description of the Portuguese Youth Participatory Budgeting, which at least in its first edition was considered successful, could inspire other public entities and serve as a benchmark to other countries.

About the Authors

Susana Jacinta Queirós Bernardino is an Adjunct Professor of Management at the Polytechnic of Porto, Institute of Accounting and Administration (ISCAP), Portugal, and researcher at CEOS.PP, Centre for Organisational and Social Studies of Porto. She completed her undergraduation degree in Management at the Faculty of Economics of the University of Porto. She got her PhD in Management in 2014 from the Portucalense University. Her research interests focus on social entrepreneurship, entrepreneurship, social and solidarity economy, crowdsourcing, and smart cities.

José de Freitas Santos is a Full Professor of International Business at the Polytechnic of Porto, Institute of Accounting and Administration (ISCAP), and researcher at CEOS.PP, Centre for Organisational and Social Studies of Porto. He obtained his PhD in 1997 from the Minho University. His research interests include tourism, regional development, social entrepreneurship, internationalisation and digital marketing. Besides presentations on these topics at various conferences in Portugal and foreign countries, his articles have been published in several domestic and international journals.

Endnotes

1. Shah, Anwar (ed.), *Participatory Budgeting,* The World Bank, Washington DC, 2007.
2. Gomez, Javier, D.R. "Insua and César Alfaro*:* A Participatory Budget Model under Uncertainty." *European Journal of Operational Research,* 249 (2014), 351–358.
3. Rexhepi, Artan, Sonja Filiposka and Vladimir Trajkovik. "Youth E-participation as a Pillar for Sustainable Societies." *Journal of Cleaner Production,* 174 (2018), 114–122.
4. Howe, Jeff. "Crowdsourcing: Why The Power of the Crowd is Driving the Future of Business." *Crown Business,* New York, 2008.
5. Shah, Anwar (ed.) (2007).
6. Coleman, Stephen and Rafael Cardoso Sampaio. "Sustaining A Democratic Innovation: A Study of Three E-Participatory Budgets in Belo Horizonte." *Information, Communication & Society,* 20(5) (2017), 754–769.
7. Shah, Anwar (ed.) (2007).
8. Coleman, Stephen and Rafael Cardoso Sampaio (2017).
9. Shah, Anwar (ed.) (2007).
10. Mcnulty, Stephanie L. "Barriers to Participation: Exploring Gender in Peru's Participatory Budget Process." *The Journal of Development Studies,* 51(11) (2015), 1429–1443.
11. Sampaio, Rafael Cardoso. "E-Participatory Budgeting as an Initiative of E-requests: Prospecting for Leading Cases and Reflections on E-Participation." *Revista de Administração Pública,* 50(6) (2016), 937–958.
12. Shah, Anwar (ed.) (2007).
13. Abreu, Júlio. "Participação *democrática em ambientes digitais: o desenho institucional do orçamento participativo digital.*" *Cadernos EBAPE.BR,* 14(3) (2016), 794–820.
14. Walczak, Dariusz and Aleksandra Rutkowska. "Project Rankings for Participatory Budget based on the Fuzzy TOPSIS Method." *European Journal of Operational Research,* 260 (2017), 260, 706–714.
15. Sampaio, Rafael Cardoso (2016).
16. https://opjovem.gov.pt.
17. Used by permission.
18. Shah, Anwar (ed.) (2007).

Chapter 18

Crowdfunding—The Indonesian Experience

Anton Root

Abstract

Crowdfunding has grown rapidly in the global North, developing into a multi-billion dollar industry annually. In emerging markets, crowdfunding has tremendous potential to close the multi-trillion MSME credit gap, thereby promoting job growth and stimulating economies.

One country where crowdfunding is seeing accelerated growth is Indonesia. The country's ministers set up friendly peer-to-peer (P2P) lending (lending-based crowdfunding) regulations that have encouraged new platforms to enter the market and borrowers to tap into the opportunity to raise capital from the crowd. From December 2016 to June 2018, P2P lending grew from having raised a total of US$17 million to US$486 million.

Since the original P2P laws were released, the government has made tweaks to the regulations as well as worked with other bodies to encourage additional guidelines. These additions have not had the same effect in promoting P2P as the initial regulations.

This chapter examines Indonesia's P2P regulatory regime, focusing on Islamic finance and P2P. Indonesia's initial success in promoting P2P can serve as a blueprint for other emerging market economies and showcases how crowdfunding can be adopted to country-specific circumstances. Its missteps can serve as learning opportunities to other governments.

This chapter will be useful to policymakers considering crowdfunding, academia researching the topic, and development organisations looking to promote private capital flows to SMEs in emerging markets.

1. Introduction

Access to credit remains poor in many emerging markets, as banks are unable or unwilling to serve micro-, small-, and medium-sized enterprises (MSMEs) or individuals with a short or non-existent credit history. The International Finance Corporation estimates that formal MSMEs face a finance gap of over US$5 trillion annually; if informal businesses were taken into account, that gap would surely widen.[1]

Multiple studies[2] have found that small- and medium-sized businesses (SMEs) are key to economic growth and job creation; the World Bank states that formal SMEs contribute as much as 60% of employment and up to 40% of gross domestic product (GDP) in emerging countries. Additionally, SEAF has found[3] that SMEs tend to create jobs that are appropriate to low-skilled workers and tend to promote from within, increasing employment levels in an economy.

This makes the case for SMEs as a key focus area for enabling countries to transition from developing to developed status. Closing the finance gap, therefore, is an important issue for developing countries.

The reasons for poor access to credit vary from country to country; likewise, the opportunities to narrow the gap are also different. This chapter examines the case of Indonesia, a rapidly developing nation with a sizeable population. Specifically, this contribution focuses on peer-to-peer (P2P) lending, a variation of crowdfunding that allows individuals and businesses to borrow money from the "crowd," typically via an online platform.

In over a year, P2P lending has grown into a US$450+ million industry in Indonesia, helping thousands of businesses and entrepreneurs to gain access to funding they may not have otherwise. This chapter examines the conditions that have enabled P2P to grow and how Indonesia has used regulations to promote the industry's growth.

2. Access to Credit in Indonesia

According to the World Bank's Findex, only 36% of Indonesian adults had an account in 2014, either at a financial institution (35.9%) or a mobile (0.4%) account. That's significantly lower than the rest of the East Asian and Pacific region (69% account penetration) and below Indonesia's lower middle income peers (42.7%).[4]

At the same time, Indonesians are credit-hungry. Over half of the adult population (56.6%) borrowed money in the previous year, more than in other countries in the region (41.2%), as well as lower middle income peers (47.2%). Most of those who borrowed (41.5%) turned to friends or family members to do so.[5]

Indonesian small businesses are crucially important to the economy— even more so than in other developing markets. According to Bank Indonesia (the country's central bank), 55 million SMEs absorb 97.2% of the country's labour force and contribute 57.5% to its GDP.[6]

Against this backdrop, increased access to credit has become crucial for the country's SMEs to continue growing. The country's regulators have sought to promote SMEs' access to finance.

This has taken various forms. One is the *Kredit Usaha Rakyat* (KUR) scheme, implemented by the Ministry of Finance. KUR is a credit guarantee programme that encourages banks to work with MSMEs by guaranteeing up to 70% of the loan. The government has committed over US$8 billion for 2018; in order to promote uptake, it has also lowered the annual interest cap at 7% (from 9% in 2017).[7] Furthermore, the Bank of Indonesia (BI) has mandated banks to lend 20% of their portfolio to SMEs (see Table 1).[8,9]

Despite these efforts, Indonesia still experiences a "missing middle"—lack of available capital for SMEs who are too big for micro-finance but not big enough to be attractive to traditional banks that prefer to work with larger corporates with an established track record and collateral.[10]

Table 1:   *Definition of MSMEs in Indonesia.*

	Assets (IDR)	Assets (US$)	Annual turnover (IDR)	Annual turnover (USD)
Micro	0–50,000,000	0–3,500	0–300,000,000	0–21,000
Small	50,000,000–500,000,000	3,500–35,000	300,000,000–2,500,000,000	21,000–175,000
Medium	500,000,000–10,000,000,000	35,000–700,000	2,500,000,000–50,000,000,000	175,000–3,500,000

There are several reasons for this financing gap. On the demand side, SMEs lack records and assets that make it easier for banks to rate their creditworthiness. They often have no credit score, little if any financial track record, and no collateral. This means banks need to spend more time working with a company to determine its ability to pay back a loan. Often, they need to hire or train staff to be able to evaluate SMEs, which adds overhead costs.[11]

On the supply side, the credit that banks offer does not meet the needs of entrepreneurs. For example, they may have a cap or floor for loan sizes. More often, the turnaround time for loan disbursement is several weeks, which does not fit with the schedules of fast-paced SMEs. Additionally, the loan tenor may be much longer than an SME needs.[12] For these reasons, there exists a breakdown in the market for SME credit.

In this context, crowdfunding, particularly P2P lending, has become a viable solution for some SMEs to fill the financing gap.

3. Government Steps In

Seeing the potential of P2P (and financial technology, or fintech, more broadly), Indonesian authorities started looking into issuing industry-specific regulations in 2015.[13]

The *Otoritas Jasa Keuangan* (OJK), the country's financial services authority, put together a dedicated task force that drafted regulations around the fintech sector by October 2016.[14]

After consulting industry actors, OJK released Regulation No. 77/POJK.01/2016,[15] summarised below. There were two key changes made compared with the draft regulation: lowered capital requirements and a maximum loan amount.[16]

3.1. P2P regulations—Key Sections

Enterprise Requirements

- Human resources expertise in information technology; risk mitigation measures via credit insurance, data collection centres located in Indonesia.
- Registration process.
- Register business with OJK.
- Apply for a business license within 1 year of registration.
- Submit company reports every 3 months after registration, containing the following:
  o the number of lenders and borrowers;
  o quality of the loan received by the beneficiary;
  o business activities conducted since registration.

Capital Requirements

- Working capital of IDR 1,000,000,000 (US$ ~70,000) at the point of applying for business registration; working capital of IDR 2,500,000,000 (US$ ~175,000) at the point of applying for a business license.
- Other restrictions.
- *Maximum interest rate*: The maximum interest rate charged on lending activities cannot exceed seven times that of the Bank of Indonesia's (the central bank of Indonesia) 7 days' repo rate.[17]
- *Lending limit*: The P2P lender cannot provide loans to a borrower exceeding 20% of the funds owned by the company; it cannot lend more than IDR 2,000,000,000 (US$ ~150,000) to any single borrower.
- *Foreign investors*: Foreign individuals and legal entities may lend through P2P lending platforms (in practice, they tend to receive lower interest rates).

- *General prohibitions*: A P2P lender is prohibited from the following:
  - o receiving funds from users of the platform (funds must go into an escrow account);
  - o collecting funds directly from the public in the form of cheques or savings;
  - o providing investment advice or recommendations.

4. Effect of Regulations

Lending legitimacy to the industry, the regulations have allowed P2P lending to take off in the country.

Since OJK introduced P2P regulations, lending-based crowdfunding activity has grown at a rapid pace (see Figure 1). There are now dozens of platforms operating in the country, and over 300,000 people have used crowdfunding of various types (i.e. donations, reward, equity, lending).[18] As of June 2018, 64 P2P platforms were registered in the country.[19]

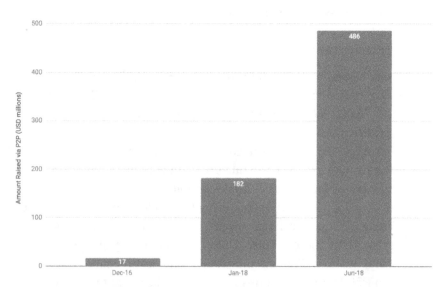

Figure 1:   *Growth of P2P amount raised in Indonesia (December 2016–June 2018).*[20,21]

The early results appear to be positive. As Figure 2 shows, the growth rate of loan volume in the country has accelerated over the previous year. It is unclear how much of the growth can be attributed to P2P lending, but Indonesians do appear to have started borrowing more, just as P2P lending has grown in popularity.

The growth has not been without adverse effects. One of the most prominent has been the incidence of fraudulent platforms. Recently, for example, OJK announced that it identified 227 platforms that were operating without a license, the vast majority being based in China.[23] It is not yet clear how much activity these unregistered platforms have experienced from borrowers or lenders.

Following the implementation of initial P2P regulations in the country at the end of 2016, OJK began to deliberate on further regulations for P2P firms, meant to continue to advance the industry's growth and create a more complete regulatory regime.

One of the key additions that OJK members discussed was around creating regulation specific to Islamic finance and P2P.[24] Despite being home to the world's largest Muslim population, Indonesia

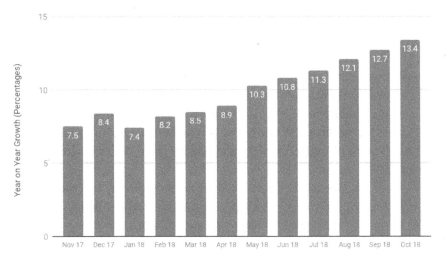

Figure 2: *Growth of year-on-year loan volume (November 2017–October 2018).[22]*

Assets ($ billion)

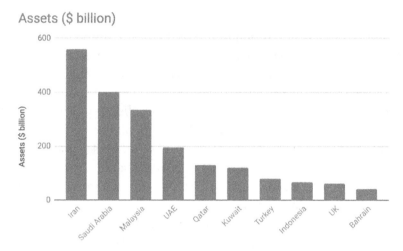

Figure 3: *Islamic financial assets in selected countries (2016).*[25]

lags behind other countries in Islamic finance in terms of total assets (see Figure 3).

The Islamic finance industry, however, is growing rapidly in Indonesia. Between 2007 and 2016, the average annual growth in assets was 56%, well above that of other countries.[26] Furthermore, the assets of Islamic banks in the country are outpacing those of traditional banks: 19% against 9.8% in 2017.[27]

The government has been looking to promote Islamic finance over the last several years, creating a "roadmap" in 2015 that aims to provide "more varied choices of banking instruments and services," promote "financial market deepening and financial system stability," and encourages financial inclusion.[28] Government ministers have also spoken about the importance of integrating P2P and other types of fintech with Islamic finance.[29,30]

Indeed, Indonesia is already making some headway in growing the Islamic finance fintech sector—it is home to 15 Islamic fintech start-ups, third in the world behind Malaysia (18) and the UK (16).[31]

Given the government's desire to promote Islamic finance and legitimise the fintech industry, it was not surprising that OJK sought to release further regulations to promote both.

5. Additional Regulations

Following months of discussions and comments to the press, OJK released further regulations (Regulation No. 13/POJK.02/2018) on fintech firms in August 2018. This provided additional guidance to entrepreneurs in the fintech space, including crowdfunding and P2P lending.[32]

In the regulation, OJK defines fintech firms as possessing the following qualities[33]:

(i) innovative and future-oriented;
(ii) involving the use of information and communication technology as their primary means of providing services to consumers in the financial services sector;
(iii) supporting financial inclusion and literacy;
(iv) beneficial and accessible to the public;
(v) compatible for integration into the existing financial services;
(vi) adopting a collaborative approach; and
(vii) complying with consumer and data protection requirement.

This regulation put into place an explicit regulatory sandbox regime for all fintech firms in the country. All start-ups must record their activity with OJK in order to be able to operate in the country. Following this, OJK will evaluate whether a company fits within the mandate of the sandbox, defined as a "a testing mechanism established by the OJK to assess the reliability of the business processes and models, financial instruments and management processes of a provider."[34]

The sandbox period is meant to last for a year, after which it may be extended by a further 6 months. During the sandbox period, platforms must submit quarterly risk assessment reports, outlining their exposure to the following risks: strategic risk; systemic operational risk; individual operational risk; risk of money laundering and financing of terrorism; and liquidity risk, among others.

After the sandbox period, OJK will evaluate the status of the business. It will then recommend it, asking to extend the sandbox period for 6 months to improve its business model, or not recommend it. If the decision is to not recommend, the company will effectively be no longer legally compliant. If the decision is to recommend, the company will have 6 months to register with OJK.

In addition to the regulations, P2P platforms are required to register with Fintech Indonesia, a self-regulating industry association.

In the summer of 2018, Fintech Indonesia released a code of conduct for P2P platforms.[35] The code is centred on three basic principles: transparence, prevention of excessive loans, and good faith practices.[36] As all P2P platforms must register with the association and accept the code of conduct, the code is effectively an extension of OJK regulation of the platforms.

Regarding fintech and Islamic finance, OJK worked with the Indonesian Ulema Council (*Majelis Ulama Indonesia* or MUI), the top clerical body in the country, to help provide guidance to start-ups.[37]

In February 2018, the MUI issued a fatwa regarding fintech.[38] The broad-ranging fatwa covers a range of issues related to fintech, including e-commerce, invoice factoring, and online payments. The fatwa proclaims such activity to be permissible under Islamic law, as long as it does not fall foul of basic tenets. For example, the funding cannot go towards supporting activity that is considered *haram*. Additionally, the fintech firms are unable to charge interest on any money lent out.

6. Analysis

The original P2P lending regulation, released in late 2016, was a boon for the industry. With their issue, OJK signalled to investors and borrowers that the industry was legitimate. By consulting industry stakeholders prior to releasing the laws, the regulatory body also ensured that the regulations allowed start-ups the freedom to operate without falling foul of unnecessarily stringent laws. That freedom, however, necessitated additional regulations to be released after a period of time to respond to market developments and answer questions that industry stakeholders developed.

OJK's additional regulation (as well as guidance from MUI and Fintech Indonesia) has not provided the level of specificity that platforms require to be clear on what is allowed and what is forbidden. Furthermore, some aspects of the regulation have been overly onerous to comply with, especially for fledgling P2P platforms.

Regarding the lack of specificity, some legal experts lamented the lack of consistency among terminology used by OJK, which can lead to confusion among companies.[39]

More importantly, by imposing a regulatory sandbox regime, OJK has delayed implementing clear, comprehensive regulation. As nearly 2 years have passed between Regulation No. 77/POJK.01/2016 and Regulation No. 13/POJK.02/2018, OJK had opportunity to observe existing P2P business models and issue regulation based on their findings. Instead, the regulatory body elected to continue observing the fintech sector, leaving P2P platforms in regulatory limbo, and subjecting them to intrusive checks.

Perhaps, OJK staff felt a continued regulatory sandbox regime was made necessary by the high number of fraudulent platforms in the country.[40] While that is an unfortunate development, the solution to this problem is more vigilant oversight of the platforms in the space, rather than submitting registered platforms to more checks.

There is also lack of clarity around how MUI's fintech fatwa interacts with existing OJK regulation. For example, there is some confusion around whether Islamic finance fintech firms need to register with both OJK and MUI, or just OJK, prior to launching.[41]

There are also aspects of regulation that have been difficult for companies to comply with. Among the most onerous is the requirement to promote financial literacy. To be sure, it is important for P2P platforms to make clear to potential users the risks and fees associated with over-borrowing and missing payments. However, this can be done online, when potential users are applying for a loan. Instead, OJK has asked platforms to hold financial literacy classes in remote regions;[42] for small start-ups, this can be prohibitively expensive.

The OJK's regulations can serve as an example for neighbouring countries, as well as other emerging markets. Regulators should note the OJK's willingness to have a flexible regulatory regime in the beginning, requiring P2P lenders to register with the OJK as well as a membership association that promotes self-policing among the platforms. OJK also did well to have a close working relationship with the association, making it easy for the platforms to interact with the government agency. However, regulators from outside countries should also note that it is important to create a set of comprehensive, rules-based regulations for the industry, instead of continued oversight of registered platforms, which can be onerous and time intensive.

7. Conclusion

P2P's growth in Indonesia has been driven by natural factors, including poor access to finance and a high number of SMEs across the country. P2P platforms offer an attractive, convenient solution for credit-starved Indonesians, who have embraced the industry so far. OJK's initial regulations at the end of 2016 legitimised the industry, promoting further growth. Since then, however, the regulator's efforts have been unhelpful.

Despite the regulation's shortcomings, Indonesia is still a good example for policymakers looking to promote the P2P lending and crowdfunding industry. Especially early on, the regulations served as a catalyst for start-ups to enter the market and test various business models, with the supervision of OJK in a controlled environment. Since then, however, the regulator has not taken the steps necessary to create a regulatory environment that promotes further innovation.

Efforts like working with MUI to issue a fintech fatwa do and will help to promote growth in the country and can be a blueprint for regulators for other countries to consider. However, OJK's regulations should be coordinated better and less onerous for start-ups if they are to promote continued growth in the industry.

Regulators should continue to keep track of P2P activity and tailor existing regulations to the country's needs. However, they should also move towards a regulatory regime that is easy for platforms to adhere to, without intrusive checks that will stymie the industry's growth. Instead of focusing its attention on oversight of platforms that have already registered with the OJK, regulators should focus on identifying bad actors.

About the Author

Anton Root is the Head of Research at AlliedCrowds, a technology and advisory firm focused on alternative finance in emerging markets. He was previously the Senior Editor at Crowdsourcing.org, a news organisation focusing on the crowdfunding and crowdsourcing industries. Since 2011, he has written industry-leading reports on crowdfunding in emerging markets, consulted on campaigns that have raised more than US$100,000, spoken at international forums, and interviewed industry leaders from around the world.

Endnotes

1. "MSME Finance Gap." *SME Finance Forum*, IFC, www.smefinanceforum. org/data-sites/msme-finance-gap.
2. For example, "Report on Support to SMEs in Developing Countries Through Financial Intermediaries." *Dalberg*, November 2011, www.eib.org/attach-ments/dalberg_sme-briefing-paper.pdf, "Scaling-Up SME Access to Financial Services in the Developing World." *IFC*, October 2010, www. enterprise-development.org/wp-content/uploads/ScalingUp_SME_Access_ to_Financial_Services.pdf, | Ndiaye, Ndeye, *et al.* "Demystifying Small and Medium Enterprises' (SMEs) Performance in Emerging and Developing Economies." *Science Direct*, Borsa Istanbul Review, April 26, 2018, www. sciencedirect.com/science/article/pii/S2214845018300280
3. "From Poverty to Prosperity: Understanding the Impact of Investing in Small and Medium Enterprises." SEAF, October 2007, www.seaf.com/wp-content/ uploads/2014/10/From-Poverty-to-Prosperity1.pdf
4. "Indonesia Financial Inclusion Data / Global Findex." *Data on Statistical Capacity*, The World Bank, www.datatopics.worldbank.org/financialinclusion/ country/indonesia
5. *Ibid.*
6. *Financing SMEs: Sharing Ideas for Effective Policies.* Bank Indonesia, 2014, https://www.bi.go.id/id/umkm/penelitian/nasional/kajian/Documents/ Financing%20SMEs%20Sharing%20Ideas%20For%20Effective%20 Policies.pdf?Mobile=1
7. "Indonesia to Distribute Rp 120t to Small, Micro Businesses." *The Jakarta Post*, January 4, 2018, www.thejakartapost.com/news/2018/01/04/govt-to-distribute-rp-120t-to-small-micro-businesses.html
8. Al Azhari, Muhamad. "Bank Indonesia's New SME Loan Rules Spark Concern." *Jakarta Globe*, November 27, 2012, www.jakartaglobe.id/archive/ bank-indonesias-new-sme-loan-rules-spark-concern/
9. Dipta, Wayan. "Indonesia SME Strategy." ILO, June 13, 2017, www.ilo.org/ wcmsp5/groups/public/---asia/---ro-bangkok/---ilo-jakarta/documents/ presentation/wcms_564690.pdf

10. "Caught in the Middle." *The Economist*, June 2, 2016, www.economist.com/finance-and-economics/2016/06/02/caught-in-the-middle

11. *Time for Marketplace Lending.* Oliver Wyman, 2016, www.oliverwyman.com/content/dam/oliver-wyman/global/en/2016/apr/Time_For_Marketplace_Lending.pdf

12. *Ibid.*

13. NLS, Vincencia. "Indonesia OJK to Regulate Crowdfunding." *DealStreetAsia*, September 30, 2015, www.dealstreetasia.com/stories/indonesia-ojk-to-regulate-crowdfunding-14068/

14. "Press Release: OJK Drafts Regulations on Fintech Development." *OJK*, October 6, 2016, www.ojk.go.id/en/berita-dan-kegiatan/siaran-pers/Pages/Press-Release-OJK-Drafts-Regulations-on-Fintech-Development.aspx

15. "OJK Issues Regulation on IT-Based Lending Services." *OJK*, January 10, 2017, www.ojk.go.id/en/berita-dan-kegiatan/siaran-pers/Documents/Pages/Press-Release-OJK-Issues-Regulation-on-It-Based-Lending-Services/SIARAN%20PERS%20POJK%20%20%20%20FIntech-ENGLISH.pdf

16. Dawborn, David, *et al.* "Regulatory Developments On Fintech In Indonesia (Part 1)—OJK Finalising Fintech Lending Regulations." *Lexology*, November 28, 2016, www.lexology.com/library/detail.aspx?g=61bd60a4-8306-4d10-8ec5-3a722f5f2915

17. Over the last two years, the repo rate has fluctuated between 4.25 percent and 5.25 percent, meaning the maximum interest rate has fluctuated between 29.75 percent and 36.75 percent. You can find the repo rates on the Bank Indonesia website: https://www.bi.go.id/en/moneter/bi-7day-RR/data/Contents/Default.aspx

18. "Indonesia Eyes Fintech Regulation to Avoid 'Loan Shark-like' Practices." *Reuters*, March 13, 2018, www.reuters.com/article/us-indonesia-fintech-regulator/indonesia-eyes-fintech-regulation-to-avoid-loan-shark-like-practices-idUSKCN1GP10B

19. Jakarta Post. "Fintech Regulation Coming Soon: OJK." *The Jakarta Post*, July 16, 2018, www.thejakartapost.com/news/2018/07/16/fintech-regulation-coming-soon-ojk.html

20. Nangoy, Fransiska. "Indonesia's Fintech Lending Boom Exploits Shortfall in Bank Loans." *Reuters*, Thomson Reuters, January 30, 2018, www.reuters.com/article/us-indonesia-fintech/indonesias-fintech-lending-boom-exploits-shortfall-in-bank-loans-idUSKBN1FJ0F4

21. Muna, Fauzul. "Indonesia's Fintech P2P Firms Disburse $486M Loans by June 2018." *The Insiders Stories*, July 13, 2018, theinsiderstories.com/indonesias-fintech-p2p-firms-disburse-486m-loans-by-june-2018/

22. Indonesia Loan Growth. TRADING ECONOMICS, www.tradingeconomics. com/indonesia/loan-growth

23. Wirdana, Ardi. "Indonesia Identifies 227 Illegal P2P Lending Startups, Mostly from China." *DealStreetAsia*, July 30, 2018, www.dealstreetasia. com/stories/indonesia-identifies-227-illegal-p2p-lending-startups-mostly-from-china-103546/

24. Winosa, Yosi. "Indonesia to Revise P2P Lending Regulation to Include Shariah-Compliant Financing." *Salaam Gateway*, January 31, 2018, www. salaamgateway.com/en/story/indonesia_to_revise_p2p_lending_regulation_ to_include_shariahcompliant_financing-SALAAM31012018062112/

25. "Overview of the Global Islamic Finance Industry." *GIFR*, 2017, www.gifr. net/publications/gifr2017/intro.pdf

26. *Ibid.*

27. "The World's Biggest Muslim Country Wants to Boost Sharia Finance." *The Economist*, May 10, 2018, www.economist.com/finance-and-economics/ 2018/05/10/the-worlds-biggest-muslim-country-wants-to-boost-sharia-finance

28. Tampubolon, Nelson. "Roadmap of Indonesian Islamic Banking 2015 - 2019." *OJK*, November 12, 2015. www.ojk.go.id/en/Documents/Pages/ Islamic-Finance-OJK-2015/1.nelson.pdf

29. United States, Congress, Maierbrugger, Arno. "A new term is born: Shariah fintech, and it has quite some potential." *Gulf Times*, July 10, 2018. www. gulf-times.com/story/599050/A-new-term-is-born-Shariah-fintech-and-it-has-quit

30. Winosa, Yosi. "Indonesia to Revise P2P Lending Regulation to Include Shariah-Compliant Financing." *Salaam Gateway - Global Islamic Economy Gateway*, January 31, 2018, www.salaamgateway.com/en/story/indonesia_ to_revise_p2p_lending_regulation_to_include_shariahcompliant_financing-SALAAM31012018062112/

31. Raconteur. "The Race to Become the World's Leading Islamic FinTech Hub." *Middle East | Bahrain FinTech Bay*, April 2, 2018, www.bahrainfintechbay. com/single-post/2018/04/02/The-race-to-become-the-world%E2% 80%99s-leading-leading-Islamic-FinTech-hub

32. Inovasi Keuangan Digital Di Sektor Jasa Keuangan. 2018. https://www.ojk. go.id/id/regulasi/Documents/Pages/Inovasi-Keuangan-Digital-di-Sektor-Jasa-Keuangan/pojk%2013-2018.pdf

33. "Highlight Provision of OJK Regulation No. 13/POJK.02/2018 on Digital Financial Innovations." *Walalangi & Partners*, September 17, 2018, www. wplaws.com/news/highlight-provision-ojk-regulation-no-13-pojk-02-2018-digital-financial-innovations/

34. Hakim, Elsie F. and Rully Hidayat. "OJK Issues Umbrella Regulation for Fintech Development, Establishes Regulatory Sandbox Regime | Lexology." *The Impact of Fake News: Economy—Part 2—Lexology*, September 20, 2018, www.lexology.com/library/detail.aspx?g=d3e294e3-c24f-4560-8908-b2b68bd01462

35. "Association Launches Code of Conduct for Fintech." *The Jakarta Post*, August 23, 2018, www.thejakartapost.com/news/2018/08/23/association-launches-code-of-conduct-for-fintech.html

36. *Code of Conduct Accountable Provision of Information Technology-Based Peer-to-Peer Lending Services*. Dewi Negara Fachri & Partners, July 2018, f.datasrvr.com/fr1/618/70127/4255v3-COC_Fintech_AG-JAKLIB01_Final.pdf

37. Winosa, Yosi. "Indonesia to Revise P2P Lending Regulation to Include Shariah-Compliant Financing." *Salaam Gateway*, January 31, 2018, www.salaamgateway.com/en/story/indonesia_to_revise_p2p_lending_regulation_to_include_shariahcompliant_financing-SALAAM31012018062112/

38. "Layanan Pembiayaan Berbasis Teknologi Infor}Iasi Berdasarkan Prinsip Syariah." *Dewan Syariah Nasional. Majelis Ulama Indonesia*, February 22, 2018. https://drive.google.com/file/d/1qCu2X6MTIFnYqK22eMx7uFL53AXBT9Vo/view

39. Hakim, Elsie F. and Rully Hidayat. "OJK Issues Umbrella Regulation for Fintech Development, Establishes Regulatory Sandbox Regime | Lexology." *Lexology*, September 20, 2018, www.lexology.com/library/detail.aspx?g=d3e294e3-c24f-4560-8908-b2b68bd01462

40. Wirdana, Ardi. "Indonesia Identifies 227 Illegal P2P Lending Startups, Mostly from China." *DealStreetAsia*, July 30, 2018, www.dealstreetasia.com/stories/indonesia-identifies-227-illegal-p2p-lending-startups-mostly-from-china-103546/

41. "OJK Belum Berencana Buat Aturan Khusus Fintech Syariah." *Republika Online*, September 1, 2018, republika.co.id/berita/ekonomi/syariah-ekonomi/18/09/01/pecs84349-ojk-belum-berencana-buat-aturan-khusus-fintech-syariah.

42. "Labuan Bajo Ready to Embrace P2P Lending Investment with Mekar." *Mekar News*, September 27, 2018, blog.mekar.id/en/labuan-bajo-ready-to-embrace-p2p-lending-investment-with-mekar/

Chapter 19

Intrastate Crowdfunding: Achieving Its Full Potential as an Economic Development Tool

Amy Cortese

Abstract

The federated structure of the USA provides considerable local regulatory autonomy within state boundaries, extending even to the operation of a wide range of financial activities. With the somewhat protracted implementation of the JOBS act delaying considerably the formal introduction of equity crowdfunding for retail investors on a national scale, many states sought to enact regulatory regimes to allow for it at a state level—the so-called intrastate equity crowdfunding. The variety of models adopted is instructive and of interest in itself but it is the powerful mechanism of binding investment into local communities with all of the additional benefits that such focused funding can bring which perhaps holds greater promise.

1. Introduction

Kyle DeWitt and Tim Schmidt spent 3 years trying to start a brewery in Tecumseh, a small Michigan town that was known as the "Refrigeration

Capital of the World" until Tecumseh Products closed its refrigeration compressor plant there in 2008.

"We went through bank after bank. They couldn't justify giving out a loan to two guys starting a business," said DeWitt, General Manager and Co-founder of Tecumseh Brewing Co., to Detroit Free Press.[1] In the wake of the Great Recession, credit was especially tight. Nor were the pair plugged into any angel investing networks.

Then in 2013, they found out about a new state law that was working its way through the Michigan legislature that might make their task easier. Called the Michigan Invests Locally Exemption, or MILE[2] for short, it allowed Michigan-based small businesses to raise money from any state resident via crowdfunding.

The would-be brewers launched a MILE campaign on Localstake, a Midwest-based crowdfunding portal and within 4 months had raised $175,000 from 21 Michigan investors. With that capital in hand, they were able to obtain a $200,000 bank loan from a local bank.

Today their Tecumseh Brewing Co. is a community gathering place that serves up craft beer and food out of two cheery storefronts in downtown Tecumseh. The brewpub has created 21 jobs, a number that is set to double as the company expands.

It's exactly the kind of success story that Michigan state leaders had in mind when they proposed MILE in the fall of 2013 and signed it into law before the year was out.

While much of the public attention in the US has been on the federal JOBS Act of 2012, which created a new, nationwide framework of exemptions for investment crowdfunding, a quiet revolution has been brewing at the state level.

Starting with Kansas—which passed its Invest Kansas Exemption in 2011, a year ahead of the JOBS Act—35 states and the District of Columbia have passed or enacted laws that allow investment crowdfunding within their borders.

The state laws grew out of frustration with the slow pace and cumbersome nature of federal reforms. While the JOBS Act was signed into law by President Obama in April 2012, it was not fully implemented until four years later, when Title III, also known as Regulation Crowdfunding, went into effect. Reg CF, as it's called, is considered the central provision of the

JOBS Act and allows small businesses to raise up to $1 million (recently increased to $1.07 million to reflect inflation) from ordinary investors.

Reg CF was also seen by many as overly burdensome. Legal and accounting costs, along with fees charged by crowdfunding intermediaries, can easily amount to 10% or more of the capital raised.

Local leaders were also motivated by a desire to support home-grown ventures, create jobs, and strengthen local economies by crafting investment crowdfunding laws tailored to their needs. The laws vary by state, but they allow small businesses based in a state to reach out to residents of that state—including customers, neighbours, and other supporters—for investment capital. For policymakers, it's a way to support local businesses—without costing the government a dime.

Most states mimic Title III's $1 million limit and caps on individual investments, but some are more liberal. Georgia, for example, allows businesses to raise up to $5 million.

Maryland, on the contrary, limits capital raises to $2,500.

There are other differences. Vermont and Oregon bar extractive industries such as mining and oil, while California has proposed bills targeted at cooperatives and social enterprises.

In addition, there is generally much less red tape involved in state crowdfunding. In some cases, businesses simply file a one- or two-page form with state regulators to notify them of the offering. About half of the 36 states do not even require a crowdfunding portal, meaning that fundraising and investment can take place offline.

"From a regulatory perspective, it's really simple," says Ryan Germany, Assistant Commissioner of Securities for the State of Georgia, of the Invest Georgia Exemption (IGE). "The downside is, you're limited to one state, so that's a much more limited crowd. But if a business is focused on its existing customers and they're all based in Georgia, then it makes sense."

What's more, he says, from an economic development perspective, "it costs the state no money." So, unlike expensive subsidies to 'attract and retain' big corporations, crowdfunding exemptions are a cost-effective way to support businesses and create jobs.

Germany cites Atlanta-based Groundfloor, which moved to Georgia from North Carolina in 2015, to take advantage of IGE. It has since created 30 local jobs and has plans to double that number.

Indeed, intrastate investment crowdfunding offers several potential advantages over federal frameworks as follows:

- It is uniquely aimed at community-scale entrepreneurs—which is to say, the majority of the US small businesses.
- The proximity and familiarity that investors have with the local market and businesses—they may even be customers—allow them to make more informed investment decisions.
- The relationships, reputations, and social capital at stake in intrastate deals can mitigate fraud and bad actors (as it has been shown to do in microfinance).
- Rather than purely speculative motives (e.g. getting in on the next Google or Facebook), intrastate crowdfunding taps into social and civic motives and can help channel capital to deserving enterprises that otherwise have limited funding options.
- It can help build local financial ecosystems in capital deserts.
- Many intrastate crowdfunding issuers choose innovative financing options, such as revenue-sharing agreements, which are more suitable for novice investors than equity—especially future equity vehicles such as SAFEs that are prevalent in Reg CF.
- States are more nimble than the federal government and can tailor and tweak their investment crowdfunding laws as they see fit. Therefore, states are laboratories that allow for experimentation and refinements that may ultimately trickle up to the federal level.
- Last but not least, intrastate crowdfunding appeals to a state's sense of civic pride and, as such, is a rare area of bipartisan co-operation in a divided age.

In other words, intrastate crowdfunding is as American as apple pie.

Representative Nancy Jenkins of Michigan, who championed Michigan's MILE Act, could have been speaking for any state when she commented on the law's appeal:

> "By investing in our communities and local small businesses, we are investing in Michigan's future. America was built by people with a vision for what could be, and the courage to step out and give their

ideas a try. That American spirit lives on in each of us. Together we can make our state a showcase of economic success for our residents, our businesses and our communities."[3]

No wonder intrastate crowdfunding has been embraced by states large and small, red and blue, urban and rural.

2. Unrealised Potential

Yet, state-based investment crowdfunding has been slow to gain traction. As of September 21, 2017, a total of 267 businesses had filed with state regulators, signalling their intent to use an intrastate crowdfunding law, according to the latest data from the North American Securities Administrators Association (NASAA), an organisation of state and local securities regulators that periodically surveys its members on their intrastate crowdfunding activity. Of those filings, 230 were cleared (filings may not be approved if the notice filing is incomplete).

In comparison, 910 businesses launched crowdfunding offerings under Regulation Crowdfunding in its first 2 years, with more than 100,000 individuals investing roughly $120 million.

In other words, there was nearly four times as much activity in the first 2 years of Regulation Crowdfunding than all of the states combined over a 7-year period (although many of the state laws are relatively recent).

The gap is likely much worse. Most states don't track what happens once a potential offering is filed. But evidence suggests that many of these businesses don't follow through with an offering—they may find the process too daunting and drop their plans, or a big investor may materialise. And those that do go ahead may not be successful in raising funds. So the number of offerings is likely fewer than the state filings suggest.

What is holding back this promising economic development tool? A number of factors seem to be contributing to its underuse.

While there is a general lack of awareness and knowledge around crowdinvesting, that's particularly true with intrastate. Entrepreneurs, investors, and the broader ecosystem of service providers (including lawyers, small business development centres, Chambers of Commerce, etc.) are often unaware that these state laws exist at all.

In the latest NASAA survey, state regulators cited a lack of awareness of the rules by companies and a lack of awareness and/or interest by investors as major factors contributing to the low usage of state crowdfunding rules. That dovetails with anecdotal data and other studies, including a survey of intrastate crowdfunding portals in 2015.

Other factors cited by NASAA survey included the availability of other exemptions, especially Regulation Crowdfunding and Rule 506 (for private offerings to wealthy investors), a lack of crowdfunding portals in some states, and limits in some states on public advertising of offerings.

3. Leaders of the Pack

That said, a handful of states are bucking the trend—and much can be learned from them.

Five states account for 66% of the intrastate filings that were allowed to proceed, according to Faith L. Anderson, Esq. Chief of Registration & Regulatory Affairs Department of Financial Institutions, Securities Division for the state of Washington, and a member of NASAA's corporate finance committee. They are Georgia (50 offerings), Texas (43), Oregon (26), Michigan (14), and Vermont (18).

What accounts for their high levels of activity? Some of these states tend to have some of the least restrictive crowdfunding laws, observes Anderson. That could contribute to a large number of filings, but as noted, it's unclear whether that equates to successful offerings.

Most of the top states also have active crowdfunding portals that have created a focal point for crowdfunding activity, such as Milk Money in Vermont, Hatch in Oregon, and NextSeed in Texas (NextSeed has since switched its focus to Regulation Crowdfunding).

In some states, such as Vermont, a strong "Buy Local" culture laid the groundwork for local investing. It doesn't hurt that local heroes Ben Cohen & Jerry Greenfield, ice cream makers of Ben & Jerry's, famously funded their early growth through a Vermont-only direct public offering, or DPO, a precursor to today's investment crowdfunding.

The most significant factor seems to be the presence of individuals and organisations that champion the law and help create an active and supportive community around it. Sometimes that is a crowdfunding portal, an incubator, or an economic development group that works to increase awareness about a state's law and educate users.

In many cases, it starts with a single individual.

In Oregon, Amy Pearl, the founder of a non-profit social enterprise incubator and co-working space called Hatch Innovation, led the charge. After neighbouring Washington State passed an intrastate crowdfunding law in March, 2014, Pearl reached out to the local legislators, lawyers, and community leaders to see what could be accomplished in Oregon. She immersed herself in the details of other state laws and helped put together a group of stakeholders that drafted Oregon's intrastate exemption, nicknamed the Community Public Offering, or CPO.

Hatch became the centre of gravity for the state's crowdfunding efforts. Before the rules were even finalised, Hatch launched a boot camp for entrepreneurs that walked them through the CPO process, from structuring their investment offering and creating a prospectus to marketing their offerings.

It created a crowdfunding portal, Hatch Oregon, to act as a clearinghouse for capital-raising campaigns and educational information. And it organised local investing meet-ups around the state, including rural towns in eastern Oregon, and dreamed up buzzy promotional events. At one rally on downtown Portland, fake dollars printed with "Community Public Offering" and the Hatch Oregon url rained from the sky.

The day the law went into effect in early January 2015, nine businesses were live on Hatch Oregon. They included an artisan ice cream company, a maker of all-terrain farm vehicles, a healthcare start-up, and, this being Portland, a taproom-cum-barbershop.

"The CPO creates a real stakeholder economy," Pearl commented at the time.[4] By her estimates, if Oregonians shift just 1% of their savings into local crowdfunding, they would inject nearly $1 billion into the state's economy.

4. "Ask Me About Crowdfunding"

More than 200 miles away, a similar scenario unfolded in Michigan 3 years earlier. It was the autumn of 2013 in tiny Adrian, Michigan, when

a run-of-the-mill economic development event launched a small revolution. It was there that a luncheon speaker (this author) described new
state crowdfunding laws that had been passed by Kansas and Georgia as
part of a broader local investing movement. Crowdfunding was still a
new notion and, as such, easy to dismiss. Yet, one attendee grasped the
idea and its implications immediately. "It struck me as something that
could potentially be amazing," says Chris Miller, the local downtown
development administrator and economic development coordinator for
Adrian.

Like much of Michigan and the US, Adrian in 2013 was still
attempting to climb its way out of a four-year recession, and credit for
any type of project was tight. "The cavalry is not going to ride into town
and save you," he recalls thinking. "You're going to have to do it on
your own."

A month later, Miller bumped into his local state representative,
Nancy Jenkins, and introduced her to this great idea he heard about. Could
Michigan pass its own intrastate crowdfunding law? Jenkins, a Republican,
took up the idea, and within 4 months, the Michigan legislature passed the
Michigan Invests Local Exemption (MILE) with bipartisan support and on
December 31, 2013, it was signed into law by the governor. Michigan
became the fourth state to have an intrastate law.

Many states might leave it at that—and many have. Yet, Miller
who wears a giant button that says "Ask me about crowdfunding," was not
done.

He talked up MILE every chance he got. The Michigan Municipal
League, a group of municipal leaders that represents all Michigan cities
and villages, also latched on. The agency created a website, Crowdfund
Michigan, explaining the new rules, and with Miller and others, criss-
crossed the state with workshops and other educational events.

One key constituency was local banks, says Miller. "Every local bank
in our area sent a rep and got trained on what MILE is and how it could
be paired with local capital." That helped Tecumseh Brewing to raise
capital via crowdfunding and unlock a larger bank loan.

The Michigan Economic Development Corp., the state's economic
development agency, also embraced the concept of crowdfunding. Led by
Katharine Czarnecki, Senior Vice President, Community Development,

the MEDC launched an innovative "crowdgranting" programme that matched, dollar for dollar, the money that Michiganers donated to local civic projects, such as new parks, maker spaces, or community gardens, via a crowdfunding site called Patronicity

The programme, called *Public Spaces, Community Places*, has been a huge success. And while it involves donations rather than investment crowdfunding, it has established the concept of crowdfunding in the state and is even seen as a model that could be expanded to MILE.

Even with all of that support, activity has been somewhat less than anticipated. "We all thought that in five years we would be banging out deals left and right," says Miller. "None of us saw the resistance—institutional and otherwise—that we'd face."

Yet, there has been a distinct change, says Miller. "Five years ago, 95% of the population—especially policymakers—didn't know what crowdfunding was." Now when there's a new project on the table, he says, "the first thing out of people's mouths is, can we crowdfund it?"

It doesn't matter whether its crowdgranting or crowdinvesting, says Miller. "The point is that they're thinking about activating and deploying local capital. The culture has shifted."

5. Forward

States like Michigan and Oregon may be ahead of the curve. But other states and countries can learn from their experience. As these leaders have demonstrated, a strong ecosystem is necessary to nurture a culture of crowdfunding, and ecosystems do not generally develop on their own. It is also very clear that much more education is needed—for entrepreneurs, the people who advise them and the general public—before this new fundraising and economic development tool can reach its full potential.

As Miller notes, the specific exemptions—federal or state, DPO or CPO—matter less that the fact that there are new options for local investors and entrepreneurs that can help grow local economies and spread wealth.

To fill a void in the national market, Miller, Pearl, and other local economy leaders (including this author) have formed a non-profit, the National Coalition for Community Capital (NC3), to help raise awareness, educate the public, and advocate for good policy around crowdfunding and other community capital tools.[5]

In particular, the group is focused on expanding crowdfunding to underserved communities that can most benefit from it. Their effort dovetails with a growing recognition within the broader impact investing world that, to truly address issues such as the wealth gap and lack of economic mobility, investment decision-making and profits must be shared more broadly rather than "hoarded" by an elite class.

New models that include and engage institutional stakeholders in the process, akin to Michigan's *Public Places, Community Spaces*, may be helpful in low-income areas, where improved access to capital is most needed, yet available capital is limited.

6. Recommendations for Policymakers

So what can policymakers do to encourage responsible use of crowdfunding tools in their regions?

Bring together a broad stakeholder group to create or improve local laws around crowdfunding and other community capital initiatives. That might include lawyers, entrepreneurs, incubators, existing crowdfunding portals or industry associations, investor advocates, and members of underserved communities. Involving a range of stakeholders creates buy-in and ensures a well-designed policy. In particular, those local crowdfunding champions and advocacy groups such as NC3 can be valuable resources and allies.

Be flexible. One stumbling block that early crowdfunding states encountered was a lack of escrow agents—the financial institutions that are entrusted to hold onto investors' money, while the company they invested in reaches its minimum funding target. Some states required that role to be played by in-state banks. Yet, given the conservative nature of banks, few stepped up and many state laws languished for lack of an escrow agent. More recently, states have addressed this gap by allowing organisations other than banks to act as escrow agents. That might include law firms (as in North Carolina) or credit unions (as in Vermont).

Give it a name. Let's face it: law is not poetry. But as Oregon's Community Public Offering, or CPO, demonstrates, a catchy name helps to rally the public and stakeholders around a local crowdfunding initiative and convey its purpose without devolving into securities jargon.

Be inclusive. Crowdfunding has found a ready niche with white, relatively affluent communities. But to reach all constituents—in particular, underserved communities that most need increased access to capital and the economic opportunity it can bring—thoughtful and deliberate outreach and education is necessary. In addition, models that pair institutional investment alongside community investment can help amplify small-dollar community investments, mitigate risk, and help crowdfunding succeed in lower income communities.

Give portals a chance. In creating policy, consider the vital role that crowdfunding portals and other intermediary platforms play. They provide a focal point for local investment, conduct basic screening of deals, manage the investment process for businesses, and offer high-touch support to investors and entrepreneurs. Still, many portals struggle to make the economics work, in part because of laws that have sharply limited the ways they can be compensated. Intermediaries need to have a sustainable business model, and laws and regulations should not make that prospect unduly difficult.

Educate, educate, educate. The single biggest factor holding back the adoption of crowdfunding is a lack of awareness and understanding on the part of investors, entrepreneurs, and the professionals that advise them. Policymakers should have a clear outreach and partnership plan to develop a network of experts who can help educate and advise entrepreneurs and investors. And where possible, they should allocate budget to support these initiatives.

Assess your local ecosystem for potential educational partners. For example, organisations such as the US Small Business Administration's network of Small Business Development Centers (SBDCs) provide free mentoring and consulting to business owners and are vital links in the entrepreneurial ecosystem. At a minimum, these centres should be well-versed on crowdfunding laws so that they can advise small business owners on their appropriate use. (In reality, however, that is often not the case). Privately run incubators and accelerators, such as Oregon's Hatch Innovation, have created successful programmes that walk entrepreneurs through the basics of structuring and launching a crowdfunding raise. At the same time, Hatch has hosted workshops for potential investors that cover the basics of due diligence and crowdinvesting. These public and private organisations can be harnessed to create a supportive ecosystem for localised crowdfunding to succeed.

Finally, consider a financial innovation sandbox. Financial technology is disrupting traditional finance and that has set up a cat-and-mouse-type dynamic between "fintech" innovators and regulators. Countries including the UK and Australia have created regulatory sandboxes designed to encourage controlled innovation and collaboration between regulators and private sector innovators. A sandbox is a safe harbour of sorts that allows financial-tech start-ups to test out new products, services, and business models without getting bogged down in costly regulations—but under the watchful eyes of regulators. State and local jurisdictions, such as Arizona, have also created localised regulatory sandboxes.[6]

About the Author

Amy Cortese is an award-winning journalist who writes about topics spanning business, finance, food, wine and travel. Her work has appeared in *the New York Times Magazine*, New York, *Business Week*, *the New York Times*, the *Daily News*, *Portfolio*, *Mother Jones*, *Afar*, *The American*, *the Daily Beast*, *Talk*, *Business 2.0*, and *Wired*, among other publications. Her recently published book, *Locavesting*: *The Revolution in Local Investing and How to Profit From it* (John Wiley & Sons, 2011), draws upon her experience covering these diverse realms to explore how a small shift in investment away from multinationals towards locally-owned enterprises can reap enormous economic and social benefits for individuals, their communities and the country.

Endnotes

1. https://www.freep.com/story/money/personal-finance/susan-tompor/2015/05/09/crowd-funding-mile-michigan-tecumseh-brewing/26861897/
2. https://www.michigan.gov/lara/0,4601,7-154-89334_61343_32915-332094--,00.html

3. https://www.adrianareachamber.com/transcript-of-nancy-jenkins-crowdfunding-workshop-speech/

4. https://www.locavesting.com/crowdfunding/crowdfunding-oregon-style-the-states-new-community-pubic-offering-is-off-to-a-fast-start/

5. http://nc3.comcap.us/

6. https://www.locavesting.com/crowdfunding/arizona-creates-fintech-sandbox-encourage-innovation/

Part 3.4

Pan National

Chapter 20

Will the Proposed European Crowdfunding Regulation Lead to a "True" European Market for Crowdfunding?

Sebastiaan Niels Hooghiemstra

Abstract

On March 8, 2018, the European Commission published a proposal for a crowdfunding regulation (CFR)[1] as well as a proposal to amend MiFID II.[2]

The CFR proposal seeks to facilitate the scaling-up of crowdfunding services across the internal market to increase access to finance for entrepreneurs, start-ups, scale-ups, and SMEs in general. Till now, crowdfunding services have not been subject to EU action. Crowdfunding service providers were therefore subject to the different national frameworks implementing existing EU law that hindered the emergence of a European market for crowdfunding. The EU regime seeks to introduce a level playing field for these service providers provided they comply with harmonised EU legislation.

The contribution seeks to assess whether and to what extent the CFR proposal will lead to a true European market for investment- and lending-based crowdfunding. To that end, this contribution evaluates several relevant aspects of the CFR. Including its scope, "platform," sales, and marketing regulation that target CSPs are reviewed in order to answer the question whether the CFR leads to a "true" European crowdfunding market. The main conclusion of this contribution is that the introduction of a European CSP passport certainly will bring benefits in terms of economies of scale and scope to platforms, project owners, and investors who are engaged in equity- and lending-based crowdfunding, but its scope will need to be reviewed in order to establish a "true" European crowdfunding market.

1. Introduction

On March 8, 2018, the European Commission published a proposal for a crowdfunding regulation CFR)[3] as well as a proposal to amend MiFID II.[4]

The CFR proposal seeks to facilitate the scaling-up of crowdfunding services across the internal market to increase access to finance for entrepreneurs, start-ups, scale-ups, and SMEs in general. Till now, crowdfunding services have not been subject to EU action.[5] Crowdfunding service providers (CSPs) were therefore subject to the different national frameworks implementing the existing EU law that hindered the emergence of a European market for crowdfunding.[6] The EU regime seeks to introduce a level playing field for these service providers provided these providers comply with harmonised EU legislation.

This contribution seeks to assess whether and to what extent the CFR proposal will lead to a "true" European market for (investment- and lending-based) crowdfunding. To that end, this contribution evaluates several relevant aspects of the CFR, including its scope, "platform," sales, and marketing regulation that target CSPs.

2. Scope of the European Crowdfunding Regulation

The CFR addresses the provision of crowdfunding services, the organisational and operational requirements of CSPs in the EEA.

Crowdfunding Services Providers and the provision of Crowdfunding Services.

The CFR applies to legal persons that seek authorisation as a CSP in relation to the provision of crowdfunding services.

Under Article 3(1)(c) CFR, a "CSP" means

"a legal person who provides crowdfunding services and has been authorised for that purpose by ESMA."

2.1. Crowdfunding Service

Under Article 3(1)(a) CFR, a "crowdfunding service" is

"the matching of business funding interest of *investors* and *project owners* through the use of a *crowdfunding platform* and which consist of any of the following:

- the facilitation of granting of loans;
- the *placing without firm commitment*, as referred to in point 7 of Section A of Annex I MiFID II of transferable securities issued by project owners and the *reception and transmission of client orders*, as referred to in point 1 of Section A to Annex I to MiFID II, with regard to those *transferable securities* (emphasis added by author)."

2.2. Investors, Project Owners and Crowdfunding Platforms

The provision of crowdfunding services (generally) relies on three types of actors[7]: the project owner[8] that proposes the project to be funded, investors[9] who fund the proposed project, generally by limited investments, and an intermediating organisation in the form of a CSP[10] that brings together project owners and investors through an online platform.

2.3. Lending-Based Crowdfunding

The meaning of "the facilitation of granting of loans" has not been defined under the CFR. Recital 10 CFR, however, states that the facilitation of granting of loans, such as presenting crowdfunding offers to clients or rating the creditworthiness of project owners, accommodates different business models enabling a loan agreement to be concluded through a crowdfunding platform between one or more clients and project owners. In this respect, "clients" are any (prospective) investors or project owners to whom CSPs (may) provide crowdfunding services.[11] Furthermore, "crowdfunding offers" include any communication by CSPs that contains information which enables (prospective) investors to decide on the merits of entering into a crowdfunding transaction.[12]

2.4. Investment-Based Crowdfunding

Investment-based crowdfunding under the CFR may take place in the form of "placing without firm commitment" and/or "the reception and transmission of client orders."

2.5. Placing without Firm Commitment

National crowdfunding regimes currently show an inconsistent approach as to what is to be regarded under Annex I, s. A.7 MiFID II as the "placing of financial instruments without a firm commitment basis."

"Placing" is the service of finding investors for securities on behalf of a seller and may or may not involve a commitment to take up those transferable securities where others do not acquire them.[13] The placing of financial instruments involves situations where a project owner wishes to raise capital for commercial purposes and, in particular, with primary market activity.[14]

The provision of crowdfunding services by CSPs is restricted to merely the placing of financial instruments *without a firm commitment basis* as the CFR under Article 7(1) CFR prohibits CSPs to have a financial participation in any crowdfunding offer on their crowdfunding platforms.

The "firm commitment" aspect of the placing service relates to the CSP arranging the placing, as opposed to the person who has agreed to purchase any instruments as part of the placing. Thus, placing by a CSP on a firm commitment basis occurs where a CSP undertakes to arrange the placing of financial instruments and to purchase some or all the instruments that it may not succeed in placing with third parties. In other words, the placing element under the CFR requires the same person to arrange the placing and provide a firm commitment that some or all of the instruments will be purchased. Consequently, CSPs are under the CFR restricted to arranging a placing in which they do not undertake to purchase those transferable securities the CSP fails to place with investors.[15]

2.6. The Reception and Transmission of Client Orders

National crowdfunding regimes currently also show an inconsistent approach as to what is to be regarded under Annex I, s. A.1 MiFID II as the "reception and transmission of client orders in relation to one or more financial instruments."

In this regard, Recital 44 MiFID II clarifies that

"the business of reception and transmission of orders should also include bringing together two or more investors, thereby bringing about a transaction between those investors."

De facto, a CFR can only provide this service in relation to equity-based crowdfunding if it is both receiving and transmitting orders.[16,17] If the CSP is a party to a transaction as an agent for a client or commits a client to it, a CSP does more than receiving and transmitting orders and will need to consider whether the CSP provides the investment service of "executing orders on behalf of clients."[18]

2.7. Transferable Securities

Crowdfunding services in relation to investment-based crowdfunding are limited to "transferable securities" within the meaning of Article 4(1)(44) MiFID II.[19] The CFR contains this limit as the European Commission considers under Recital 11 CFR the transferability of a security

"as an important safeguard for investors to be able to exist their investment since it provides them with the legal possibility to dispose of their interest on the capital markets."

The CFR only covers and permits investment-based crowdfunding services in relation to transferable securities.[20] Financial instruments other than transferable are, therefore, excluded from the scope of the CFR because those securities entail risks for investors that cannot be properly managed within the legal framework of the CFR.[21]

2.8. Exemptions Under the European Crowdfunding Regulation

Several crowdfunding services that will now be discussed are exempt from the scope of the CFR.

2.9. Crowdfunding Services Provided to Project Owners who are Consumers

Crowdfunding services provided to "project owners," i.e. persons who seek to fund their crowdfunding project[22] through a crowdfunding platform[23] to "consumers"[24] within the meaning of the Consumer Credit Directive, are excluded from the scope of the CFR.

 The CFR proposal explains that, the CFR does not include consumer lending for consumption purposes as this is not business lending and it (partially) falls within the scope of existing EU legislations, such as the Consumer Credit Directive and, in case of a consumer receiving a loan to purchase immovable property, the Mortgage Credit Directive.[25,26]

 In a similar vein, Recital 8 CFR explains that the objective of the CFR is to address the obstacles to the functioning of the internal market in crowdfunding services in order to *foster cross-border business funding*. Crowdfunding services in relation to lending to consumers, such as P2P consumer lending, therefore do not fall within the scope of the CFR.

2.10. Crowdfunding Services Provided by MiFID II Investment Firms

In order to avoid that the same activity is subject to different authorisations within the EU, crowdfunding services provided by investment firms that have been authorised under MiFID II are excluded from the scope of the CFR.[27]

2.11. Crowdfunding Services that are Provided by Natural or Legal Persons in accordance with National Law

Under the CFR, CSPs providing crowdfunding services have the possibility either to opt for "EU regulatory treatment" under the CFR or "national regulatory treatment," i.e. the national crowdfunding regulatory framework, if available, in the Member State of residence of the CSP. For crowdfunding services provided to consumers, as defined under the Consumer Credit Directive, or investment-based crowdfunding related to securities other than "transferable securities," CSPs have no other choice than to comply with "national regulatory treatment," if available. Recital 9 CFR, similar as for investment firms, however, prohibits CSPs to obtain an authorisation under both the CFR and the national regime in place. This is to avoid that the same activity is subject to different authorisations.[28]

2.12. Crowdfunding Offers Subject to the Prospectus Regulation

Article 2(1)(d) CFR excludes from the scope of the CFR offers with a consideration of more than €1,000,000 per crowdfunding offer, which shall be calculated over a period of 12 months in relation to a particular crowdfunding project. In this regard, Recital 12 CFR explains that given the risks associated with crowdfunding investments, it is appropriate and in the interest of the effective protection of investors to impose a threshold for a maximum consideration for each crowdfunding offer. Recital 12 CFR continues to explain that the threshold should be set at €1,000,000, because that threshold corresponds to the threshold for the mandatory drawing up and approval of a prospectus above that threshold as set out in the Prospectus Regulation[29] that will in 2019 replace the current Prospectus Directive.[30]

3. "Platform Regulation"

European financial law requires licensing to promote fairness, honesty, and professionalism by those who provide financial services, on the one hand, while ensuring that intermediaries remain financially solvent on the other hand. In the crowdfunding context, crowdfunding services shall under the CFR only be provided by legal persons who have an effective and stable establishment in an EEA Member State and who comply with the licensing and corresponding business organisational requirements under the CFR.[31]

3.1. Licensing Requirements

Some of the licensing requirements to which CSPs are subject, *inter alia*, include the following:

- fit and properness requirements for senior management;[32]
- a business plan;[33]
- an adequate risk organisation;[34]
- an adequate and appropriate business organisation;[35]
- reliable significant shareholders[36] and
- penalties if the platforms no longer meet the requirements of the applicable legislation.[37]

3.2. Business Organisational Requirements

The commitment to fairness, honesty, and acting in the investor's best interest is the overarching duty of loyalty/care the CSPs need to comply with.[38] To maintain a high standard of investor protection, to reduce the risks associated with crowdfunding, and to ensure fair treatment of all clients, CSPs, for example, have to ensure a policy designed to ensure that projects are selected in a professional, fair, and transparent way and that crowdfunding services are provided in the same manner.[39]

The overarching principle of duty of loyalty/care, as is common in European financial law, has been elaborated in a number of more detailed business organisational requirements for CSPs.

3.3. Effective and Prudent Management

Ensuring an effective system of governance is a common principle in European financial legislation that seeks to ensure the proper management of risk and prevent conflict of interests.[40] For that purpose, the management of CSPs shall establish, and oversee the implementation of, adequate policies and procedures to ensure effective and prudent management, including the segregation of duties, business continuity, and the prevention of conflicts of interest, in a manner that promotes the integrity of the market and the interest of their clients.[41]

3.4. Complaints' Handling

CSPs have to establish and maintain effective and transparent procedures for the prompt, fair, and consistent handling of complaints received from clients.[42] Clients may be filing complaints with the CSP free of charge.[43] CSPs have to keep a record of all complaints received and the measures taken.[44]

3.5. Conflicts of Interest Rules

To ensure that CSPs operate as "neutral intermediaries" between clients on their crowdfunding platform, CSPs have to comply with certain conflict of interest rules related to the CSPs, managers and employees, or any person directly or indirectly controlling them.[45] In particular, CSPs are not allowed to have any financial participation ("skin in the game") in any crowdfunding offer on their crowdfunding platforms.[46] Furthermore, shareholders holding 20% or more of share capital or voting rights, managers and employees, or any person directly or indirectly controlling crowdfunding platforms are not allowed to be clients in relation to the crowdfunding services offered on that platform.[47] For that reason, CSPs shall maintain and operate effective internal rules to prevent conflicts of interest.[48] In particular, CSPs shall take all appropriate steps to prevent, identify, manage, and disclose conflicts of interest between the CSPs themselves, their shareholders, their managers and employees, or any person directly or indirectly linked to them by control,[49] and their clients, or between one client and another client.

CSPs shall disclose to their (potential) clients the general nature and sources of conflicts of interest and the steps taken to mitigate those risks when they consider that this is necessary for the measures taken in accordance with their internal rules.[50]

3.6. "Best Execution Rules"

CSPs have under the CFR to comply with specific "best execution rules." To improve the service to their clients, CSPs have discretion on behalf of clients with respect to the parameters of the clients' orders, provided they take all necessary steps to obtain the best possible result for their clients and that they disclose the exact method and parameters of the discretion.[51]

3.7. Rules on CSP Remuneration

To ensure that prospective investors are offered investment opportunities on a neutral basis, CSPs may not pay or accept any remuneration, discount, or non-monetary benefit for routing investors' orders to a particular offer provided on their platform or on a third-party platform.[52]

3.8. Rules on the use of SPVs

In practice, CSPs use "special purpose vehicles" (SPVs)[53] or similar structures for various reasons.[54] First, CSPs negotiate and/or enter into investment agreements and related documents with project owners for the benefit of investors. By doing so, CSPs ensure that investors receive the same investor rights as "non-crowdfunding investors." Second, the use of SPVs allows project owners to raise further capital and an easier "exit." Indirect investments in a company through an SPV allows for easier corporate transactions, such as further financings or sales. CSPs may exercise discretions on behalf of the joint investors to sign transactions and other (corporate) decisions to support further capital raising or an exit. Finally, SPVs ease the administration of investments for both investors and project owners.

The CFR has, however, as its objective to facilitate direct investments and to avoid creating regulatory arbitrage opportunities for financial intermediaries, such as asset managers.[55] For that reason, the use of legal structures, including SPVs, to be interposed between the crowdfunding project and investors, is regulated under the CFR and is only permitted where it is considered to be justified.[56] Under Article 4(5) CFR, the use of SPVs for the provision of crowdfunding is allowed, provided that the CSP only transfers one asset to an SPV to enable investors to take exposure to that asset by means of acquiring securities. The decision to take exposure to that underlying asset shall exclusively lie with investors.[57]

3.9. Rules on Money Laundering and Terrorist Financing

Crowdfunding services may be exposed to money laundering and terrorist financing risks.[58] The CFR, therefore, considers that safeguards should be envisaged when meeting conditions for authorisation, assessing the good repute of the management, providing payment services only through PSD II[59] authorised entities subject to anti-money laundering and terrorist financing requirements.[60] Although the CFR does not impose AMLD IV/V[61] requirements to CSPs, the European Commission will review the necessity and proportionality of compliance with the standards laid down in these directives.[62]

3.10. Outsourcing

To facilitate an efficient and smooth provision of crowdfunding services, CSPs are allowed to entrust any operational function, in whole or in part, to third-party service providers.[63] CSPs may outsource to the extent that it does not materially impair the quality of the CSP's internal controls and effective supervision, i.e. operation functions are outsourced to the extent that the CSP is a "letter-box entity."[64] CSPs are, therefore, fully responsible for compliance with the CFR and, to that end, take all reasonable steps to avoid additional operational risks.

3.11. Client Asset Safekeeping, Holding of Funds, and the Provision of Payment Services

CSPs have to inform their clients whether[65]

- they provide asset safekeeping services and on which terms and conditions, including references to applicable national law;
- asset safekeeping services are provided by them or by a third party;
- payment services and the holding and safeguarding of funds are provided by the CSP or through a third party acting on their behalf.

The holding of clients' fund and the provision of payment service require an authorisation as a payment service provider in accordance with Article 4(11) PSD II.[66] That authorisation requirement cannot be satisfied by a mere authorisation as a CSP under the CFR. For that reason, CSPs or third-party providers acting on their behalf need, in addition, to be authorised as a payment institution in accordance with PSD II if payment services are carried out in relation to crowdfunding services.[67] Where CSPs or third parties on their behalf are not providing payment services or the holding and safeguarding of funds in relation to crowdfunding services, such CSPs have to put in place and maintain arrangements to ensure that project owners accept funding of crowdfunding offers or any payment only by means of a payment service provider.[68]

4. Sales and Marketing Regulation

Under the CFR, CSPs are required to comply not only with "platform regulation" but also with various sales and marketing requirements.

4.1. Information to Clients

All information, including marketing communication, from CSPs to (potential) clients such as the costs and charges related to crowdfunding services or investments, the crowdfunding conditions, including crowdfunding project selection criteria, and the nature of risks associated with the services, shall be clear, compensable, complete, and correct.[69] Information, including investor information and marketing communication, shall be provided to potential clients before they enter into a crowdfunding transaction.[70] Such information shall be available to all (potential) clients on a clearly identified section of the website of the crowdfunding platform and in a non-discriminatory manner.[71]

4.2. KIIS Obligation

The CFR is limited to crowdfunding offerings that fall outside of the scope of the Prospectus Regulation.[72] In order for prospective investors to have a clear understanding of the nature, risks, costs, and charges of crowdfunding services, CSPs should require prospective investors with a key investment information sheet (KIIS).

The obligation to provide a KIIS for each crowdfunding offer to prospective investors lies with the CSPs,[73] whereas the document itself shall be drawn up by the project owner. The CFR foresees this division in responsibilities as the project owner concerned is in the best position to provide the relevant information.[74] Since the CSP is, however, responsible for informing their prospective investors, they are required to ensure that the KIIS is complete.[75]

Along the trend for short form investor disclosure, such as the "key investor document" under the PRIIPR,[76] the KIIS is a document of maximum six pages that should warn prospective investors that the investing environment they have entered into entails risks and is covered neither by a deposit compensation scheme nor by investor compensation guarantees.[77]

The KIIS should also take into account the specific features and risks associated with early-stage companies and focus on material information about the project owners, the investors' rights and fees, the type of securities offered, and loan agreements.[78] To that end, the KIIS contains general information related to the following[79]:

- the project owner(s) and the crowdfunding project, including information about the identity of the project owner(s), the principles activities, products, or services offered, and a description of the crowdfunding project;
- main features of the crowdfunding process and conditions for the capital raising or funds borrowing, including, among others, the minimum target capital to be raised, the deadline for reaching the target, and the maximum offering amount;
- presentation of the main risks associated with financing the crowdfunding project, with the sector, the project, the project owner, and the investment instrument, including geographic risks, where relevant;
- fees, where additional information can be obtained and legal redress.

The KIIS contains additional specific information depending upon whether the project concerns investment- or loan-based crowdfunding.

For equity-based crowdfunding, the KIIS contains disclosure related to the loan agreement, including[80]:

- information related to the offering of securities, such as, among others, the total amount and type of investment instruments to be offered, the terms of subscription and payment, and the custody and delivery of investment instrument to investors;
- issuer's information, where the issuer is different from the project owner and therefore is an SPV;
- investor rights, such as, among others, key rights attached to the securities, restrictions to which securities are subject and opportunities for exit.

For loan-based crowdfunding, the KIIS contains disclosure related to the loan agreement, including[81]:

- the nature and duration of the credit agreement;
- applicable interest rates or, where applicable, other compensation to the investor;
- risk mitigation measures, such as whether credit is secured;
- an amortisation schedule of the principle and repayment of interest.

When a CSP identifies a material omission, mistake, or inaccuracy, the project owner shall complement or amend that information.[82] If that is not possible, the CSP shall not make the crowdfunding offer or cancel the existing offer until the KIIS complies with the requirements laid down in the CFR.[83] *De facto*, the CSP has the obligation to keep the KIIS updated during the period in which the project is open for funding and the CSP has, thus, a "duty of care" to ensure that investors are not misled by the KIIS.

To ensure seamless and expedient access to capital markets for start-ups and SMEs, to reduce the costs of financing, and to avoid delays and costs for CSPs, the KIIS is not subject to approval by a Competent Authority.[84]

4.3. KYC: The Crowdfunding Knowledge Test

Investments in products marketed on crowdfunding platforms are not comparable to traditional investment or savings products and should not be marketed as such.[85] To ensure that prospective investors, however, understand the level of risk associated with crowdfunding investments, CSPs are required to run an entry knowledge test of their prospective investors to establish their knowledge and understanding of risk in relation to investing in general and the types of investments offered on a crowdfunding platform.[86]

For the purpose of the assessment, CSPs shall request information about the prospective investor in relation to the following:

- the prospective investor's past investments in transferable securities or loan agreements, including in early- or expansion-stage businesses;
- any relevant knowledge or professional experience in relation to crowdfunding investments.

The knowledge test by CSPs shall be carried out for each investor once every 2 years.[87] CSPs should explicitly warn prospective investors whenever the crowdfunding services provided are deemed as inappropriate for them.[88] That information or risk warning shall, however, not prevent prospective investors from investing in crowdfunding projects.[89]

Simulating the "ability to absorb losses" as a Substitute for Investment Limits.

Other than is the case under various national crowdfunding legislations, the CFR does not impose an investment limit per investor per project. Instead, CSPs shall, in connection with the knowledge test, simulate the ability of prospective investors to bear loss, calculated as 10% of their net worth.[90] The simulation by the CSP takes place based upon the following information[91]:

- regular income and total income, and whether the income is earned on a permanent or temporary basis;
- assets, including financial investments, personal and investment property, pension funds, and any cash deposits;

- financial commitments, including regular, existing, or future commitments.

Irrespective of the results of the simulation, (prospective) investors shall, however, not be prevented from investing in crowdfunding projects.[92]

4.4. Bulletin Board

CSPs may under Article 17 CFR allow their investors to interact directly with each other to buy and sell loan agreements or transferable securities which were originally crowdfunded on their platforms on a so-called "bulletin board." This possibility is foreseen by the CFR as to open up opportunities for secondary market trading. Investments of retail investors are, generally, limited in value and volume. A bulletin board allows a retail investor to make his investment liquid again at any moment in time and for any reason.[93] This fits in the objectives of the Capital Markets Union to allow investors, wherever residing within the EEA, to continuously invest and trade their investments in start-ups and SMEs through CSPs.

When offering the possibility of a "bulletin board," CSPs shall inform their clients that they do not operate a trading system and that such buying and selling activity on their platforms is at the client's own discretion and responsibility.[94]

Recital 35 CFR explicitly confirms that CSPs should not be able to provide any discretionary or non-discretionary matching of buying and selling interest, because that activity requires an authorisation as an investment firm[95] for operating an MTF or OTF, or as a regulated marketed in accordance with Article 44 MiFID II.[96] CSPs should, in the interest of transparency and flow of information, be able to allow investors who have made investments through their platform to contact and transact with each other over their platforms in relation to investments originally made on their platform.

CSPs that suggest a reference price for the buying and selling shall inform their clients that the suggested reference price is non-binding and substantiate the suggested reference price.[97]

4.5. Marketing Communications

To ensure fair and non-discriminatory treatment, CSPs that promote their services through marketing shall not treat any particular project more favourably by singling it out from other projects offered on their platform.[98] For that purpose, CSPs shall ensure that all marketing communications to investors are clearly identifiable as such.[99] Planned projects should therefore not feature in marketing communications of a crowdfunding platform.[100] CSPs may, however, mention successfully closed offers in which investments through the platform are no longer possible. Competent authorities shall, however, not require an *ex ante* notification and approval of marketing communications.

5. Does the CFR Lead to a "True" European Crowdfunding Market?

The question to be answered in this contribution is whether the CFR will lead to a "true" European crowdfunding market?

The introduction of an "European passport" under which CSPs that are authorised for providing crowdfunding services in a "home Member State" under the CFR may market their services throughout the internal market without incurring further conditions in "host Member States" enhances market efficiency and leads to economies of scale and scope.[101]

Authorisation and notification requirements under CSP European passport arrangements, for example, only need to be fulfilled in one Member State while having a large market to offer their financial products and services without worrying about the establishment of subsidiaries and multiple authorisation applications that would have led to a duplication of legal costs. In particular, CSPs that intend to offer their services on a cross-border basis do not have to legally "engineer" their business models to comply with various fragmented national legal frameworks. The introduction of a European passport with a uniform set of legislation will, thus, lead to a convergence of the legal structuring of crowdfunding platforms. Apart from this, the introduction of the CFR leads to a decline of prices in the internal market for crowdfunding services as fixed costs are shared among a larger client base. CSPs may organise themselves anywhere in the EEA, in small and big Member States and, as a result, have the potential to maximise their efficiency and become more competitive. In addition to reducing transaction costs, CSPs from either smaller or bigger EEA Member States may market their crowdfunding services to all of the EEA's 500 million citizens resulting in more revenue.

It can, however, not be said that the introduction of a European passport for CSPs leads to a "true" European crowdfunding market as its scope is limited to equity- and lending-based crowdfunding projects offered by SMEs.[102] Other crowdfunding forms, such as civic, reward- and donation-based crowdfunding are excluded from its scope and, thus, will need to rely upon the national legal framework restricting access for consumers of (small) EEA Member States to offer their crowdfunding projects in their Member States and under high costs. In addition, the focus of the Capital Markets Union on SMEs and the exclusion of crowdfunding offers made by consumers ensure that a significant number of crowdfunding projects are excluded from obtaining crowdfunding on a cross-border basis in the EEA.

Overall, the CFR is a good start for a "true" European crowdfunding market. After taking effect, it will need to be assessed whether and to what extent its limitations in scope pose problems in practice that then can be effectively remedied to unleash the potential of a "true" European crowdfunding market.

6. Conclusion

In this chapter, several relevant aspects of the proposed CFR, including its scope, "platform," and sales and marketing regulation that target CSPs, were reviewed in order to answer the question whether the CFR leads to a "true" European crowdfunding market.

The CFR applies to CSPs in relation to the provision of crowdfunding services. In this regard, "crowdfunding services" only include lending- and equity-based crowdfunding. In line with European financial law, authorised CSPs are required to comply with "intermediary regulation" in the form of licensing and business organisational requirements, including, among others, specific crowdfunding rules on the use of SPVs, client asset safekeeping, the holding of funds, and the provision of payment services.

In addition, CSPs have to comply with sales and marketing regulation, including but not limited to, information to be provided to clients, the obligation to provide a KIIS, and a crowdfunding KYC knowledge test.

The CSP European passport to be introduced allows platforms to exploit economies of scope and scale and to offer crowdfunding services to project owners and investors throughout the entire EEA. Nevertheless, its scope is limited. The CFR, for instance, only applies to equity- and lending-based crowdfunding. Other types of crowdfunding, such as civic and consumer crowdfunding projects, are left out of the scope. Although the introduction of a European CSP passport certainly will bring benefits, its scope will need to be reviewed to establish a "true" European crowd-funding market.

About the Author

Sebastiaan Niels Hooghiemstra is an Associate in the investment funds practice of NautaDutilh Luxembourg. He assists asset managers, funds promoters, depositary banks, and institutional investors in the structuring and setting up of Luxembourg investment fund structures. Prior to joining NautaDutilh Avocats Luxembourg, he was a Doctoral Candidate at the University of Utrecht and worked as a Research Associate at the University of Liechtenstein. He recently obtained a PhD in Financial Law from Utrecht University. He was a member of the working groups implementing UCITSD IV/V, CRD IV, and the AIFMD in Liechtenstein. He also advised the European Commission and EIOPA on a regulatory framework for PEPPs and the introduction of a European depositary passport.

Endnotes

1. Proposal for a Regulation of the European Parliament and of the Council on European Crowdfunding Service Providers (ECPS) for Business {SWD(2018) 56 final} {SWD(2018) 57}, March 8, 2018 COM (2018) 113 final.
2. Proposal for a Directive of the European Parliament and of the Council amending Directive 2014/65/EU on markets in financial instruments {SWD(2018) 56} {SWED(2018) 57}, March 8, 2018 COM (2018) 99 final.

3. Proposal for a Regulation of the European Parliament and of the Council on European Crowdfunding Service Providers (ECPS) for Business {SWD(2018) 56 final} {SWD(2018) 57}, March 8, 2018 COM (2018) 113 final.

4. Proposal for a Directive of the European Parliament and of the Council amending Directive 2014/65/EU on markets in financial instruments {SWD(2018) 56} {SWED(2018) 57}, March 8, 2018 COM (2018) 99 final.

5. Ferrarini, Guido and Eugenia Machiavello: FinTech and Alternative Finance in the CMU: The Regulation of Marketplace Investing. Busch, Danny, Avgouleas, Emilios and Guido Ferrarini (eds.), Capital Markets Union in Europe. Oxford University Press, 2018.

6. Hooghiemstra, Sebastiaan and Kristof de Buysere, The Perfect Regulation of Crowdfunding: What Should the European Regulator Do? Gajda, Oliver and Dennis Brüntje (eds.), Crowdfunding in Europe—State of the Art in Theory and Practice, Springer 2015.

7. Recital 3 CFR.

8. Art. 3(1)(f) CFR.

9. Art. 3(1)(g) CFR.

10. Art. 3(1)(b) and (f) CFR.

11. Art. 3(1)(e) CFR.

12. Art. 3(1)(d) CFR.

13. *See*, for example, in the UK: PERG 13.3 Investment Services and Activities, Q22. What is underwriting of financial instruments and/or placing of financial instruments on a firm commitment basis (A6)?

14. *Ibid.*

15. *Supra* note 12, Q23. When might placing of financial instruments without a firm commitment basis arise (A7)?

16. *See* Art. 4 Commission Delegated Regulation (EU) 2017/590 of July 28, 2016 supplementing Regulation (EU) No. 600/2014 of the European Parliament and of the Council with regard to regulatory technical standards for the reporting of transactions to competent authorities.

17. Supra note 12, Q13. When might we be receiving and transmitting orders in relation to one or more financial instruments? (A1 and recital 44)?

18. Art. 3(1)(a) CFR in its original proposal does not include this service in the 'crowdfunding service' definition. The European Crowdfunding Network proposes to include this in the final CFR draft. *See* European Crowdfunding Network, Support for—and Proposed Improvements to—The European Commission *Proposal for a Regulation on European Crowdfunding*

Service Providers (ECPS) for business, Brussels, March 19, 2018, p. 12, available at https://eurocrowd.org/2018/03/19/crowdfunding-service-provider-regulation/.
19. *See* Art. 3(1)(i) CFR.
20. Recital 11 CFR.
21. *Ibid.*
22. Art. 3(1)(h) CFR.
23. Art. 3(1)(b) CFR.
24. Art. 3(a) Directive 2008/48/EC of the European Parliament and of the Council of April 23, 2008 on credit agreements for consumers and repealing Council Directive 87/102/EEC (hereafter: Consumer Credit Directive).
25. Directive 2014/17/EU of the European Parliament and of the Council of February 4, 2014 on credit agreements for consumers relating to residential immovable property and amending Directives 2008/48/EC and 2013/36/EU and Regulation (EU) No 1093/2010.
26. CFR, p. 2.
27. Recital 4 Proposal for a Directive of the European Parliament and of the Council amending Directive 2014/65/EU on markets in financial instruments 2018/0047(COD).
28. Recital 9 CFR.
29. Regulation (EU) 2017/1129 of the European Parliament and of the Council of June 14, 2017 on the prospectus to be published when securities are offered to the public or admitted to trading on a regulated market and repealing Directive 2003/71/EC.
30. Directive 2003/71/EC of the European Parliament and of the Council of November 4, 2003 on the prospectus to be published when securities are offered to the public or admitted to trading and amending Directive 2001/34/EC.
31. Art. 4(1) CFR.
32. Art. 10(2)(h) and (i), (3) CFR.
33. Art. 10(2)(g) CFR.
34. Art. 5, 10(2)(e) CFR.
35. Art. 10(2)(d), (e), (f), (h), (k) CFR.
36. Art. 10(2)(j) CFR.
37. Recital 28, Art. 13 CFR.
38. Art. 4(2) CFR.
39. Recital 15 CFR.
40. Recital 18 CFR.

41. Art. 5 CFR.
42. Art. 6(1) CFR.
43. Art. 6(2) CFR.
44. Art. 6(3) CFR.
45. Recital 19 CFR.
46. Art. 7(1) CFR.
47. Recital 19, Art. 7(2) CFR.
48. Art. 7(3) CFR.
49. 'Control' is defined in Art. 4(1)(35)(b) MiFID II.
50. Art. 7(3) and (5) CFR.
51. Recital 16, Art. 4(4) CFR.
52. Recital 16, 4(3) CFR.
53. SPVs are under Article 3(1)(l) CFR defined as 'entities whose sole purpose it to carry on a securitisation within the meaning of Art. 1(2) of Regulation (EU) No. 1075/2013 of the European Central Bank.'
54. These three examples are adapted from *supra* note 17, p. 12.
55. Recital 17 CFR.
56. *Ibid.*
57. Art. 4(5) CFR.
58. COM(2017) 340 final, Report from the Commission to the European Parliament and the Council on the assessment of the risks of money laundering and terrorist financing affecting the internal market and relating to cross-border activities.
59. Directive (EU) 2015/2366 of the European Parliament and of the Council of November 25, 2015 on payment services in the internal market, amending Directives 2002/65/EC, 2009/110/EC and 2013/36/EU and Regulation (EU) No. 1093/2010, and repealing Directive 2007/64/EC.
60. Recital 24 CFR.
61. Directive (EU) 2018/843 of the European Parliament and of the Council of May 30, 2018 amending Directive (EU) 2015/849 on the prevention of the use of the financial system for the purposes of money laundering or terrorist financing, and amending Directives 2009/138/EC and 2013/36/EU.
62. *Ibid.*
63. Recital 20, Article 8 CFR.
64. *Ibid.*
65. Art. 9(1) CFR.
66. Recital 21 CFR.
67. Recital 21 CFR.

68. Art. 9(4) CFR.
69. Art. 14(1) CFR.
70. Art. 14(2) CFR.
71. *Ibid.*
72. *See* Recital 12, Art. 2(2)(d) CFR.
73. Art. 16(1) CFR.
74. Recital 32 CFR.
75. Recital 32, Art. 16(5) CFR.
76. Regulation (EU) No 1286/2014 of the European Parliament and of the Council of November 26, 2014 on key information documents for packaged retail and insurance-based investment products (hereafter: PRIIPR).
77. *See* Recital 30, 32, Art. 2(b) and (c) CFR.
78. Recital 32 CFR.
79. European Commission Brussels, 8.3.2018 COM(2018) 113 final Annex— Annex to the Proposal for a Regulation of the European Parliament and of the Council on European Crowdfunding Service Providers (ECSP) for Business {SWD(2018) 56 final}–{SWD(2018)57 final} (hereafter: KIIS Annex).
80. Part G: Disclosure related to the loan agreement KIIS Annex.
81. *Ibid.*
82. Art. 16(6) CFR.
83. *Ibid.*
84. Recital 33, Art. 16(6) CFR.
85. Recital 30 CFR.
86. Art. 15(1) CFR.
87. Art. 15 CFR.
88. Art. 15(4) CFR.
89. Art. 15(4) CFR.
90. Art. 15(5) sub-para. 1 CFR.
91. *Ibid.*
92. Art. 15(5) sub-para. 2 CFR.
93. Hakvoort, Anne: FinTech—een worsteling voor wetsgevers en toezich- thouders. Brengt het FinTech actieplan verduidelijking? *Tijdschrift voor Financieel Recht* 4 (2018).
94. Art. 17(1) CFR.
95. Art. 5 MiFID II.
96. Recital 35 CFR; Van Praag, Emanuel: Wat is een handelsplatform onder MiFID I(I)? *Ondernemingsrecht* 41(2018), 250–255.

97. Art. 17(2) CFR.
98. Recital 37 CFR.
99. Recital 19(1) CFR.
100. Recital 37, Art. 19(2) CFR.
101. *See* Zetzsche, Dirk and Christina D. Preiner, "Cross-Border Crowdfunding: Towards a Single Crowdlending and Crowdinvesting Market for Europe". *European Business Organization Law Review* 2 (2018), 217–251.
102. Macchiavello, Eugenia, "Financial-return Crowdfunding and Regulatory Approaches in the Shadow Banking, FinTech and Collaborative Finance Era." *European Company and Financial Law Review* 4 (2018), 662–721.

Chapter 21

Crowdfunding for Civic and Development Projects

Robert Pasicko and Marina Petrovic

Abstract

UNDP recognises that to help cities to implement the 2030 Agenda, be more resilient, and their inhabitants less vulnerable, it needs to act as a middleman and a support platform that integrates a multi-sectoral action and promotes a greater collaboration across sectors and partners to deliver impacts at scale and utilise limited resources efficiently.

Crowdfunding goes in accordance with UNDP multi-stakeholder "whole-of-society" approaches by introducing the new opportunities for innovative win–win collaborations between citizens and municipalities. UNDP sees crowdfunding as not just a new marketplace for accessing funds but a tool to build community engagement; not just a PR fad but a powerful outreach and advocacy mechanisms; and not merely a support system for an initiative but an effort to catalyse new business models, technologies, and solutions for social issues.

UNDP AltFInLab crowdfunding pilots have demonstrated that the innovative initiatives can be catalytic if they are linked to larger policymaking or reform processes (e.g. the crowdfunding campaign we did in Moldova that was scaled from 350 kids to 350,000 kids was linked to the broader reform of implementing new eating habits in school kids), or opening up the civic space and making the business of governing cheaper and more effective (e.g. a civic crowdfunding campaign for the construction of an all-inclusive playground in Kombinat, Tirana, we co-developed and co-founded together with Municipality of Tirana).

The quote "If you have \$1 to invest in knowledge management, invest 0.1 in content, and 0.99 in connecting people" best reflects UNDP AltFInLab work and points out that governments ought to focus on enabling its citizens to support their community development, which can result in absorbing complexity vs. reducing it.

As innovative finance model, crowdfunding also has immense potential to fund municipal projects, but it should be no alternative for traditional financing structures used in the urban area development, rather an addition. It can be used as additional funding mechanism during any phase of the project, and as a tool to finance small, citizen-led projects to build social resilience and rejuvenate urban areas and for getting support for the local community. Civic crowdfunding should be a way of "widening the funnel" of funding. Most civic projects are related to the use of public spaces or public services, so it is important that local authorities have a role in these projects. Local authorities often tend to stick to their traditional way of doing things, rules, and procedures and are not used to the "creative and innovative way of thinking." In fact, local authorities have to obey the rules and procedures and finding the way to test some innovation is not that straightforward. In some cases, the reason is that local authorities don't have enough human resources to get into new projects and operations. For all these reasons, it is suggested that local NGOs (group of architects, designers, volunteers, students, green activists, etc.) or community groups run the crowdfunding campaign, while municipality would be a partner and supporter, but not a lead role player. In order to do that, the municipality needs to sign an agreement with the local organisation, showing that the key stakeholders are supportive during the campaign and project idea implementation. Besides providing match-funding to the crowdfunded amount, local authorities can be supportive by speeding up permitting procedures, connecting the team with other local partners, using their channels of communication and media marketing, etc.

The quote "If you have $1 to invest in crowdfunding knowledge management, invest 0.1 in content, and 0.99 in connecting people" points out that governments ought to focus on enabling its citizens to support their community development, which can result in absorbing complexity vs. reducing it. Governments need to move away from internal, traditional funding management systems towards people-oriented approaches that seek to connect and create knowledge just in time and on demand.

1. UNDP Alternative Finance Lab—Track Record of Crowdfunding Development Projects

Decreasing public budgets is an unfortunate reality and core services cannot be compromised. Crowdfunding is a strong tool to motivate citizens to be a part of the transformation of the urban environment and for keeping the momentum behind the urban regeneration projects. UNDP Alternative Finance Lab (AltFinLab) crowdfunding pilots have demonstrated that the innovative initiatives can be catalytic if they are linked to larger policy-making or reform processes (e.g. the crowdfunding campaign we did in Moldova was scaled from 350 kids to 350,000 kids and is linked to the broader reform of implementing new eating habits in school kids). The AltFinLab has been working and experimenting with crowdfunding for more than 5 years, by developing campaigns, workshops, booklets, and observing how this new way of financing can help communities to become more resilient and empowered. By trialling numerous crowdfunding campaigns and tracking the results after one, two, or more years, we have recognised that crowdfunding is not a trend but is here to stay and disrupt the market as no business model has ever done before.

What the application of crowdfunding showed was that, once successfully tested, new ways of doing things expand quickly through the region and easily build up in various more complex models. The following shows what we have learned in the last 5 years working on numerous crowdfunding projects and campaigns:

- *Community building instrument*—By capitalising social power, we are enabling broader visibility and scaling-up of a given project, as well as catalysing social action and policy-level impact. We are using crowdfunding not only to collect the money, for example, our Tajik campaign attracted donations of their Australian diaspora. A similar sidekick happened with a campaign for STEM spider-robots that attracted attention of 30+ schools in Croatia and made them implement it in their official education curriculum.
- *Marketing instrument*—Our Taste of Home campaign for a restaurant run by asylum seekers in Croatia was featured on Yahoo news, hugely increasing their visibility; Brlog, a cooperative brewery run by women, appeared as a featured story in WizzAir inflight magazine and their Indiegogo webpage attracted more than 20,000 visitors within the 40-day campaign; contributions came from 25 countries, have 50+ media appearances, and were often promoted by celebrities such as Jamie Oliver and Mashrafe Bin Mortaza.
- *Political instrument*—We have examples of governments using crowdfunding as a supplement to their traditional financing models, and on the contrary, we can see crowdfunding giving voice to citizens who otherwise would remain unheard. Our campaign for the Subversive Festival in Zagreb, for example, received prompt support by the Croatian Ministry of Culture but only after public interest was confirmed through campaign contributions; another campaign for building an energy-independent school in Croatia helped to trigger a pilot project from the Ministry of Construction for energy refurbishment of schools all over Croatia, worth €7 million.
- *Innovation instrument*—Our campaign for healthy snacks in primary schools in Moldova got an interest from a large national network of supermarkets afterwards and was upscaled by the government from 3,500 to 350,000 kids. Through crowdfunding, we've supported lots of UNDP tech-based platforms, like LiveLebanon.org, GreenCrowds in Ecuador, or YemenOurHome. We're working with them to move away from online donations, towards building communities that can deliver long-term impact.

- *Powerful tool*—The first 5 years of experience with crowdfunding taught us that the approach is not just a new market place for accessing funds but a tool to build community engagement; not just a PR fad but a powerful outreach and advocacy mechanism; and not merely a support system for an initiative but an effort to catalyse new business models, technologies, and solutions for social issues as well as re-inventing the business model, building an ecosystem of supporters and ambassadors, and being a powerful leapfrog.

Crowdfunding is so much more than about getting the required financing for your project. Here are our takeaway points:

- *Link up with tech platforms*: Through crowdfunding we've supported lots of UNDP tech-based platforms, like LiveLebanon.org, GreenCrowds in Ecuador, or YemenOurHome. We're working with them to move away from online donations, towards building communities that can deliver long-term impact. It's not enough just putting a "donate" button on your page. These online communities can be powerful starting points for continuing to rally investors and partners.
- *Partner with cities to achieve quick wins*: London is creating a city for all Londoners through crowdfunding. Madrid has its own crowdsourcing platform. We're transferring these practices to some unlikely places like Somalia, working with the diaspora to help people in Mogadishu create revenue and withstand violence and disaster.
- *Use Diaspora power to rebuild cities*: Use the crowdsourcing digital platform to enable participation and inclusion of all segments of the city in making decisions that impact their lives, including returning IDPs and refugees, and provide a channel for proposals that will make the city more resilient and speed up its recovery. In Aleppo, Syria, we are planning to leverage Alepines diaspora engagement to reverse the outflow of financial and human capital from Aleppo in the last few years and support the recovery of Aleppo. Live Lebanon is a UNDP donation-based crowdfunding platform bridging the Lebanese Diaspora and the development efforts in Lebanon. Since its launch in 2009, it has implemented 67 projects by raising funds from the Lebanese communities abroad which resulted in the collection of

donations worth US\$4.2 million. By connecting with various crowd-investing platforms, we are converting Live Lebanon crowdfunding platform to become an impact investment platform with sustainable impact investments in reforestation/nature conservation, solar energy, social entrepreneurship, health, and development projects.

- *Go solar*: It's the ultimate platform initiative. It's not development if it isn't green. Funding solar is the ultimate way to hit multiple SDG targets. We teamed up with over 30 crowdfunding platforms to deliver Citizenergy, which helped invest €40 million into clean energy. With UNDP Moldova and Sun Exchange, we are also developing a US\$1 million solar plant using cryptocurrencies.

- *Support small businesses*: It's hard for SMEs to get and repay their credit. But our experience in Ukraine, Bosnia and Herzegovina, and Morocco and Turkey shows that crowdfunding can be a great way to get new businesses off the ground. By the same token, development organisations should intensify their work to bring crowd-funders, businesses, third-party verifiers, and others together to make business more inclusive.

- *Use new sources of financing*: In the Muslim world, Zakat (donations) are worth €200 billion to €1 trillion. We're now working with the Islamic Development Bank on a proposal to fund NGOs. In Indonesia, UNDP is designing a brand-new platform that will—among other things—use Islamic finance to help the country achieve the SDGs.

- *Build networks to help people recover from disaster*: Campaigns such as GoFundMe and YouCaring are putting a face on individuals affected by disasters and mobilising global funding. The Connecting BusinessInitiative is taking that approach further, mobilising business networks so they too can get involved.

Going forward within UNDP, we are now aiming at **creating platforms for collaboration between civic actors, the private sector, and government** to harness capacity, creativity, knowledge, and expertise from citizens, especially young and women technologists, to strengthen responsiveness, openness, and efficiency of institutions, as well as to stimulate innovation and cross-sectoral collaboration with the private sector. Crowdfunding is the originator of all modern development platforms. When we turn to platforms, we direct money where it is most needed. People can crowdsource the best ideas and vote for them. Governments and donors can match the funds collected, thereby financing projects citizens actually support.

About the Authors

Robert Pasicko co-founded UNDP Alternative Finance Lab (www. AltFinLab.org) through which he supports development on alternative finance mechanisms in over 30 countries globally. He also cofounded UNDP Crowdfunding Academy, a training programme for crowdfunding which has been implemented in over 20 countries worldwide and has created over 50 successful campaigns. As a co-founder of Green Energy Cooperative, Robert is providing alternative finance and low-carbon development expertise in Croatia and Western Balkans. In parallel, he works as Assistant Professor at the Faculty of Geotechnical Engineering, University of Zagreb, focusing on the low-carbon development and alternative finance topics. He defended his PhD dissertation in 2014 at the University of Zagreb and holds diplomas in Electrical Engineering and Management.

Marina Petrovic is super excited when talking about innovation and connecting different dots from a more creative and interdisciplinary perspective. Her main focus are alternative financing models, mostly blockchain, crowdfunding, and investment crowdfunding projects that can assist in addressing economic, societal, and environmental facets of community growth. She is one of the Alternative Finance Lab and Crowdfunding Academy founders, which has just entered into its second year. She has led and co-developed few UNDP blockchain projects, from which the last one—blockchain for supporting small farmers in Ecuador—has just been launched.

Part 3.5

Blockchain and Cryptocurrency

Chapter 22

Relationships among Crowdassets, Crowdfunding, and ICOs for Civic Engagement

Angelo Miglietta and Emanuele Mario Parisi

Abstract

Since the infrastructure of Blockchain first appeared on the Internet, the number of digital assets such as Tokens and cryptocurrencies has grown dramatically, often facilitated through ICOs. Tokens and Tokenisation are digital assets used to transfer value from one wallet to another, which is a transaction between peers in the crowd. The relevance of this topic is crucial not only for private institutions but also for the public. Policymakers, civic bodies, and governments are starting to assess the crowd-based tools and see the crowd as an asset to pursue economic and social goals. In this chapter, we will analyse and assess what potentials ICOs and Blockchain allow for social engagement and value creation.

1. Introduction

Blockchain first appeared on the Internet mostly from programmers and developers as an alternative to the traditional *World Wide Web* infrastructure. Since then, the number of digital resources that were created

447

leveraging on Blockchain infrastructure has grown dramatically.[1] The birth of BitCoin in 2009 represented the moment when first glimpses and fundamentals of what would soon become "digital assets" became evident.[2] BitCoins were among the first—and with time among the most—valuable assets to leverage on Blockchain infrastructure, but at the same time, other competing platforms and schemes emerged in parallel. Ethereum, for example, was originally instituted to set smart contracts through the blockchain distributed ledger, but soon it has also become a cryptocurrency and a tool to create additional value out of the block infrastructure.

Tokens and Tokenisation are the digital values' or digital assets' creation through blockchain schemes.[3] Tokens are not currencies—they represent the symbols of a contract between the buyer and the embedded digital value. Tokens can be issued in two ways:

(1) through a new blockchain scheme—this method is less convenient because it is accompanied by the difficulties in reaching network effect that represents the basics of value creation;
(2) through the issuing of tokens on top of the existing infrastructure—it is also difficult considering that the raw bitcoin transactions leave a very small space for additional information encoding.

To overcome these problems, Ethereum, in 2015, introduced the concept of decentralised smart contracts, building the infrastructure for both digital assets transactions and creation and for the management of secondary digital tokens with full disintermediation.

Through Ethereum, tokens can be produced, distributed, and traded easily. When tokens are created and distributed as digital assets reflecting the overall network value, it is called Initial Coin Offering (ICOs). This phenomenon reshapes and fosters the way the companies use Blockchain for creating and distributing digital assets. Ethereum represented the best infrastructure solution to issue tokens, as it originally operated as a decentralised smart contract enabler. As for the extent of digital assets, BitCoin is a Blockchain scheme allowing payments (so users transfer money to one another); while Ethereum allows users to write wallet-based programmes.

The competitive advantages of Blockchain-based tokenisation remain and can be synthetised as follows:

- *The underlying smart contract and automatic execution that these programmes entail*: Such infrastructures can be designed to receive digital assets from wallets automatically, as long as it fulfils two conditions: firstly, the programme applies the same conditions to all users, and secondly, it complies with known principles to be predictable, fair, clear, and unalterable. When these principles are applied, they are automatically executed, posing the basis of a smart contract or self executing contract. Being written on the Blockchain, it can also be stored and executed permanently to produce *erga omnes* effects.
- *The network effect*: The exchange of currencies is the transfer of value from one wallet to another, which is a transaction between peers. Blockchain networks add two other benefits to this possibility: (1) the peers (o users) involved in the transaction gather maximum consensus in the shortest time span to increase the value of the first-hand digital asset; (2) key parties of the network (i.e. the miners, and not merely of the transactions) compete and validate network transactions promptly, creating positive spillovers for the whole infrastructure. In both cases, the reward for these tasks is valuable for both miners and users either as increased asset value or in fees paid in the very same network token. This process is built on trust, because the safer the scheme becomes, the more difficult and expensive it is for violators to alternate or violate it.

A second-hand token, created atop a Blockchain network, is a formalisation of the property right; in other words, while Blockchain chain represents a kind of register, the token is the "legal" title that the network uses to prove native rights of the title.

ICOs, similarly to IPOs,[4] are schemas the start-ups use to raise finance through the distribution of tokens issued via Blockchain platforms on the secondary market combining Smart Contracts and Network effects as described earlier. The first ICO was conducted in July 2013 by Mastercoin, a digital currency built on Bitcoin's blockchain, that collected a total of 5,000 Bitcoins ($500,000 at the time) from 500 investors.[5] Since then, hundreds of ICO platforms came by following them. CoinSchedule (www.coinschedule.com), one of the major ICO-tracking sites, reports 269 ICOs in 2017 that raised a combined US$4.8 billion. In contrast, the first crowdfunding platform Kickstarter raised a total of US$3.5 billion in 8 years, since its inception in 2009.[6] Every month, a growing number of companies leverage on ICOs to explore ways to connect their product with the token and leverage smart contracts to add more features to these tokens, creating great potentials in that way. In other words, start-ups continue to leverage on the network effect embedded on Blockchain technologies to issue tokens on the secondary market via ICOs.

It is necessary to observe that the ICOs not only create the impact on the traditional financial schemas of shares, securities, etc., but also decentralise the creation of value and the trading leveraging on assets whose value is determined by the network usage, or, in other words, by the crowd.

The relevance of this topic is crucial not only for private institutions but also for the public. Policymakers, civic bodies, and governments are starting to assess crowd-based tools to pursue economic and social goals, recognising the crowd as an important asset. Consequently, ICOs represent an approach to crowdassets, and, perhaps, nowadays it is also the most common one.

In this chapter, we will analyse and assess the potentials of ICOs and Blockchain for social engagement and value creation, starting with the literature on the topic, identifying the evidence from the market and from the public, in particular assessing the value brought by ICOs and further applications and models to civic engagement and the initiatives of public or policy engagement in Blockchain.

2. Literature Review

The topic of ICOs has started gaining consistent scientific interest only during these last few years, due mostly to its innovativeness and disruptive approach to traditional funding schemas. First examples belong to Christian Fisch that, in 2016 and 2017, empirically explored the capital raised in 238 ICOs and deduced that several characteristics of the ICO campaign and the underlying technology determine the amount raised, while the venture characteristics were less important to investors.[7]

The topic obtained a higher interest in 2017, due to the fact that, however new, the market size of ICOs has reached dramatic values. Only in

2017, the overall volume of funds raised through ICOs reached \$5.3 billion and the success rate of 81%.[8] This amount is surprisingly high when paralleled with overall financing raised in initial public offerings (IPOs) in the United States.[9] Nevertheless, since then only a few research reports, and almost no paper published in scientific journals, have dealt with this subject.[10] As shown by Talk on the logarithmic regression chart of 2017,[11] the global tendency of bitcoin's performance has always been positive: the first time it overcame the Market Cap indicator in 2011 with US\$100 Million result, and the second time in 2013 with US\$1 billion result, after that remaining permanently at the top. In 2017, its weekly price exceeded US\$100 billion. Also, the time flows of ICO proposed by Giudici *et al.*[12] confirm the tendency: after three Initial Coins Offering completed in 2015, in 2016 they were 27, together with six failures. In the first 8 months of 2018, their numbers already reached 174, with a minor rate of failures (42, 19.4% of total number).

The number of categories of projects that actually use ICO is also important and shows the wide recognition of this method[13]: finance (12.2%), payments/wallets (6.7%), commerce/retail (5.6%), blockchain platform (5.5%), asset management (5.4%), betting/gambling (4.3%), gaming (4.9%), data/computing/AI (4.9%), prediction market (3.9%), IoT (3.8%), entertainment (3.8%), media/content (3.9%), health (3.3%), security/identity (3.3%), funding/venture philanthropy (3%), job market (2.7%), content/advertising (2.7%), etc.

As for the total funds raised by categories,[14] on comparing with 2017 the situation is very changeable. The leadership belonged to blockchain platforms 2 years ago with 38.2%, in 2019 belonging to network/communications (59.3%, raised up from 16.6%), while blockchain platforms gathered 19.3% of funds. All other categories are at a relevant distance, starting from finance (5.2% from 3.2% in 2017), payments/wallets (with 3%, lowered from 5.3%), and health (3.2% from 4.8%), followed by others that don't arrive to 2%.

This rather sudden growth of the phenomenon—considering that this funding schema exists only since 2013—has led practitioners and scholars to increasingly assess ICO variables and mechanisms either quantitatively and qualitatively. As for the state of the literature topic, it is focused either on the more legal aspects of it[15] or on the socio-economical roots and impacts,[16] mostly extending financial and economical schemas to this nascent market.

In 2018, Fenu *et al.* and Giudici *et al.*[17] claimed that the probability of success of an ICO is strongly and positively affected by the presence of a set of codes for the blockchain project, being unaffected by the availability of a "white paper." They have also underlined that the conditions of the cryptocurrency markets underlying the ICOs—that do not create *ex novo* blockchains of their own—are not taken into account by investors and, thus, do not affect the probability of success of the ICO, being measured by average return and volatility.

ICOs raise a number of issues that the literature is starting to address. The first issue is the optimal technical design of an ICO that permits to prevent any type of fallacy in the code and/or to attract contributors.[18] The second issue is the relationship between ICO activity and regulations[19]: normally, companies and entrepreneurs must propose a public prospectus approved by market authorities whenever they want to tap public retail investors for funds in exchange for securities.

In terms of legal perspective, as discussed in Bohme *et al.*[20] and Yermack[21], academics have concentrated their attention on the risks embedded in the ICO financing. The short history of ICOs is indeed constellated with various scams: fraudulent projects that lacked underlying technological or economical value, like Tezos,[22] or anonymous phishing activities to major companies, like hacks in BRIG or Titanium.[23] A certain number of organisations are dedicated to restore broken ICOs, such as CoinJanitor, but many of them have a dubious nature and a short time of activity.[24] The information about the failed crypto projects is collected on sites such as Coinopsy and DeadCoins. The sensitivity of ICO market to adverse industry events is higher than IPOs. To name a few, the biggest hacks of virtual currency projects such as Bitfinex in the August of 2016 or MtGox in 2011 and 2014,[25] the most severe regulatory bans by the Chinese and the South Korean governments in 2017,[26] and the recent Facebook announcement to ban ICO ads arrived at the beginning of 2018,[27] are problems that have posed serious questions about the legitimacy and opportunity to ICOs funding in the long term. ICOs issue tokens that are essentially currencies that reflect value for a specific platform. The amount of tokens is often predetermined, consequently, with the higher demand on the platform with constant amount of tokens, the token price also becomes higher.

In the short term indeed, consistent and genuine projects tend to provide strong incentives to reward investors, usually by underpricing the ICO value to generate the necessary market liquidity first. At the same time, the fraudulent ICOs aim at inflating value only in the short term. This creates an adverse selection where the most-valued token might not reflect any actual value beyond a well-organised marketing and communication strategy to solicit investors.

For these reasons, the literature has concentrated on customers' protection above all, identifying risks and uncertainties that investors face with respect to ICOs in the markets. More precisely, it is possible to cite factors such as the following:

- lack of transparency;
- a shorter settlement period compared to traditional currencies such as Euro or US dollar;
- high volatility;
- high amount of credit, liquidity, legal, and operational risks.[28]

As an instrument to evaluate risks and the actual value of an ICO, an empirical approach has been proposed that analyses the correlation between the network activities.[29]

The literature agrees a playing field where ICOs operate and offer no or little protection for customers mainly because of the following reasons:

Firstly, ICOs' lack of regulatory standards (although some jurisdictions are starting to pose basis for legal standards).

Secondly, because cryptocurrency projects that solicit investors lack corporate governance, ICOs occur without any underwriting activity.

Lastly, because they intervene on the web, they often leverage on regulatory perimeters that are more tolerant or absent. According to Lewis *et al.*[30] in countries like Italy, Belgium, France, the Netherlands, and Spain, there is no specific regulatory framework or legislation with regards to either ICOs or digital tokens. The financial regulator of Hong Kong, the Securities and Futures Commission, has opined that typical digital tokens and ICOs will be considered a form of "virtual commodity" and as such will not be subject to regulation. On the contrary, countries like Germany, the United Kingdom, and Singapore have already made efforts in this direction by introducing some elements of classification and regulatory structure for the phenomenon: In those cases, it is not correct to say that all ICOs are unregulated.

Notwithstanding these uncertainties, other authors have emphasised positive or uncommon features of the ICO phenomenon that differ from IPOs. For example, the value of rights and information. Where the corporate governance law and finance literatures[31] affirm that rights and guarantees are the key factors that determine a higher share of wealth allocated for any asset, ICO investors appear to commit substantial amounts of wealth to ICOs without claiming any legal right to get a fair return on their investment.[32] Another uncommon feature, in line with the literature on the value of soliciting information and transparency of IPOs, is that traditional tools that aim at protecting customers appear to decrease the value of ICOs instead of increasing it.[33] Empirical research conducted on the topic identifies, in fact, the value of ICO funding as decreasing when it involves a know-your-customer process, in which the project team gets to know its investors and, hence, can better gauge its true value.

As for Tokens in general, the literature agrees in defining a token as "a virtual currency that entitles the owner to pay for the specific services of the project that has issued the token,"[34] in other words, *token-for-service* value. This definition is surely true reflecting the origins of tokens. However, the emergence of various token exchange platforms has lead tokens to become speculative asset, thus expanding their initial token-for-service value. The speculative schema of BitCoins is so evident today that it was possible to illustrate by the empirical evidence that most transactional Bitcoin volume comes from trading activities on token exchange platforms.[35]

On the same line, the literature also concentrates on the general public's contribution in increasing the value of tokens—when compared to traditional IPOs. Catalini and Gans[36] argue that ICOs add economic value to the tokens because they establish buyer competition, thus, strongly signalling consumer value in a way that is superior to IPOs. Similarly, Li and Mann[37] demonstrate that the platform is able to make the value generated inside the platform grow. It becomes higher because ICOs consolidate the fragmented and dispersed information about platform quality/value and resolve the coordination dilemma, typical in many peer-to-peer (P2P) platforms. By adding dynamics to a platform launch, ICOs can also solve a coordination failure inherent of networks.

The salient feature of a platform is that its value is largely driven by the interactions among its users who benefit from each other's participation. Consequently, there are two ulterior ways as to how an ICO increases platform value that can be presented as two related channels based on this insight that both lend value to an ICO:

- platform users are directly benefiting from each other's participation, generating a strategic complementarity: a user's gain from joining a platform increases with the number of other users;
- the information about the platform quality is dispersed within the user base incentivising each user to learn the "wisdom of the crowd," so as to make more informed participation decisions.[38]

These traits are particularly interesting for what concerns crowdassets. ICOs strengthen the value creation of tokens by relying on information diffusion and consolidation and create coordination among peers at higher pace and with stronger magnitude—considering this fact, their underlying schema could be leveraged in crowdassets too. Chapter 23 analyses facts and figures of solutions that leveraged on ICOs dynamics to increase the value of digital assets or crowdassets.

3. Crowdassets and the Role of Platforms

In the digital realm, a new type of business models is emerging, particularly favourable for digital companies. It is defined "platform" meaning the P2P online spaces that use tokens as an internal medium of exchange.[39]

This business model entails a common infrastructure where multiple parties are pivoted to exchange goods, services, and information, creating added value for the platform itself at each transaction performed in its perimeter (i.e. Amazon, Apple store, to name a few). The customer acquisition and retention costs in such cases are higher than other business models and, however, they offer the owner of the platform access to profits at each transaction performed by third parties upon the platform. The crowd, or in other words, the users, are the main assets of these platforms. Examples such as N26—German digital bank offering free standard services to gather customers to offer third-party services—illustrate the value of having exclusive customer relation to motivate the company's value, despite their revenues.

Digital companies are commonly targeting to establish such kind of companies, while the emerging "platforms" business models in organisations foster, precisely, the use of the crowd as an asset. This approach is to intend as *Crowdassets*.[40] It is the result of the cryptocurrency market explosion, which has led to the spring of ICOs—merging crowdfunding[41] with crowd assets and creating new potentials for both entrepreneurs and investors.[42]

For the situations and circumstances illustrated above, the value brought by crowd assets is increasingly recognised in several financial, entrepreneurial, and administrative domains, also enabled by distributed ledger technologies and blockchain to pursue tokenisation goals.[43]

Considering that BitCoin is a Blockchain scheme to allow payments where users transfer money to one another and that Ethereum adds to it the possibility of users to write wallet-based programmes, the competitive advantages of Blockchain-based tokenisation is mainly the underlying smart contract that can be stored and executed permanently to produce *erga omnes* effects, together with the network that increases the value of the underlying platform. There are several advantages as follows:

Firstly, peers involved in the tokenisation creation and acquisition (users) aim to pursue maximum consensus in the shortest time span to increase the value of the first-hand digital asset.

Secondly, miners of the network will compete and validate network transactions promptly, creating positive spillovers for the whole infrastructure. In both cases, the reward for these tasks is valuable for both miners and users either as increased asset value or in fees paid in the same network token.

Thirdly and last, trust is embedded in the process as the safer the scheme becomes, the more expensive it is for violators to alternate or violate it.

However, when it comes to ICOs, policymakers are polarised: China has banned ICOs while the US has been regulating them aggressively. Other nations claim on the contrary as supportive to ICOs, hoping to involve a growing community entrepreneur, start-ups, and the potential economic windfall, also in the shape of Tokens.[44]

What is interesting, however, it is to analyse the approach of governments to Blockchain and, potentially, ICOs as a standard for public engagement. In particular, the literature on the subject assesses Blockchains as technology with the potential to revolutionise the activities of government (i.e. healthcare; national identity management systems; tax and internal revenue monitoring; voting; secure banking services, etc.). Dubai, for example, plans to leverage the power of blockchain technology in facilitating license renewals, payment of bills, and visa applications; Gibraltar issued a ruling that effectively grants licenses which allow blockchains to be used as conduits for the storage and transfer of digital assets; United States, and in particular Illinois, launched a trial of their proposed birth registry and identification system that will be powered by blockchain technology.

More recently (September 19, 2018), France announced the release of a legal ICO framework that targets to provide safeguards and guarantees to investors. Following a provision of the Business Growth and Transformation bill (PACTE) dedicated to Initial Coin Offerings (ICO), French Authorities issue allowances to start-ups if they want to leverage on ICOs to raise capital. It represents a part of a wider strategy that shall bring France to become a major financial centre: the legislation provides standards by which the authority shall verify white paper contents prior to capital raise, through the release of a particular Visa.

The assessment performed by the authority shall lead to draft and certify a white paper description, a kind of project analysis of the ICO and its development roadmap, together with the rights conferred by the token, the legislative court in the case of disputes and the economic purpose and use of funds collected during the ICO. The white paper would allow potential investors to undertake full due diligence before investing.[45]

The country that has proved to be most advanced in the extents of ICOs is Estonia. In 2018, its government has publicly declared the target to offer the best technological and legal architecture with world-wide legitimation for regulated ICOs. Surely, its target is not the only one but it follows the path of the well-defined vision to become the world's most digital-friendly ecosystem. Indeed, Estonia has established an innovative e-residency scheme for foreign nationals and has approached a fully digital bureaucracy schema upon the digital ID system.

Unsurprisingly, Estonia was also planning to launch the EstCoin, the first example of crypto token issued from a country, establishing a digital token at fixed value to avoid volatility. It would initially serve only e-residents in three different fashions: as a reward for supporting the e-residency architecture, as a form of digital signature for smart contracts, and lastly, as a digital token to allow free transactions through blockchain. This last aspect could be particularly interesting for the sake of this research, but in June 2018, the Estonian government denied the idea of implementation of the EstCoin.[46]

4. Potential Applications in Public Administration

Gathering the state-of-the-art on the subject, the academic insights and the best practice examples assessed in Chapter 21, we can envisage further applications of ICOs and Tokenisation for public administrations, with the goal to establish platforms based on blockchain technology that would allow the broader participation of individuals to specific programmes and projects, leveraging on distributed ledger registration. These approaches entail the engagement of the public as an enabling factor.

Firstly, as mentioned earlier, blockchain and distributed ledgers strongly enhance the digitisation of any financial activity, but prior to that, any online activity shall also foresee a thorough customer identification (i.e. KYC). As experienced in Estonia, a digital identity registry is a necessity for all governments, pushed by private sectors too such as Financial and Health. Administrations need to set adequate standard for online identification because it will limit the risk of fraud and enhance access to online services. This registry must be public, mostly because it needs constant exchange and extraction information from public data-bases, often secured and inaccessible for individuals (i.e. fiscal, public security, health data). A decentralised registry of national IDs would strongly enhance identification online for both ICO managers and issuers and investors, granting updated and verified information on both categories.

Secondly, the engagement of the public is needed to perform a so-called "enhanced due diligence" of any ICO campaign. In a previous research, we emphasised the role of the public as a tool to enhance the consistency of due diligence activities due to the network of founders who share information. This assumption fits ICO assumptions as well, because there is no difference in the two models—crowdfunding and ICOs. In other words, online collection of capital requires investors to read and collect information on the investment prior to investing. It is recognised that investors tend to collect more information on their own in case the authority does not provide a regulatory informative paper.[47] Moreover, authorities are currently obliged to perform project assessment for IPOs that despite their overall growth are still in a limited amount. Asking the regulator to perform assessments on a wide, uncontrolled quantity of start-ups targeting ICOs would be extremely difficult and time-consuming. Indeed, the participation of many investors from the "crowd," posing a higher risk on themselves, would also require them to perform a due diligence and share results with the community, together with targeting a higher control on the legitimacy of the ICO.

Lastly, the exit strategy. One of the limitations of the crowdfunding industry, indicated by both practitioners and academics, is the difficulty to monetise investments due to the lack of a secondary market. ICOs partially solve this issue as tokens are cryptocurrencies that can be traded on the many platforms that exist currently and permit an immediate monetisation.

This gives rise to the combination of several benefits. Firstly, ICOs for public ventures create added value due to a pivotal use of existing assets. Investors (i.e. tax payers) can allocate their resources in favour of projects financed by them with an enhanced screening and monitoring potential through the platform. Such kind of sharing returns sell-off in the second market.[48]

Similarly, Civic Crowdfunding fosters efficiency of assets otherwise unused or misused—in the same way as the marginal benefit involves more taxpayers than decisions adopted at the top. When taxpayers pool their resources in favour of a project, it is more likely that the venture satisfies the needs of the highest possible number of citizens.[49]

A potential application could be, to name a few, on project financing for public structures and projects, i.e. administrations facing budget cuts or governments searching how to promote public engagement both at entrepreneurial and economical extents, leveraging on proven project financing and crowdfunding experiments worldwide.[50] Tokenisation can further strengthen public engagement due to its full disintermediation of issuing process and speed of tokens emission for government projects, such as physical infrastructures, i.e. research indicate that investors are unlikely to invest in public projects due to unstable regulatory settings, political interference, and lack of information about issuer exit strategies.[51] In case of governmental projects using blockchain—if adequately regulated—token acquirers (i.e. investors) would increase their knowledge on both project feasibility and investment performances and returns, due to a stronger commitment in the project that leverages on the network effect. Participants would not only be investors but also be owners of tokens built upon specific platforms whose usage and distribution increases the token value, with easier exit strategies due to tokens second market. This scenario entails governments to create *ad hoc* a technological infrastructure capable of soliciting the general public to invest in specific projects through the creation of a project token. This token would behave like a share, hence through an ICO it would be distributed to the public. Clearly, a threshold for feasibility shall be set, and so is the token value. This allows to assess which projects would be feasible and economically realisable through the public engagement. Hence, projects that do not collect enough resources would be aborted with clear benefits for public budgets. Instead, projects that reach the given threshold would be implemented. When managers are also paid with tokens, there is a mutual interest for both managers and investors to accomplish higher returns.

Also, ICO schemas can be scaled on bonds and in particular on issuing mini-bonds. In Europe, for example, the mini-bonds market is still small compared to large-size bonds. In theory, SMEs may access the bond market, but this is less frequent due to their small size. The bigger the company (and the bond issuing), the larger the investor base and therefore the potential participants also in the second market. Moreover, the majority of bond indexes demands a minimum issuance size in order to embed bonds in their portfolio.

To this extent, private institutions and particularly banks—attracted by the huge and untapped SME potential opportunity—started approaching blockchain and ICOs as tools to finance SMEs. BNP Paribas has revealed that its securities services division is working on a blockchain platform that would enable retail investors to lend money to businesses via an instrument known as a mini-bond. ICOs can be scaled in the same way to grant easier issuing of securities for SMEs through blockchain applications.

The combination of ICOs schemas and the value of crowdassets originated indicate the potential for further application of its functioning for public governments.

5. Conclusions

This analysis can lead to three conclusions, keeping in mind that the pace of change in two tumultuous areas (finance and tech), allow limited space for consistent forecasting. Based on occurring events and potentials, we identify the following:

   I. Potential implementations of Blockchain technologies and in particular the tokenisation as crowdassets enablement are recognised and consistent in various areas and governments.

  II. The widespread governments' interest in leveraging on blockchain technology is subordinated to the success of these actions. It will most probably depend on the ability of governments—and by reflection of the private sector—to promote this growth by also limiting and reducing reputational risks that surround ICOs globally and in that way protecting the investors.

 III. Technological evolution is proceeding at a very fast pace, followed by social and behavioural paradigms. Variables in these areas can hinder—or foster—the success of tokenisation as a crowdassets enabler according to changes in people behaviours and expectations. We refer here to the rise of sharing economy as described by Jeremy Rifkin in "marginal society."

Firstly, there are already several examples globally proving the governmental interest in Blockchain and ICOs growth. In the years to come, and in the aftermath of the global enthusiasm for the speculation around blockchain and ICOs,[52] governments globally will enter the playing field attracted by the opportunities that this schema can offer: the approach will be by regulating this nascent industry in mainly two approaches:

- a supportive approach, thus, leveraging on technological developments and the role of the crowd as enablers, interpreting tech and crowds as facilitators for capital collection from start-ups and companies with higher transparency due to the wider public participation and distributed ledgers for data collection and conservation;
- neglecting as a whole its functioning and inhibiting it to develop nationally—leaving local investors exposed to online dangers not regulated locally.

A most promising third approach is indeed cooperation. Reckoning that governments today need to cooperate in the financial sector as much as they do for more sensitive areas (illegal trafficking, for example), ICOs regulation is most effective when participated—hence the mentioned examples coming from local governments are extremely positive but will need a global adoption to become fully applicable in the era of Internet where investors are offered products and services from virtually any legislation (and often with no legislative protection at all).

Secondly, the governments most appealed by tokenisation for crowdassets enablement are correctly supporting this growth by limiting and reduce reputational risks surrounding ICOs globally. This occurs first by protecting investors: most diffused governmental conducts are aimed indeed at protecting investors and the general public from potential dangers, often curbing technologies and business models that are hard to detect through the Internet at an early stage. Online financial frauds, where investors are often granted no protection from governments as issuers and solicitors, are based on non-regulated legal entities and therefore cannot be persecuted, governments will do their best to detect these misconducts promptly, hence the control of the online activities considered dangerous follows necessarily their network, and as identified in

Chapter 1, the greater is the network effect of misconducts, the higher are the chances of prompt identification and prevention (China or South Korea). However, some governments are positioning themselves at the edge of the innovation curve, hence identifying potentials of these emerging technologies for the benefit of their communities. Regulatory Sandboxes, for examples, can strongly enhance the development on new technology applications in compliance with the law, where the regulator can also understand, discover, and promote development via a frictionless testing and prototyping IT infrastructure.

Lastly: the technological evolution is happening at a very rapid pace, together with social and behavioural paradigms. We do not refer here to the technological evolution of blockchain that will most likely proceed at growing speed. We refer here only to its implementation as the enabler for the creation of crowdassets, i.e. in the changing context of sharing economics and marginal costs societies. In a thorough definition by Antonia Hyman,[53] main building blocks of the shared economy style businesses shall include: a technology-enabled platform, a preference for access over ownership, a P2P sharing of personal assets, granted ease of access, enhanced social interaction, collaborative consumption, and openly shared user feedback. The author also reckons that, given each of these do not apply to all business, they are predominantly exiting dominant shared-economy companies—Uber and AirBnb. Both companies allow individuals to share personal assets with consumers who incrementally use the asset, resulting in increased utilisation rates. What tokenisation can do to crowdassets is changing this paradigm, shifting from personal assets to crowdassets whose ownership is distributed and also registered through ICOs. One might argue that even with sharing economies, the existence of a last resort beneficial owner is needed (i.e. the one that bears the ownership risks and benefits at first); however, it might be imagined as a "public" sharing economy ownership whose governance is defined precisely by token-holders, similar to what occurs in public companies and shareholders. In this scenario, it becomes automatic that the action of governments shall be more and more oriented towards the protection of investors or token-holders and customers—to ensure that governing bodies act in accordance with the law and in the benefit of the stakeholders. If this scenario occurs, steered by the transparency of available information that crowdassets embed, by the consistency in authentication and registration of rights and possession, together with the sound storing mechanism underlying its functions, in such a case, it would be eventually realised as a "efficient-market hypothesis (EMH)," or in other words, a theory in financial economics that states that asset prices fully reflect all available information.

About the Authors

Angelo Miglietta is a Professor of Economy and Corporate Governance at IULM University of Milan, where he also coordinated the PhD course in Economics Management and Communication for Creativity. At Bergamo University, he coordinated the PhD course in Marketing and Corporate Governance. Throughout his academic career, he taught at Cattolica University and at the universities of Pavia, Bergamo and Turin. For nearly 30 years he has carried out professional activities for numerous enterprises and entities with global standing in the area of company evaluations, mergers and transfers, acquisitions and other corporate finance operations (leveraged buyout, financial planning, project financing). He is also Statutory Auditor since 1995, and as such he carries out technical advisory activities for the Court of Milan and for the Arbitration Board of Milan's Chamber of Commerce. Currently he is member of the Board of Directors of the Italian sub holdings of important foreign organisations, as Aviva, Bain and Eon.

Emanuele Mario Parisi is currently Business Manager to Head of Central Eastern Europe Retail at Unicredit Bank and Professor at IULM of Entrepreneurship & Innovation. Previously, he has worked as Investment Team Member of Anthemis UniCredit EVO, the Group's Corporate Venture Fund focused on financial technologies. He is also lecturer at the Luiss Master on "Fintech and Innovation Management" and Professor at the Master in Fintech at Lumsa University in Rome. Since 2016, he is Head of Business Planning of IULM Innovation LAB in Milan. He works with various public and private companies on innovation, digitalisation, corporate venturing and micro-venturing issues. He is a mentor at "Startup Direct UK;" Country Representative for Italy of the Fintech Holding

Company "Planet-N;" judge in several international contests and hackathons including CosmoPharma and Awesome Foundation.

Endnotes

1. Chesbrough, H.W. "Open Innovation: The New Imperative for Creating and Profiting from Technology." *Harvard Business Review* (2003).

   Gregory, T., Berente, N. and Howison, J. "Digital Technologies and Patterns of Distributed Innovation." In: *Americas Conference on Information Systems*, AMCIS, 2015.

   Johansen, S. A Comprehensive Literature Review on the Blockchain Technology as an Technological Enabler for Innovation, Mannheim University, Department of Information Systems, 2016.

   Manyika, J. and Roxburgh, C. *The Great Transformer: The Impact of the Internet on Economic Growth and Prosperity*, McKinsey Global Institute (2011).

   Shuradze, G. and Wagner, D. "Technological Platforms and Innovation: Review, Integration, and Extension." In: *Twenty-first Americas Conference on Information Systems*, 2011, pp. 1–13.

2. Chatterjee, Rishav and Chatterjee, Rajdeep. *An Overview of the Emerging Technology: Blockchain, International Conference on Computational Intelligence and Networks*, 2017, doi: 10.1109/CINE.2017.33

   Christidis, K. and Devetsikiotis M. "Blockchains and Smart Contracts for the Internet of Things." *IEEE Access*, 4 (2016) 2292–2303.

   Crosby, M., Pattanayak, P., Verma, S. and Kalyanaraman, V. "Blockchain Technology: Beyond Bitcoin." *Applied Innovation*, 2 (2016), 6–10.

   Davidson, S., Filippi, P.D. and Potts, J. *Economics of Blockchain, Social Science Research Network* (2016).

   Nofer, M., Gomber, P., Hinz, O. and Schiereck, D. *Blockchain*, Springer, 2017, doi: 10.1007/s12599-017-0467-3

3. Chen, Y. "Blockchain Tokens and the Potential Democratization of Entrepreneurship and Innovation." *Business Horizons*, 61 (2018), 567–575, doi: 10.1016/j.bushor.2018.03.006

   Conley, J. Blockchain and the Economics of Crypto-tokens and Initial Coin Offerings, Vanderbilt University Department of Economics Working Papers 17-00008, 2017, http://www.accessecon.com/Pubs/VUECON/VUECON-17-00008.pdf

4. An initial public offering, or IPO, is the very first sale of stock issued by a company to the public. Prior to an IPO, the company is considered private,

with a relatively small number of shareholders made up primarily of early investors (such as the founders, their families, and friends) and professional investors (such as venture capitalists or angel investors). The public, on the contrary, consists of everybody else—any individual or institutional investor who wasn't involved in the early days of the company and who is interested in buying shares of the company. Until a company's stock is offered for sale to the public, the public is unable to invest in it. You can potentially approach the owners of a private company about investing, but they're not obligated to sell you anything. Public companies, on the contrary, have sold at least a portion of their shares to the public to be traded on a stock exchange. This is why an IPO is also referred to as "going public." (*Source*: Investopedia.com, https://www.investopedia.com/university/ipo/ipo.asp).

5. Shin, L. "Here's The Man Who Created ICOs and this is the New Token He's Backing." *Forbes*, 21/09/2017.

6. Fisch, Christian. Initial Coin Offerings (ICOs) to Finance New Ventures: An Exploratory Study. SSRN Working Paper, 2018, doi: 10.13140/RG.2.2. 17731.91683

7. *Ibid.*

8. Giudici, G., Adhami, S. and Martinazzi, S. "Why Do Businesses Go Crypto? An Empirical Analysis of Initial Coin Offerings." *Journal of Economics and Business* (2018), doi: 10.1016/j.jeconbus.2018.04.001

9. Talk, F. "Are ICOs Replacing IPOs?" In: *US Global Investors,* 2017, http://www.usfunds.com/investor-library/frank-talk/are-icos-replacing-ipos/#. W6DTwfbOPIU

10. Fenu, G., Marchesi, L., Marchesi, M. and Tonelli, R. "The ICO phenomenon and its Relationships with Ethereum Smart Contract Environment." In: *Proceedings of the SANER 2018 Conference*, IWBOSE, 2018.

11. Talk (2017).

12. Giudici (2018).

13. *Source*: icowatchlist.com, data 2019.

14. *Ibid.*

15. Figuet, J.-M. "ICO: *un nouveau mode de financement de l'innovation?*" *Journal of Economics and Business* (2018), doi: 10.13140/RG.2.2. 21583.18080

16. Berryhill, J., Bourgery T., Hanson A. Blockchains Unchained: Blockchain Technology and its Use in the Public Sector. OECD Working Papers on *Public Governance* 28 (2018), OECD Publishing, Paris, http://dx.doi. org/10.1787/3c32c429-en

Conley, J. Blockchain and the Economics of Crypto-tokens and Initial Coin Offerings, Vanderbilt University Department of Economics Working Papers 17-00008 (2017), http://www.accessecon.com/Pubs/VUECON/VUECON-17-00008.pdf

Chen, Y. "Blockchain Tokens and the Potential Democratization of Entrepreneurship and Innovation." *Business Horizons*, 61 (2018), 567–575, doi: 10.1016/j.bushor.2018.03.006

17. Fenu, G., Marchesi, L., Marchesi, M. and Tonelli, R. "The ICO Phenomenon and its Relationships with Ethereum Smart Contract Environment." In: *Proceedings of the SANER 2018 Conference*, IWBOSE, 2018.

Giudici, G., Adhami, S. and Martinazzi, S. "Why Do Businesses Go Crypto? An Empirical Analysis of Initial Coin Offerings." *Journal of Economics and Business* (2018), doi: 10.1016/j.jeconbus.2018.04.001

18. Conley, J. Blockchain and the Economics of Crypto-tokens and Initial Coin Offerings. Vanderbilt University Department of Economics Working Papers 17-00008 (2017), http://www.accessecon.com/Pubs/VUECON/VUECON-17-00008.pdf

Yadav, M. Exploring signals for investing in an Initial Coin Offering (ICO). SSRN Working Paper (2017), DOI: 10.2139/ssrn.3037106

19. Enyi, Jin and Le, Ngoc. "The Legal Nature of Cryptocurrencies in the US and the Applicable Rules." *SSRN Electronic Journal* (2017), doi: 10.2139/ssrn.2995784

20. Böhme, R., Christin, N., Edelman, B. and Moore, T. "Bitcoin: Economics, Technology, and Governance." *Journal of Economic Perspectives*, 29(2) (2015), 213–238.

21. Yermack, D. "Is Bitcoin a Real Currency?" In: David, K.C. Lee (ed.), *The Handbook of Digital Currency*, Elsevier, 2015, pp. 31–44.

22. Pollock, D. "From $2.9 Billion in a Month to Hundreds Dead: Trends of the Rollercoaster ICO Market in 18 Months." In: *Cointelegraph. The Future of Money*, July 7, 2018, https://cointelegraph.com/news/from-2-9-billion-in-a-month-to-hundreds-dead-trends-of-the-rollercoaster-ico-market-in-18-months

23. Biggs, J. "Thousands of Cryptocurrency Projects are already Dead." In: Techcruch.com (2018), https://techcrunch.com/2018/06/29/thousands-of-cryptocurrency-projects-are-already-dead/?guccounter=1

24. *Ibid.*

25. Price, D. "The Worst Cryptocurrency Hacks Everyone Needs to Know About." In: MUO, April 2018, https://www.makeuseof.com/tag/cryptocurrency-hacks/

26. Russell, J. "First China, Now South Korea has Banned ICOs." In: Techcrunch. com, September 2017, https://techcrunch.com/2017/09/28/south-korea-has-banned-icos/

27. Dickey, M.R: "Facebook is Banning Cryptocurrency and ICO Ads." In: Techcrunch.com, January 2018, https://techcrunch.com/2018/01/30/facebook-is-banning-cryptocurrency-and-ico-ads/

28. Fridgen, G., Regner, F., Schweizer, A. and Urbach, N. "Don't Slip on the Initial Coin Offering (ICO)—A Taxonomy for a Blockchain-enabled Form of Crowdfunding." In: *Twenty-Sixth European Conference on Information Systems* (*ECIS 2018*), Portsmouth, UK.

29. Venegas, P. "Initial Coin Offering (ICO) Risk, Value and Cost in Blockchain Trustless Crypto Markets." *SSRN Electronic Journal* (2017), doi: 10.2139/ssrn.3012238

30. Lewis, M., Anning, P. and Perry, A. "Interpretations of Existing Regulation Concerning ICOs in selected European and Asian Countries." Osborne Clarke, 2018, Publication number 38991821, http://www.osborneclarke.com/wp-content/uploads/2018/06/ICO-Booklet.pdf

31. Gompers. P.A., Ishii, J.L. and Metrick, A. "Corporate Governance and Equity Prices." *Quarterly Journal of Economics*, 118(1) (2003), 107–155; La Porta, R., Lopez-de-Silanes, F., Shleifer, A. and Vishny, R.W. "Law and Finance." *Journal of Political Economy*, 106(6) (1998), 1113–1155.

32. Romero, T. "Why Your ICO Investment Is Going To Zero." In: *Forbes*, 2018, https://www.forbes.com/sites/tromero/2018/01/09/why-your-ico-investment-is-going-to-zero/#45b5ab833922

33. Ljungqvist, A., Richardson, M. and Wolfenzon, D. The Investment Behavior of Buyout Funds: Theory and Evidence. NYU Working Paper SC-AM-03-12 (2007), https://papers.ssrn.com/sol3/papers.cfm?abstract_id=1295177

34. Chen, Y. "Blockchain Tokens and the Potential Democratization of Entrepreneurship and Innovation." *Business Horizons*, 61 (2018), 567–575, doi: 10.1016/j.bushor.2018.03.006

    Fenu, G., Marchesi, L., Marchesi, M. and Tonelli, R. "The ICO Pheno-menon and Its Relationships with Ethereum Smart Contract Environment." In: *Proceedings of the SANER 2018 Conference*, IWBOSE, 2018.

    Fisch, Christian. "Initial Coin Offerings (ICOs) to Finance New Ventures: An Exploratory Study." In: *SSRN Working Paper* (2018), doi: 10.13140/RG.2.2.17731.91683

35. Athey, S., Parashkevov, I., Sarukkai, V. and Xia, J. "Bitcoin Pricing, Adoption, and Usage: Theory and Evidence." In: *Stanford Business*, Working

Paper 3469 (2016), https://www.gsb.stanford.edu/faculty-research/working-papers/bitcoin-pricing-adoption-usage-theory-evidence

36. Catalini, Ch. and Gans, J.S. Initial Coin Offerings and the Value of Crypto Tokens. MIT Sloan Research Paper 5347-18, Rotman School of Management Working Paper 3137213 (2018), http://dx.doi.org/10.2139/ssrn.3137213

37. Li, J. and Mann, W. Initial Coin Offering and Platform Building. UCLA Working Paper, 2018, https://papers.ssrn.com/sol3/papers.cfm?abstract_id=3159528

38. *Ibid.*

39. *Ibid.*

40. Taeihagh, A. "Crowdsourcing, Sharing Economies and Development." *Journal of Developing Societies*, 33(2) (2017), 191–222, doi: 10.1177/0169796X17710072

41. Already existed practice of getting a large number of people to each give small amounts of money in order to provide the finance for a business project, typically using the internet (Source: Cambridge Dictionary online).

42. Fisch, Christian (2018).

43. Giudici G., Adhami S., Martinazzi S. Why Do Businesses Go Crypto? An Empirical Analysis of Initial Coin Offerings. *Journal of Economics and Business* (2018), doi: 10.1016/j.jeconbus.2018.04.001.

44. Wu, K., Wheatley, S. and Sornette, D. *Classification of Cryptocurrency Coins and Tokens by the Dynamics of Their Market Capitalizations*, The Royal Society, Online ISSN 2054-5703 (2018), https://doi.org/10.1098/rsos.180381.

45. Rud, D. "France Introduces New ICO Framework to Become Europe's ICO Hub." In: Coinspeaker, 2018, https://www.coinspeaker.com/2018/09/19/france-introduces-new-ico-framework-to-become-europes-ico-hub/

46. https://www.cnbc.com/2018/06/04/estonia-wont-issue-national-cryptocurrency-estcoin-never-planned-to.html

47. Hartmann, F., Xiaofeng, W. and Lunesu, M.I. "Evaluation of initial Cryptoasset Offerings: The State of the Practice." In: *International Workshop on Blockchain Oriented Software Engineering* (*IWBOSE*), 2018, IEEE.

48. Ordanini, A., Miceli, L., Pizzetti, M. and Parasuraman, A.P. "Crowd-funding: Transforming Customers into Investors through Innovative Service Platforms." *Journal of Service Management*, 22 (2011), 443–470, doi: 10.1108/09564231111155079

49. Miglietta, A. and Parisi, E.M. "Civic CrowdFunding: Sharing Economy Financial Opportunity to Smart Cities." In: Riva, Sanseverino E., Riva,

Sanseverino R. and Vaccaro, V. (eds.) *Smart Cities Atlas: Western and Eastern Intelligent Communities*, Springer, 2016.

50. Miglietta, A. and Parisi, E.M. "Means and Roles of Crowdsourcing vis-à-vis CrowdFunding for the Creation of Stakeholders Collective Benefits." In: Aiello, L.M. and McFarland, D. (eds.), *Social Informatics*, SocInfo 2014 International Workshops in Barcelona, Springer, 2015.

Miglietta, A. and Parisi, E.M. *"L'equity crowdfunding in Italia: opportunità, normative e regolamenti. Investitori e piattaforme di equity Crowdfunding: ruoli e opportunità."* In: *L'equity crowdfunding per lo start up d'impresa e lo sviluppo del Territorio*, Conference in Lecce, March 28, 2014.

51. Data OECD 2014.

52. That is now approaching a more balanced phase with less volatility, i.e. Ripple has lost and as of September 2018 is below 0.50% bps.

53. Hyman, A. "Inclusion of Blockchain in Course of Distributed Systems at the School of Computer Science." In: *ITiCSE 2018 Proceedings of the 23rd Annual ACM Conference on Innovation and Technology in Computer Science Education*, pp. 390–390.

Chapter 23

The Berkeley Blockchain Initiative— Crypto Municipal Bonds

Addressing the Challenge of Homelessness and a Whole Lot More

Tim Wright

Abstract

The Vice Mayor of Berkeley, California, is a man on a mission. Troubled by the knowledge that up to 1,400 people are homeless in Berkeley and the death of two homeless people in the past year, Ben Bartlett has made tackling this widespread problem a particular area of focus for his public service. Drawing on novel schemes from elsewhere and driving innovation to explore the capability of technology to address this issue, Berkeley is pioneering a blockchain secured municipal bond or cryptobond which could offer a blue print for there and elsewhere to tackle funding gaps and inclusivity and create an asset base for a generation that otherwise might not have one.

Ben Bartlett is a lawyer by day and the Vice Mayor[1] of Berkeley, California, by night. He is also a man on a mission to address the plight of the homeless of Berkeley. Not only does he hope to address this pressing issue through the use of emerging finance models underpinned by blockchain, but he also hopes to simultaneously address a number of other

issues that challenge his electorate and the wider community. His desire
to address what he sees as the systemic challenges of the future that are
visibly emerging today is what in large part inspired him to enter public
service. The City of Berkeley[2] is in California, across the bay from San
Francisco, with a campus of the University of California as a notable part
of the city infrastructure. With a tradition of embracing counter culture
and being a location in the forefront of the hippie movement in the 1960s,
Berkeley City proudly describes its modest size as belying its considerable credentials as being "famous around the globe as a center for academic achievement, scientific exploration, free speech, and the arts."

For all its credentials as a beacon of liberalism and free thought, it has
a problem that is not uncommon across much of the developed world,
namely a shortage of affordable housing. One of the manifestations of this
is that among its population of some 112,000 people, it has been estimated
that around 1,400 are homeless. This surprisingly high percentage of
homelessness has resulted in unfortunate and uncomfortable outcomes
with two deaths occurring among the homeless population last winter and
a series of fires occurring at a tent encampment in the city which had
served as a base to a number of the homeless. Bartlett knows that something needs to be done, "We have people in their 60s homeless which is
a nightmare, right?" he asserts.

Quite apart from the moral dilemma and the distress homelessness
presents to us all, Bartlett has a keen sense to most of the perils of the
homeless. As someone who encountered homelessness himself as a young
man, Bartlett is well aware of the reality of being found vulnerable and
shocked at the sudden loss of settled accommodation and this was a key
driver for him to act. "Many of us are just a paycheck or personal crisis
away from being homeless" he reflects as retells a tale of having to spend
several months in a women's refuge with his mother and sister when
family circumstances disrupted and challenged his hitherto stable and
comfortable life experience.

The situation of burgeoning numbers of homelessness is not unique
to Berkeley. California, more widely, has an estimated number of
100,000 homeless people, and many other cities around the globe have
similar challenges. There are a range of initiatives being employed to
try to resolve the problem of homelessness, and for the motivated public

servant, it is possible to search out these attempts to resolve the challenge and consider their efficacy. And so, it was that Mayor Jesse Arreguin and Vice Mayor Bartlett began to explore a range of existing or experimental methods to address the challenges they were faced with the crisis of homelessness and consider them for use in Berkeley.

Among the most common and recurring themes that challenge those seeking to find some mechanism of providing viable homes for the homeless are the lack of availability of land for affordable housing development, cost of constructing or developing accommodation, and access to the finance required for any public intervention.

Inspired by San Francisco developer Patrick Kennedy's experimental MicroPAD Step Up[3] housing, Bartlett sought to bring forward moves to create a similar 100 prefabricated micro- homes into Berkeley. While the resulting Council recommendation stopped short of specifying the type of units to be created, it significantly moved the process of finding parcels of land where they might be located. Both pre-fabricated housing units and the relaxation of controls to permit development play a part in creating a circumstance where action can be more readily taken but the key constraint in all of these initiatives is the availability of finance to enable these civic projects to develop.

Historically, municipal bonds have consistently played a part in financing such public initiatives. But, with Federal corporate tax rates falling in the US, the incentive for the business community to offset their tax exposure through civic investment and charitable giving is in decline, and as a consequence, the financial constraints placed on public bodies like the City of Berkeley to raise the funds necessary to drive these novel initiatives are exacerbated.

But Berkeley has a number of advantages in seeking out solutions to this dilemma. Crowdasset resources and insight are surprisingly concentrated in the local regions. One significant advantage is its proximity to the centres of burgeoning technology-led firms with the skills and understanding of the possibilities of what new digital technologies can be harnessed to achieve. Could this offer a potential foundation for a solution?

Digital tools can let us do things more quickly, more cheaply, more inclusively, more widely, and, importantly, they can also permit us to do things differently. It is this capacity for technology to facilitate re-imagined processes and systems that is at the heart of some of the most disruptive aspects of the crowd-empowered economy.

Bartlett's own personal experience also helped him recognise the possibilities available to him through bringing together a group to resolve a pressing issue and the viability of novel finance. As a lawyer, Bartlett is engaged in the finance of renewable energy projects, which brings him into contact with emerging alternative and novel finance, which is increasingly forming a part of these principle-driven finance schemes.

Additionally, his own personal involvement of an event on a New York train demonstrated both the power of self-forming groups, emerging from the crowd, and their willingness to act for good when a crisis is brought before them. Bartlett recalls an occasion where a bloodied distressed woman boarded a train he was travelling on. The woman had been assaulted, thrown out of her home, and denied access to her children by an abusive partner. Twelve strangers on that train, including Bartlett, immediately responded as a collective presented with the need to act.

"We encircled her and protected her for three months" he recalls. "We got her a job, an apartment, a lawyer, and custody of her kids."

The result of this insight and the readily available digital expertise led to the initial tentative discussion and ideation process of developing novel finance for civic-funding projects through a partnership with the University of California Berkeley Blockchain Lab[4] and a number of local blockchain experts. The resulting project was christened the Berkeley Blockchain initiative or BBI.

The key components of the project were to prove the concept of creating a new type of cryptobond secured by the distributed ledger technology of the blockchain and was used to create finance streams for use on civic projects.

The advantages of the scheme are potentially considerable if they prove to be sustainable ones.

Key among these is the fundamental crowd principle of disintermediation. Quite apart from the advantages of building deep relationships and connections between the funders and the project by bringing them closer together, the disintermediation process has the advantage of significantly reducing the issuance cost of a civic bond. Civic bonds are not new products. They have been widely used as a source of public finance for generations, but they do, however, have a number of significant constraints which this new approach can begin to address. Chief among these is the cost of issuing bonds. The regulatory, diligence, and transactional complexity of issuing traditional bonds is such that in order to make an issuance worthwhile, they need to be of a significant size to accommodate these costs and as a result they tend to have a significant purchase value which places them in a comparatively restricted marketplace with high-worth investors who able to tie up significant sums for extended periods.

It has been estimated by the Haas Institute[5] at University of California, Berkeley, that United States State and local governments spend up to a collective $4 billion annually on issuance overheads alone. Stripping cost from this process by removing much of the human reassurance layer through the application of blockchain distributed ledger technology, which can instantaneously register ownership and so obviate a large amount of compliance and diligence steps, is clearly advantageous in reducing those lost costs of issuance. Bartlett's estimates are that using the crypto-based approach could reduce costs by as much as 42%. If born out, this permits the size of each issuance to be significantly reduced so a bond can quite reasonably be issued for relatively small sums.

Consequently, instead of a minimum denomination of a $5,000 bond, "cryptobonds" can be issued in denominations as low as $5 or $10. Over and above the lower total, it is reasonable to assume that with transaction costs as low as 0.00035 cents per transaction, these cryptobonds become readily tradeable assets bringing liquidity to a marketplace.

The features of lower ticket price and liquidity introduce a further key component of the crowd engagement proposition, that of accessibility. At these sums, and with the promise of not tying investments in for extended periods, it becomes possible for a much wider group of investors to become engaged in the project which both expands the pool of finance available and brings with it a much more diverse expectation of return. Furthermore, it provides a mechanism to create a highly targeted portfolio of micro-offerings helping investors be very specific about where their funds are directed. Berkeley has suggested one of its first issuances might be to fund a new fire truck, for example. This type of a direct linkage from investment to specified and traceable local assets significantly increases engagement, develops insight, and creates transparency.

For Bartlett, there are other significant additional and linked opportunities that could emerge from this initiative. He asserts that "the next generation may have an average asset wealth of zero assets." For him this is a significant emerging structural issue. "If we are going to address the issue of homelessness, we need to address the systemic issues of poverty and create a circumstance where people have access to wealth and can be masters of their own destiny and so not become homeless."

The proposed model of cryptobonds may well provide a mechanism for a new generation of citizens to actually create an asset base which otherwise may never happen. "By fixing this problem we can fix others. Through this model we can provide fractional ownership of assets, a building, a dam and electrical grid for example" Bartlett enthuses.

Other potential benefits from the scheme could be that any returns realised on the bonds by the investors could be tokenised and redeemable in local traders and retailers thereby creating a localised cryptocurrency, which would bring with it all the demonstrable highly focused additional value that local currencies are able to bring.

For Bartlett though this is an option for the marketplace to innovate, building upon the foundation that the BBI project is creating. In his mind, the project has local application but has a much wider potential application. Speaking of the desire to develop a broadly applicable solution, Bartlett says "this technology is sited in the first world but is perfectly suited to the third world. In Berkeley, we are a microcosm and so we wanted to develop a solution which was very flexible and widely applicable, a sort of 'As Above, So Below' model."

This ambition for the scheme plays into his wider concerns over a need to address ecological issues such as desertification and progressive degradation of the availability of drinking water over time. The need to address these looming challenges will necessitate the creation of civic assets like desalination plants to mitigate against these forces which, if left unchallenged, are likely to lead to further deprivation and dislocation for citizens.

Bartlett is unapologetic about his ambitions and vision for these types of holistic solutions and refreshingly unencumbered approach echoing the service design mind-set of constantly adding "and" to the "what if questions" which always seem to lead to the novel solutions. He cites work being undertaken by others, like mobile phones being developed as nodes to facilitate faster transactional speeds on blockchain transactions, to address the potential scalability constraints of these technologies. For him, this opens a whole new set of inclusive, engaged, and purposeful crowd-financing opportunities—Smart Path[6] as he calls it.

But for the time being, the focus is on the Berkeley project. The regulatory and administrative hurdles are progressively being overcome and the scheme is developing well. "By 2020 you and I could be owning a microbond" he happily posits with a discernible glee in his voice.

When asked what success would look like for the scheme, he reflects for a moment and responds "What does success look like? A smooth roll out with kids and homeless people actually owning bonds."

I wouldn't bet against.

Endnotes

1. http://www.benbartlett.vote/
2. https://www.cityofberkeley.info/Home.aspx
3. https://sf.curbed.com/2016/8/5/12389646/prefab-modular-micropad-panoramic-sf
4. https://scet.berkeley.edu/blockchain-lab/
5. Doubly Bound The Costs of Issuing Municipal Bonds BY Marc Joffe Haas Institute https://haasinstitute.berkeley.edu/sites/default/files/haasinstitutere-fundamerica_doublybound_cost_of_issuingbonds_publish.pdf
6. https://hackernoon.com/the-smart-path-solving-for-zero-assets-with-blockchain-technology-566ece753da0

Part 4

Conclusion

Chapter 24

Making It Happen

Tim Wright, Oliver Gajda and Dan Marom

The opportunity and responsibility placed upon policymakers to be open to, and to facilitate, the radical and remarkable possibilities presented by the crowdasset opportunity is a weighty one. To not make use of the opportunity would be remiss, and to not make it more possible for others to adopt it, or at worst delay that process, would be unforgivable. But, by the same token, policymakers are bound by responsibilities and duties to a range of stakeholders and imperatives which can make this decision-making process delicate, nuanced, and problematic.

Nevertheless, the key obligation placed upon policymakers is, in our view, to act and we have set out the case for that in earlier chapters. We have also offered up a set of frameworks and thinking tools to help the policymaker undertake that process in a more managed and structured way.

But the rarefied atmosphere of conceptual models is leavened significantly by reflecting on what has already been done by policymakers in this space. To that end, we have collected a range of remarkable examples of what policymakers have already been doing in their response to the call to act. The intention of bringing this collection together is to provide evidence that it is entirely possible to act from a policymaker's standpoint and that the results are remarkable. It is also to build confidence among that community of policymakers that it can, and is, being done.

We have grouped the cases in very broad groupings, which perhaps make it easier for the more focused reader to zero in on an area or theme that perhaps most engages them. But, in truth, we could have easily arranged them into a number of groupings in order to highlight specific characteristics or themes that are of interest to each of us as collectors of these tales.

The range of cases and exemplars that we have included here is, in our estimation, an impressive one. But, in truth, it is just a selection of remarkable and exciting projects occurring across the world. These projects are helping us to re-imagine how we address long-standing and seemingly immutable problems through engaging with the crowd and unearthing the assets that sit within them that are so valuable and at last available. We fundamentally believe that the progressive recognition of the concept of crowdassets and the willingness of policymakers to explore and experiment with methods to harness this resource is a characteristic of public discourse and interaction which will continue to grow rapidly and will be shown to yield ever increasingly valuable results.

It is important to emphasise that we make no claims that the cases included here have each made use of our frameworks and approaches. It is entirely possible that the thinking of the actors who have taken the actions described here could have echoed some of our thoughts but, to our knowledge, none have explicitly used our approaches, and we thank our contributors for agreeing to share their experience and insight in the context of the book which does include such frameworks and tools.

However, we think that in your examination of these cases you will certainly detect many occasions where the principles we set out as key elements of the crowdasset opportunity are directly identifiable and that the approaches adopted by the actors in the cases collected here either knowingly or unknowingly utilised the principles which we set out. The opportunity for you, however, is to draw on the inspiration of the projects set out here and to take action using the frameworks in a structured and managed way.

The cases we have included here from our collaborators demonstrate awareness of the full range of potential returns from crowd engagement at both the planning and review stages of their projects. Similarly, you will also detect the full range of intervention options used in isolation and in combination.

In terms of the returns, it is the "communication" element which is in large part obvious through the projects and initiatives focused at a more local and civic level with enormously sophisticated conversations and engagement emerging from them. Of course, this is sometimes apparent in hindsight, and we would suggest that specifically seeking to include a targeted and strategic approach to achieving this engagement from the outset would be an approach which would yield maximum value. As we set out communication is a two-way process and opening channels for project participants to speak to the policymaker is a key and valuable return from these projects. In London and many other cities cited in the Civic Crowdfunding section, the insight derived about citizens desires and priorities are emphatically demonstrated.

Building community assets through the projects set out in Germany and elsewhere speak to the growth of an infrastructure in line with citizens' needs and wants. The growing resilience and investment in the future are writ large through the Berkeley Blockchain initiative as it seeks to address the social ill of homelessness. Examples of the networks and connections fostered between disparate groups are catalogued and mapped in many cases.

At a national level, there is a clear emphasis on developing ecosystems to facilitate the growth and operation of the sector, and at a supranational level, we can observe the importance of education and awareness raising as to the opportunity.

The growth of entrepreneurial finance driven by empowering the possibility of a wider group of investors and creating an ecosystem through enlightened regulation and partnership with technical facilitators has brought new flows of capital at a time of constraint from the traditional markets and freed up the access to inclusive funding for public and civic activities at a time of significant constraint.

Without questions, the four key returns of Finance, Insight, Communication, and Networks, all driving growth can be amply demonstrated across the cases, both specifically sought returns planned for as part of the programmes and the serendipitous returns that emerge as a result of the activity. While the emphasis may vary, the presence is indelible.

While we have sought to group the cases into similar thematic sections, we could equally have built paths through the cases to illustrate common groupings based on the intervention models we set out in the framework. This would, to our mind, have been somewhat self-serving and disingenuous. However, and perhaps more importantly, like the spread of returns that are visible across the cases so too are the interventions and so such singular set of groupings by the intervention type would have perhaps masked the complexity and subtlety of the intervention approaches employed.

In any case, you can build your own path through these cases and expand upon them by experiences you find that are not included here.

But, by way of a simple reminder of what the cases contain, we offer this simple précis of each.

1. Civic Crowdfunding, Equity, and the Role of Government

David Weinberger speaks of the Iobby experience and how policymakers at a city level have identified the power of crowdfunding to build strong and authentic relationships and to empower local organisations to take ownership to build and grow public goods and local assets.

2. The Design of Paying Publics

Ann Light and Jo Briggs challenge the notion of the use of the term "crowd," preferring the term "public" as a better expression of more profound and enduring change being imagined and enabled through crowdfunding platforms.

3. The Patronicity Crowdgranting Model

Ebrahim Varachia looks at the matching model as used by the Indiana Housing & Community Development Authority in conjunction with the Patronicity platform to drive economic development and enhance community engagement.

4. Match-Funding Calls for Open Crowdfunding: The Experience from Goteo.Org in New Policies for Crowdvocacy

In setting out its analysis of match funding in crowdfunding initiatives, the Goteo model identifies what it describes as new dynamics of institutional cooperation emerging as a result of this approach and how this can become a powerful instrument for public participation and policy innovation, which they describe as crowdvocacy.

5. Co-Creating Cities—Practical Experiences from Crowdsourcing and Crowdfunding Urban Areas in Turin, Brussels, and London

While we have endeavoured to concentrate on the crowdfunding aspect of crowdassets, we also make the point that this is a much wider dynamic and it is hard to be dogmatic about the topic. And it is perhaps inevitable that we explicitly point at how these practices bleed into and incorporate other crowd engagement models as demonstrated in co-creating cities which explicitly include open innovation models and the broader crowdsourcing alongside enabling finance.

6. Crowdfund Angus: The Impact of Public-Level Crowdfunding

Shelley Hague shows us how those tasked with developing enterprise and the infrastructure that underpins it are harnessing the crowdfunding opportunity and creating a highly targeted but sophisticated approach of integration including educational support, ecosystem development through partnership with platforms, and leverage from the Council itself.

7. Crowdfunding Social or Community-Focused Projects

Jonathan Bone sets out the role of institutions as partners in delivery and founds the item in extensive research from an organisation which has played an important role in encouraging adoption of crowdfunding by public and civic bodies in the UK and further afield.

8. Crowdfunding for Civic and Development Projects

The importance of crowdfunding to NGOs in a time of constrained funding is at the heart of the UNDPs approach, but the emphasis is clearly on the role of the NGO to empower and provide the structure for local empowerment rather than undertake the activity itself. As we point out, the emphasis for policymakers should generally be to help others do it as opposed to doing it themselves.

9. Crowdfund London: Civic Crowdfunding as a Tool for Collaborative Urban Regeneration in London

As a comprehensive exercise in driving city-wide participation, the efforts and results from Crowdfund London are hard to deny. Focused on many small projects, the initiative demonstrates how the communities running them can have a big social impact, and how policymakers can catalyse a powerful mix of public, private, and local collaboration to improve places. As a clear example of the growth, networks, and communication elements of the five key returns, it is hard to find a more telling example.

10. Lessons Learned from Civic Crowdfunding and Match-Funding Schemes in Europe

Once again, the role of the civic crowdfunding initiatives to build networks and communications between policymakers and their citizens is explained, and this chapter also demonstrates the type of insight which can be derived from such actions and offers some excellent examples of best practice.

11. Civic Crowdfunding in Milan: Between Grass-Roots Actors and Policy Opportunities

The experience of Milan sets out interesting reminders for policymakers in terms of the importance of their shaping the context within which civic crowdfunding takes place in order to ensure that the participatory aspects and the distribution of resulting benefits are maximised in the process of crowdfunding, and also emphasises that the evidence from here would not support a suggestion that the result of this type of activity causes a retreat on the part of civic bodies from being suppliers of critical services, in fact quite the contrary is apparent.

12. Initiative Comes at Cost: Russian Experience of Crowdfunding for Policymakers

The Russian experience highlights a real emphasis on a collaborative approach and the value of the crowdfunding process as a mechanism of driving communication and provides considerable insight into citizens' preferences.

13. Civic Crowdfunding in Germany—An Overview of the Developing Landscape in Germany

The German experience, as set out in this chapter, demonstrated the diversity of sectors addressed through the proactive approach and its broad applicability.

14. Crowdfunding Act: Accelerating the Growth of Crowdfunding Market in Finland

The Finish experience details a high-level governmental intervention to create an ecosystem suitable for crowdfunding and its impact on the sector. The chapter exposes the differing views associated with this intervention as it emerged over the legislative cycle and how, over time, policymakers can adjust and amend their intervention to reflect their own political objectives, values, and priorities.

15. Crowdsourcing Ideas for Public Investment: The Experience of Youth Participatory Budgeting in Portugal

The importance of participation for a specific demographic, in this case young people, is captured very effectively in the Portuguese example. Based on an analysis of a significant number of schemes, the example demonstrates the importance of the communication and insight aspect of crowd engagement.

16. Intrastate Crowdfunding: Achieving Its Full Potential as an Economic Development Tool

The experience of intrastate crowdfunding in the USA is a fascinating exploration of policymakers reacting to other policymakers. By utilising the flexibility that they have to act within a local context to bring about change, the states have created specific ecosystems to permit crowdfunding in ways that are specific to local concerns and needs. The returns in terms of empowering enterprise and growth are remarkable and inspiring. The item includes some valuable pointers and advice for policymakers.

17. Will the Proposed European Crowdfunding Regulation Lead to a "True" European Market for Crowdfunding?

The opportunity for supranational policymakers to set an expansive ecosystem to grow the crowdfunding opportunity is examined in a pan European model. The complexity and challenges of creating intervention within what are often nationally bound competencies is a fascinating case of the breadth of awareness and the imperative of the opportunity that crowdfunding is creating, so much so that bodies like the EU are now compelled to engage.

18. Crowdfunding—The Indonesian Experience

It is sometimes easy to forget that the environment we operate within is not simply constrained by governmental edict. Religious and cultural issues create important and significant norms that we adhere to and so developing an ecosystem that conforms to not just financial regulatory constraints but also the conventions and permissibility of religious and ethical considerations represents an additional complexity for policymakers. For Indonesia, abiding by the permissible actions in an Islamic context was essential and this chapter offers a fascinating experience of the process of developing that approach.

19. Relationships among Crowdassets, Crowdfunding, and ICOs for Civic Engagement

Innovation is central to the process of policymakers taking advantage of the crowdfunding opportunity, be that innovation in terms of the type of intervention or simply embracing the innovations that underpin the crowdasset phenomenon and make it possible. In this vein, we dive directly into the emerging ideas of cryptofinance and the technical architectures that underpin the blockchain or distributed ledger. This digital revolution will undoubtedly cement the move to new finance and also provide new opportunities to explore how policymakers can make it possible and drive forward their policy agendas through it. This chapter gives a sound introduction to what is already happening in this space.

20. From Crowdfunding Initiative to Fintech Hub: Lithuanian Case

Fintech is a wider term that embraces many new innovations in finance and this opening out of digital finance innovation is growing all the time, but the popularity of crowdfunding can offer a strong foundation for policymakers to both begin the process of engagement with new models and allow the ambitious to create a broader ecosystem that embraces both crowdfunding as a foundational element but actively encourages and underpins the growth of other emerging models, creating a crucible to attract innovators and entrepreneurs to an ambitious hub for fintech innovation.

21. The Berkeley Blockchain Initiative—Crypto Municipal Bonds

Rounding off our collection of examples and cases, we encounter a fascinating story of the coming together of innovation for traditional finance, which provides a visionary and holistic approach to address a range of apparently unrelated challenges by harnessing crowdassets to benefit the crowd.

These examples are to us inspiring and informative, and should you search more widely, you will undoubtedly find others not included here but of equal validity and importance to shaping the perspective and debate around how to best seize the crowdasset opportunity.

We urge you as a policymaker to consider how you will respond and act to this opportunity for the benefit of those who fall within your policy remit. Take courage and inspiration from these examples and take confidence that with these tools you can act in a properly considered and structured way, which will ensure that the returns available to your constituents are maximised and that the strategic objectives of your institution are advanced in a managed and effective way.

Good luck!

Epilogue

Crowdfunding is an important contributor to the democratization of innovation—the biggest paradigm shift in innovation since the Industrial Revolution.

— Prof. Eric von Hippel (MIT)

Crowdfunding is a global phenomenon. Each day, more and more innovators and backers are joining forces online for profit and non-profit projects and for societal impact alongside financial outcomes. The power of the crowd is being leveraged, instead of and alongside the conventional funding methods. Innovators, change makers, and entrepreneurs are finding new fascinating opportunities for growth, based not only on the direct financial benefits of this mechanism but much more than that—they are leveraging the societal superpowers of crowdfunding and harnessing it to their use. Evidence is revealing the substantial impact of crowdfunding on businesses, industries, and causes. Across the globe, millions of individuals are simultaneously conducting financial transactions worth billions of USD, using thousands of different platforms, all of which are creating trustworthy cooperation among total strangers.

Are policymakers aware of this phenomenon? Can they leverage the crowd power? Can these impactful mechanisms be instrumental for the sake of their communities?

If you reached this page, you have read our humble answers—YES, YES, and YES.

Policymakers across the globe are unleashing Crowdassets that were hidden within their communities, waiting to be leveraged to create mutually beneficial results. Pushing for a Win–Win–Win. Matching funds, educating the market, implementing various interventions models are done in order to design a better future, where resources are more aligned to the needs of the society, and by joining forces, the impact becomes much clearer and stronger.

We spent a year trying to present the best practices, frameworks, and insights. A year filled with learning new ideas, conducting research, gathering data, brainstorming, and writing. Then re-writing, focusing, and sharpening. The outcome is in your hands. Thanks to our wonderful contributors, professional publisher, and respectful reviewers, we feel extremely fortunate to gather all these wonderful experts, hopefully making your crowdfunding exploration journey a more interesting and diverse one.

As you have probably noticed, the three of us are crowdfunding enthusiasts, strongly believing that crowdfunding is changing the world of today and shaping our future. We see and feel its impact on individuals, businesses, organisations, and communities. More than that, we understand its enormous potential for policymakers, who can and should use it wisely, for the sake of their citizens, and promoting their welfare and better outcomes via crowd engagement. That was our mission. To make crowdfunding best practices accessible for policymakers from all levels. To go beyond theoretical discussions, into various case studies which will demonstrate its practical implications. To host as many stakeholders as we can, in a book which will provide all the relevant point of views.

We hope that you enjoy and benefit from this book. We hope it will serve as a practical guide, which will be instrumental for all of you. We have done our best to minimise the gap between your intention and action—by providing practical insights, tools, and guidelines. We do hope that these pages are not only inspiring but also a call for action. That was our humble goal. To push you forward. Whether you will be using our generic framework or a very specific recommendation, just do it! We deeply believe crowdfunding should be the standard for citizen engagement, democratising and pushing the boundaries of fruitful relationships between policymakers and citizens.

Now it's your turn. The biggest paradigm shift is already taking part. Now make it happen for your stakeholders and crowd empower your community!

Best of luck!

Index